4/24/89

# Particle Physics

## A Los Alamos Primer

# Particle Physics

## A Los Alamos Primer

**Edited by Necia Grant Cooper and Geoffrey B. West**

*Los Alamos National Laboratory*

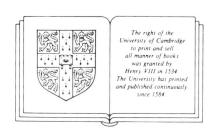

The right of the
University of Cambridge
to print and sell
all manner of books
was granted by
Henry VIII in 1534.
The University has printed
and published continuously
since 1584.

## CAMBRIDGE UNIVERSITY PRESS

*Cambridge*

*New York   New Rochelle   Melbourne   Sydney*

Published by the Press Syndicate of the University of Cambridge
The Pitt Building, Trumpington Street, Cambridge CB2 1RP
32 East 57th Street, New York, NY 10022, USA
10 Stamford Road, Oakleigh, Melbourne 3166, Australia

© Cambridge University Press 1988

First published 1988
Reprinted 1989

Printed in the United States of America

*Library of Congress Cataloging-in-Publication Data*

Particle physics.

An updated version of Los Alamos science, no. 11
(summer/fall 1984).

Includes index.
1. Particles (Nuclear physics)   I. Cooper, Necia
Grant.   II. West, Geoffrey B.
QC793.P358   1987      539.7'21         87-10858

*British Library Cataloguing in Publication Data*

Particle physics : a Los Alamos primer.
1. Particles (Nuclear physics)
I. Cooper, Necia Grant   2. West, Geoffrey B.
539.7'21      QC793.2

ISBN 0-521-34542-1 hard covers
ISBN 0-521-34780-7 paperback

| | |
|---:|:---|
| *General Editors* | Necia Grant Cooper |
| | Geoffrey B. West |
| | |
| *Editor* | Necia Grant Cooper |
| | |
| *Associate Editors* | Roger Eckhardt |
| | Nancy Shera |
| | |
| *Designer* | Gloria Sharp |
| | |
| *Illustration and Production* | Jim Cruz |
| | Anita Flores |
| | John Flower |
| | Judy Gibes |
| | Jim E. Lovato |
| | Lenny Martinez |
| | LeRoy Sanchez |
| | Mary Stovall |
| | Chris West |

# Contents

*Preface to Los Alamos Science, Number 11, Summer/Fall 1984* _____ viii

*Introduction* _____ ix

## Theoretical Framework

**Scale and Dimension—From Animals to Quarks** _____ 2
  *by Geoffrey B. West*

  Fundamental Constants and the Rayleigh-Riabouchinsky Paradox _____ 12


**Particle Physics and the Standard Model** _____ 22
  *by Stuart Raby, Richard C. Slansky, and Geoffrey B. West*

  QCD on a Cray: The Masses of Elementary Particles _____ 41
  *by Gerald Guralnik, Tony Warnock, and Charles Zemach*

  **Lecture Notes—From Simple Field Theories to the Standard Model** _____ 54
  *by Richard C. Slansky*


**Toward a Unified Theory: An Essay on the Role of Supergravity in the Search for Unification** _____ 72
  *by Richard C. Slansky*

  Fields and Spins in Higher Dimensions _____ 86


**Supersymmetry at 100 GeV** _____ 98
  *by Stuart Raby*

  Supersymmetry in Quantum Mechanics _____ 102

**The Family Problem** _____ 114
   *by T. Goldman and Michael Martin Nieto*

   Addendum: CP Violation in Heavy-Quark Systems _____ 124

# Experimental Developments

**Experiments to Test Unification Schemes** _____ 128
   *by Gary H. Sanders*

   An Experimentalist's View of the Standard Model _____ 130

   Addendum: An Experimental Update _____ 149


**The March toward Higher Energies** _____ 150
   *by S. Peter Rosen*

   Addendum: The Next Step in Energy _____ 156

   LAMPF II and the High-Intensity Frontier _____ 158
   *by Henry A. Thiessen*

   The SSC—An Engineering Challenge _____ 164
   *by Mahlon T. Wilson*


**Science Underground—The Search for Rare Events** _____ 166
   *by L. M. Simmons, Jr.*

# Personal Perspectives

**Quarks and Quirks among Friends** _____ 180
   *A round table on the history and future of particle physics with Peter A. Carruthers, Stuart Raby, Richard C. Slansky, Geoffrey B. West, and George Zweig*


*Index* _____ 196

# Preface

On the cover a mandala of the laws of physics floats in the cosmos of reality. It symbolizes the interplay between the inner world of abstract creation and the outer realms of measurable truth. The tension between these two is the magic and the challenge of fundamental physics.

According to Jung, the "squaring of the circle" (the mandala) is the archetype of wholeness, the totality of the self. Such images are sometimes created spontaneously by individuals attempting to integrate what seem to be irreconcilable differences within themselves. Here the mandala displays the modern attempt by particle physicists to bring together the basic forces of nature in one theoretical framework.

The content of this so-called standard model is summarized by the mysterious-looking symbols labeling each force: $U(1)$ for electromagnetism, $SU(2)$ for weak interactions, $SU(3)_C$ for strong interactions, and $SL(2C)$ for gravity; each symbol stands for an invariance, or symmetry, of nature. Symmetries tell us what remains constant through the changing universe. They are what give order to the world. There are many in nature, but those listed on the mandala are special. Each is a *local* symmetry, that is, it manifests independently at every space-time point and therefore implies the existence of a separate force. In other words, local symmetries determine all the forces of nature. This discovery is the culmination of physics over the last century. It is a simple idea, and it turns out to describe *all* phenomena so far observed.

Where does particle physics go from here? The major direction of present research (and a major theme of this issue) is represented by the spiral that starts at electromagnetism and turns into the center at gravity. It suggests that the separate symmetries may be encompassed in one larger symmetry that governs the entire universe—one symmetry, one principle, one theory. The spiral also suggests that including gravity in such a theory involves understanding the structure of space-time at unimaginably small distance scales.

Julian Schwinger, whose seminal idea led to the modern unification of electromagnetic and weak interactions, regards the present emphasis on unification with skepticism: "It's nothing more than another symptom of the urge that afflicts every generation of physicists—the itch to have all the fundamental questions answered in their own lifetime."* To others the goal seems tantalizingly close, an achievement that may be reached, if not this year—then maybe the next . . . .

The hope of unification depends on a second theme of this issue, symbolized by the ants and elephants walking round the mandala. These creatures are our symbol of scaling, the sizing up and sizing down of physical systems. Strength (or any other quality, for that matter) may look different on different scales. But if we look hard enough, we can find certain invariances to changes in scale that define the correct variables for describing a problem. Why do ants appear stronger than elephants? Why does the strong force look weak at high energies? How could all the forces of nature be manifestations of a single theory? These are the questions explored in "Scale and Dimension—from animals to quarks," a seductively playful article that leads us to one of the most important contributions to modern physics, the renormalization group equations of quantum field theory. The insights about scaling gained from these equations are important not only to elementary particle physics but also to phase transition theory and the dynamics of complex systems.

All the articles in this issue were written by scientists who care to tell not only about their own research but about the whole field of particle physics, its stunning achievements and its probing questions. Outsiders to this field hear the names of the latest new particles, the buzz words such as grand unification or supersymmetry, and the plans for the United States to regain its leadership in this glamorous, high tech area of big science. But what is the real progress? Why does this field continue to attract the best minds in science? Why is it a major achievement of human thought? From a distance it may be hard to tell—except that it satisfies some deep urge to understand how the world works. But if one could be given a closer look at the technical content of this field, its depth and richness would become apparent. That is the aim of the present issue.

The hardest job was defining the technical level. How could the framework of the standard model be appreciated by someone unfamiliar with symmetry principles? How could modern particle physics research, all of which builds on the standard model, be understood by someone unfamiliar with what everyone in the field takes for granted? We hope we have solved this problem by presenting some of the major concepts on several levels and in several different places. We even include our own reference material, a remarkably clear and friendly set of lecture notes prepared especially for this issue.

As one who was trained in this field, I returned to it with some trepidation—to deal with the subject matter, which had been so difficult, and with the personalitites competing in the field, who sometimes ride roughshod over each other as they battle these unruly abstractions. Much to my delight and the delight of the *Los Alamos Science* staff, the experience of preparing this issue was immensely enjoyable and rewarding. The authors were enthusiastic about explaining and re-explaining, about considering the essence of each point one more time to make sure that the readers too would be able to grasp it. Their generosity and interest made it fun for us to learn. May this presentation also be a treat for you.

---

*This quote appeared in "How the Universe Works" by Robert P. Crease and Charles C. Mann (The Atlantic Monthly, August, 1984), a fast-paced article about the history of the electoweak theory.*

Necia Grant Cooper
1984

# Introduction

Beginning with the dramatic discovery of the J/ψ particle in 1974, particle physics has gone through a remarkably productive and exciting period. Quantum field theory, developed during the 1940s and '50s but abandoned in the '60s, was re-established as the language for formulating theoretical concepts. The unification of the weak and electromagnetic interactions via a so-called non-Abelian gauge theory could only be understood in this framework. A similar theory of the strong interactions, quantum chromodynamics, was also constructed during this period, and nowadays one refers to the total package of the strong and electroweak theories as "the standard model." Over the last decade the predictions of the standard model have been spectacularly confirmed, so much so that it is now almost taken for granted as embodying all physics below about 100 GeV. The culmination of this exuberant period was the inevitable discovery in 1983 of the $W^\pm$ and $Z^0$ particles, the massive bosons predicted by the standard model to mediate the weak interactions. Although the masses of these particles were precisely those predicted by the SU(2) × U(1) electroweak theory, their discovery was almost anticlimactic, so accepting had the particle-physics community become of the standard model. Indeed, future research in particle physics is often referred to as "physics beyond the standard model," an implicit tribute to the progress of the past decade.

The development of the standard model during the 1970s brought with it a lexicon of new words and concepts—quark, gluon, charm, color, spontaneous symmetry breaking, and asymptotic freedom, to name a few. Supersymmetry, preons, strings, and worlds of ten dimensions are among the buzz words added in the '80s. While scientists, engineers, and even many lay people will recognize some subset of these words, only a few have more than a superficial understanding of the profound achievement they denote. Add to this the demand by particle physicists for several billion dollars to build a super-accelerator in order to explore "physics beyond the standard model," and one can sense the gap between the particle physicist and his "public" reaching irreparable proportions. On the other hand there remains an endless wonder and fascination in the public's eye for such speculative conceptual ideas, which are more usually associated with the literature of science fiction than with *Physical Review*.

It was with some of these thoughts in mind that a group of us at Los Alamos National Laboratory decided to put together a series of pedagogical articles explaining in relatively elementary scientific language the accomplishments, successes, and projected future of high-energy physics. The articles, intended for a wide scientific audience, originally appeared in a 1984 issue of *Los Alamos Science*, a technical publication of the Laboratory. Since that time they have been used as a teaching tool in particle-physics courses and as a reference source by experimentalists in the field.

*Particle Physics—A Los Alamos Primer* is basically an updated version of the original *Los Alamos Science* issue. We believe it will continue to help educate undergraduate and graduate students as well as bridge the gap between experimentalists and theorists. We are also confident that it will help non-experts to develop a good feel for the subject.

The text consists of eight "chapters," the first five devoted to the concepual framework of modern particle physics and the last three to experiments and accelerators. Each is written by a separate author, or group of authors, and is to a large extent self-contained. In addition, we have included a round table among several particle physicists that addresses some of the broader issues facing the field. This discussion is in some ways a unique evaluation of the present status of particle physics. It is quite personal and idiosyncratic, sometimes irreverent, and occasionally controversial. For the non-expert it is probably the place to begin!

The first article addresses the question of scaling. In its broadest sense this lies at the heart of any attempt to unify into one theory the fundamental forces of Nature—forces seemingly so very different in strength. "Scale and Dimension—From Animals to Quarks" begins by reviewing in an elementary and somewhat whimsical fashion the whole question of scale in classical physics and then introduces the more sophisticated concept of the renormalization group. The renormalization group is really no more than a generalization of classical dimensional analysis to the area of quantum field theory: it answers the seminal question of how a physical system responds to a change in scale. The concept plays a central role in the modern view of quantum field theory and has been particularly successful in elucidating the nature of phase transitions. Indeed, it is from this vantage point that the intimate relationship between particle and condensed-matter physics has developed. Clearly, the manner in which physics evolves from one energy or length scale to another is of fundamental importance.

The second article, "Particle Physics and the Standard Model," addresses the question of unification with an elementary yet comprehensive discussion of how the famous electroweak theory is constructed and works. The role of internal symmetries and their incorporation into a principle of local gauge invariance and subsequent manifestation as a non-Abelian gauge field theory are explained in a pedagogical fashion. The other component of the standard model, namely quantum chromodynamics (QCD), the theory of the strong interactions, is similarly treated in this article. Again, the discussion is rather elementary, beginning with an exposition of the "old" SU(3) of the "Eightfold Way" and finishing with the field theory of quarks and gluons. For the more ambitious reader we have included a set of "lectures" by Richard Slansky that give some of the technical details necessary in going "from simple field theories to the standard model." Crucial concepts such as local gauge invariance, spontaneous symmetry breaking, and emergence of the Higgs particles that give rise to the masses of elementary particles are expressed in the mathematical language of field theory and should be readily accessible to the serious student of the field. These lectures are very clear and provide the reader with the explicit equations embodying the physics discussed in the article on the standard model.

Following this review of accepted lore, we begin our journey into "physics beyond the standard model" with an essay on supergravity by Slansky entitled "Toward a Unified Theory." In it he discusses some of the speculative ideas that gained popularity in the late 1970s. Among them are supersymmetry (a proposed symmetry between fermions and bosons) and the embedding of our four-dimensional space-time world in a larger number of dimensions. Supergravity, a theory that encompasses both of these ideas, was the first serious attempt to include Einstein's gravity in the unification scheme. This article also includes a description of superstring theory, which has gained tremendous popularity just in the last year or so. Slansky explains how the shortcomings of the supergravity scenario are circumvented by basing a unified theory on elementary fibers, or strings, rather than on point particles. This area of research is in a state of flux at the moment, and it is still far from clear whether strings really will form the basis of the "final" theory. The problems are both conceptual and technical. Conceptually there is still no hint as to what principles are to replace the equivalence principle and general coordinate invariance, which form the bases of Einstein's gravity. Technically, the mathematics of string theory is beyond the usual expertise of the theoretical physicist; indeed it is on the forefront of mathematical research itself. This may be the first time for a hundred years or more that research in physics and mathematics has coincided. Some may view this as a bad omen, others as the dawning of a new exciting age leading to the equations of the universe! Only time will tell.

A less ambitious use of supersymmetry has been in the attempts to unify, without gravity, the electroweak and strong theories of the standard model. Stuart Raby, in his article "Supersymmetry at 100 GeV," discusses some of these efforts by concentrating on the phenomenological implications of a world in which every boson has a fermion partner and vice versa. These include a possible explanation for why proton decay, certainly one of the more dramatic predictions of grand unified theories, has not yet been seen. Supersymmetric phenomenology has served as an important guide for speculating about what can be seen at new accelerators. A special feature of this article is the self-contained section "Supersymmetry in Quantum Mechanics," in which Raby explains this novel space-time symmetry in a setting stripped of all field-theoretic baggage.

One of the more mysterious problems in particle physics is "the family problem" described in an article of that title by Terry Goldman and Michael Nieto. The apparent replication of the electron and its neutrino in at least two more families differing only in their mass scales has remained a mystery ever since the discovery of the muon. This replication, exhibited also by the quarks, can be accommodated in unified theories, though no satisfactory explanation of the family structure, nor even a prediction of the total number of families, has been advanced. The phenomenology of this problem as well as some attempts to understand it are carefully reviewed. An addendum to the original article presents a slightly more technical discussion of how experiments involving the third quark family might extend our knowledge of CP violation. This symmetry violation remains perhaps the most mysterious aspect of the known particle phenomenology.

The next three articles concern the experimental side of particle physics. Although the choice of Los Alamos experiments to illustrate certain points does reflect some parochial interests of the authors, these articles succeed in providing a broad overview of experimental methodology. In this era of elaborate detection techniques requiring extensive collaboration, it is often difficult for the uninitiated to unravel the complicated machinations that are involved in the experimental process. In "Experiments to Test Unification Schemes" Gary Sanders presents a very clear exposition of the physics input to this process. Indeed, as if to emphasize the departure from the world of theory, he has included a brief page-and-a-half précis subtitled "An Experimentalist's View of the Standard Model." For the beginner this might be read immediately following the round table! Sanders describes in some detail four experiments designed specifically to test the standard model, all being conducted at Los Alamos. Each is a "high-precision" experiment in which, say, a specific decay rate is measured and compared with the value predicted by the standard model. These experiments are prototypical of the kind that have been and will continue to be done at accelerators around the world to push the theory to its limits. Most exciting, or course, would be the observation of some deviation from the standard model that could be associated with grand unification. However, in an addendum Sanders reports that no such deviations were seen in the data from the Los Alamos experiments and others. So far the standard model has stood the test of time.

The following article by Peter Rosen, "The March toward Higher Energies," surveys the high-energy accelerator landscape beginning with a historical perspective and finishing with a glimpse into what we might expect in the not-too-distant future. The emphasis here is on tests of the standard model and searches for new and exotic particles not included in it. The traditional methodology is quite simple: go for the highest energy possible. This has certainly been successful in the past, and we have no reason to believe that it won't be successful in the future. Thus, there is a push to build a giant superconducting supercollider (SSC) that could probe mass scales in excess of 20 TeV, or $2 \times 10^{13}$ eV. We have also included a brief report by Mahlon Wilson, an accelerator physicist, on some of the problems peculiar to the gigantic scale of the SSC.

An alternative technique for probing high mass scales is to perform very accurate experiments in search of deviations from expected results, such as those described in Sanders' article. Obviously, high-intensity beams are the desired tool in this approach. A high-intensity machine has been proposed for Los Alamos, and another brief report by Henry Thiessen, also an accelerator physicist, describes that machine and some of the questions it might answer. The reports on the SSC and LAMPF II provide an idea of what is involved in designing tomorrow's accelerators.

The final article is a review by Mike Simmons on "science underground." In it he discusses what particle physics can be learned from experiments performed deep underground to isolate rare events of interest. The most famous of these is the search for proton decay. Other experiments measure the flux of neutrinos from the sun and search for exotic particles (such as magnetic monopoles) in cosmic rays. These essential fishing expeditions use "beams" from the biggest accelerator of them all, namely the universe!

*Particle Physics—A Los Alamos Primer* thus provides the reader with a comprehensive, up-to-date introduction to the field of particle physics. Our belief is that it will be a useful educational guide to both the student and professional worker in the field as well as provide the general scientist with an insight into some of the recent accomplishments in understanding the fundamental structure of the universe.

In conclusion we would like to thank the staff of *Los Alamos Science* for their invaluable help in making this primer lively and accessible to a wide audience.

Geoffrey B. West
1986

# Particle Physics

## A Los Alamos Primer

imension

— FROM ANIMALS TO QUARKS!

BY GEOFFREY B. WEST

*F*OR THOSE OVER 40 YEARS OF AGE OR FOR THOSE WHO ENJOY WATCHING 3ʳᵈ RATE MOVIES AT 2 O'CLOCK IN THE MORNING, THE SPECTER OF BEING **ATTACKED BY GIANT ANTS**, BEETLES, OR SPIDERS IS FAMILIAR AND POSSIBLY EVEN AMUSING. WHAT PERHAPS IS EVEN MORE AMUSING, AND CERTAINLY MORE PROFOUND, IS THAT THIS PARANOID FANTASY OF THE 1950's HAD ALREADY BEEN CONJURED UP AND ANALYZED ALMOST THREE HUNDRED YEARS EARLIER BY NONE OTHER THAN **GALILEO!**

CONTINUED ON NEXT PAGE.

*"I have multiplied visions and used similitudes."* — Hosea 7:10

In his marvelous book *Dialogues Concerning Two New Sciences* there is a remarkably clear discussion on the effects of scaling up the dimensions of a physical object. Galileo realized that if one simply scaled up its size, the weight of an animal would increase significantly faster than its strength, causing it ultimately to collapse. As Galileo says (in the words of Salviati during the discorso of the second day), ". . . you can plainly see the impossibility of increasing the size of structures to vast dimensions . . . if his height be increased inordinately, he will fall and be crushed under his own weight." The simple scaling up of an insect to some monstrous size is thus a physical impossibility, and we can rest assured that these old sci-fi images are no more than fiction! Clearly, to create a giant one "must either find a harder and stronger material . . . or admit a diminution of strength," a fact long known to architects.

It is remarkable that so many years before its deep significance could be appreciated, Galileo had investigated one of the most fundamental questions of nature: namely, what happens to a physical system when one changes scale? Nowadays this is the seminal question for quantum field theory, phase transition theory, the dynamics of complex systems, and attempts to unify all forces in nature. Tremendous progress has been made in these areas during the past fifteen years based upon answers to this question, and I shall try in the latter part of this article to give some flavor of what has been accomplished. However, I want first to remind the reader of the power of dimensional analysis in classical physics. Although this is stock-in-trade to all physicists, it is useful (and, more pertinently, fun) to go through several examples that explicate the basic ideas. Be warned, there are some surprises.

## Classical Scaling

Let us first re-examine Galileo's original analysis. For *similar* structures* (that is, structures having the same physical characteristics such as shape, density, or chemical composition) Galileo perceived that weight $W$ increases linearly with volume $V$, whereas strength increases only like a cross-sectional area $A$. Since for similar structures $V \propto l^3$ and $A \propto l^2$, where $l$ is some characteristic length (such as the height of the structure), we conclude that

$$\frac{\text{Strength}}{\text{Weight}} \propto \frac{A}{V} \propto \frac{1}{l} \propto \frac{1}{W^{1/3}}. \qquad (1)$$

Thus, as Galileo noted, smaller animals "appear" stronger than larger ones. (It is amusing that Jerome Siegel and Joe Shuster, the creators of Superman, implicitly appealed to such an argument in one of the first issues of their comic.[†] They rationalized his super strength by drawing a rather dubious analogy with "the lowly ant who can support weights hundreds of times its own" (sic!).) Incidentally, the above discussion can be used to understand why the bones and limbs of larger animals must be proportionately stouter than those of smaller ones, a nice example of which can be seen in Fig. 1.

Arguments of this sort were used extensively during the late 19th century to un-

4

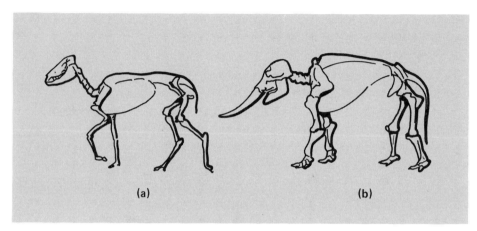

**Fig. 1. Two extinct mammals: (a)** Neohipparion, *a small American horse and (b)* Mastodon, *a large, elephant-like animal, illustrating that the bones of heavier animals are proportionately stouter and thus proportionately stronger.*

derstand the gross features of the biological world; indeed, the general size and shape of animals and plants can be viewed as nature's way of responding to the constraints of gravity, surface phenomena, viscous flow, and the like. For example, one can understand why man cannot fly under his own muscular power, why small animals leap as high as larger ones, and so on.

A classic example is the way metabolic rate varies from animal to animal. A measure $B$ of metabolic rate is simply the heat lost by a body in a steady inactive state, which can be expected to be dominated by the surface effects of sweating and radiation. Symbolically, therefore, one expects $B \propto W^{2/3}$. The data (plotted logarithmically in Fig. 2) show that metabolic rate does

indeed scale, that is, all animals lie on a single curve in spite of the fact that an elephant is neither a blown-up mouse nor a blown-up chimpanzee. However, the slope of the best-fit curve (the solid line) is closer to 3/4 than to 2/3, indicating that effects other than the pure geometry of surface dependence are at work.[‡]

It is not my purpose here to discuss why this is so but rather to emphasize the importance of a scaling curve not only for establishing the scaling phenomenon itself but for revealing deviations from some naive prediction (such as the surface law shown as the dashed line in Fig. 2). Typically, deviations from a simple geometrical or kinematical analysis reflect the dynamics of the system and can only be understood by examining it in more detail. Put slightly differently, one can view deviations from naive scaling as a probe of the dynamics.

The converse of this is also true: generally, one cannot draw conclusions concerning dynamics from naive scaling. As an illustration of this I now want to discuss some simple aspects of birds' eggs. I will focus on the question of breathing during incubation and how certain physical variables scale from bird to bird. Figure 3, adapted from a *Scientific American* article by Hermann Rahn, Amos Ar, and Charles V. Paganelli

entitled "How Bird Eggs Breathe," shows the dependence of oxygen conductance K and pore length $l$ (that is, shell thickness) on egg mass $W$. The authors, noting the smaller slope for $l$, conclude that "pore length probably increases slower because the eggshell must be thin enough for the embryo to hatch." This is clearly a dynamical conclusion! However, is it warranted?

From naive geometric scaling one expects that for similar eggs $l \propto W^{1/3}$, which is in reasonable agreement with the data: a best fit (the straight line in the figure) actually gives $l \propto W^{0.4}$. Since these data for pore length agree reasonably well with geometric scaling, no *dynamical* conclusion (such as the shell being thin enough for the egg to hatch) can be drawn. Ironically, rather than showing an anomalously slow growth with egg mass, the data for $l$ actually manifest an anomalously fast growth (0.4 versus 0.33), not so dissimilar from the example of the metabolic rate!

What about the behavior of the conductance, for which $K \propto W^{0.9}$? This relationship can also be understood on geometric grounds. Conductance is proportional to the *total* available pore area and inversely proportional to pore length. However, total pore area is made up of two factors: the total number of pores times the area of individual pores. If one assumes that the number of pores *per unit area* remains constant from bird to bird (a reasonable assumption consistent with other data), then we have two factors that scale like area and one that

*The concept of similitude is usually attributed to Newton, who first spelled it out in the* Principia *when dealing with gravitational attraction. On reading the appropriate section it is clear that this was introduced only as a passing remark and does not have the same profound content as the remarks of Galileo.*

[†] *This amusing observation was brought to my attention by Chris Llewellyn Smith.*

[‡] *This relationship with a slope of 3/4 is known as Kleiber's law (M. Kleiber, Hilgardia 6(1932):315), whereas the area law is usually attributed to Rubner (M. Rubner, Zeitschrift für Biologie (Munich) 19(1883):535).*

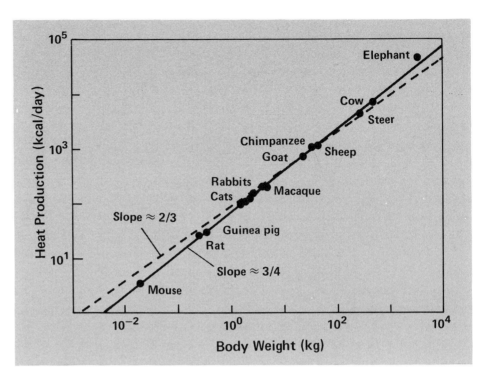

Fig. 2. *Metabolic rate, measured as heat produced by the body in a steady state, plotted logarithmically against body weight. An analysis based on a surface dependence for the rate predicts a scaling curve with slope equal to 2/3 (dashed line) whereas the actual scaling curve has a slope equal to 3/4. Such deviation from simple geometrical scaling is indicative of other effects at work. (Figure based on one by Thomas McMahon,* Science *179(1973):1201-1204 who, in turn, adapted it from M. Kleiber,* Hilgardia *6(1932):315.)*

scales inversely as length. One thus expects $K \propto (W^{2/3})^2/W^{1/3} = W$, again in reasonable agreement with the data.

**Dimensional Analysis.** The physical content of scaling is very often formulated in terms of the language of dimensional analysis. The seminal idea seems to be due to Fourier. He is, of course, most famous for the invention of "Fourier analysis," introduced in his great treatise *Theorie Analytique de la Chaleur*, first published in Paris in 1822. However, it is generally not appreciated that this same book contains another great contribution, namely, the use of dimensions for physical quantities. It is the ghost of Fourier that is the scourge of all freshman physics majors, for it was he who first realized that every physical quantity "has one *dimension proper to itself*, and that the terms of one and the same equation could not be compared, if they had not the same *exponent of dimension*." He goes on: "We have introduced this consideration . . . to verify the analysis . . . it is the equivalent of the fundamental lemmas which the Greeks have left us without proof." Indeed it is! Check the dimensions!—the rallying call of all physicists (and, hopefully, all engineers).

However, it was only much later that physicists began to use the "method of dimensions" to *solve* physical problems. In a famous paper on the subject published in

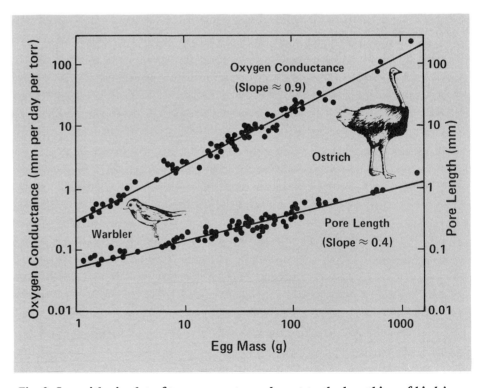

Fig. 3. *Logarithmic plot of two parameters relevant to the breathing of birds' eggs during incubation: the conductance of oxygen through the shell and the pore length (or shell thickness) as a function of egg mass. Both plots have slopes close to those predicted by simple geometrical scaling analyses. (Figure adapted from H. Rahn, A. Ar, and C. V. Paganelli,* Scientific American *240(February 1979):46-55.)*

*Nature* in 1915, Rayleigh indignantly begins: "I have often been impressed by the scanty attention paid even by original workers in the field to the great principle of similitude. It happens not infrequently that results in the form of 'laws' are put forward as novelties on the basis of elaborate experiments, which might have been predicted a priori after a few minutes consideration!" He then proceeds to set things right by giving several examples of the power of dimensional analysis. It seems to have been from about this time that the method became standard fare for the physicist. I shall illustrate it with an amusing example.

Most of us are familiar with the traditional Christmas or Thanksgiving problem of how much time to allow for cooking the turkey or goose. Many (inferior) cookbooks simply say something like "20 minutes per pound," implying a linear relationship with weight. However, there exist superior cookbooks, such as the *Better Homes and Gardens Cookbook*, that recognize the nonlinear nature of this relationship.

Figure 4 is based on a chart from this cookbook showing how cooking time $t$ varies with the weight of the bird $W$. Let us see how

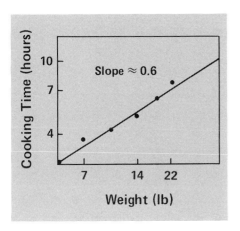

*Fig. 4. The cooking time for a turkey or goose as a logarithmic function of its weight. (Based on a table in* Better Homes and Gardens Cookbook, *Des Moines:Meridith Corp.,* Better Homes and Gardens Books, *1962, p. 272.)*

one can understand this variation using "the great principle of similitude." Let $T$ be the temperature distribution inside the turkey and $T_0$ the oven temperature (both measured relative to the outside air temperature). $T$ satisfies Fourier's heat diffusion equation: $\partial T/\partial t = \kappa \nabla^2 T$, where $\kappa$ is the diffusion coefficient. Now, in general, for the dimensional quantities in this problem, there will be a functional relationship of the form

$$T = f(T_0, W, t, \rho, \kappa) , \qquad (2)$$

where $\rho$ is the bird's density. However, Fourier's basic observation *that the physics be independent of the choice of units*, imposes a constraint on the form of the solution, which can be discerned by writing it in terms of dimension*less* quantities. Only two independent dimensionless quantities can be constructed: $T/T_0$ and $\rho(\kappa t)^{3/2}/W$. If we use the first of these as the dependent variable, the solution, whatever its form, must be expressible in terms of the other. The relationship must therefore have the structure

$$\frac{T}{T_0} = f\left( \frac{\rho(\kappa t)^{3/2}}{W} \right) . \qquad (3)$$

The important point is that, since the left-hand side is dimensionless, the "arbitrary" function $f$ must be a dimensionless function of a dimensionless variable. Equation 3, unlike the previous one, does not depend upon

the choice of units since dimensionless quantities remain invariant to changes in scale.

Let us now consider different but *geometrically similar* birds cooked to the same temperature distribution at the same oven temperature. Clearly, for all such birds there will be a scaling law

$$\frac{\rho(\kappa t)^{3/2}}{W} = \text{constant} . \qquad (4)$$

If the birds have the same physical characteristics (that is, the same $\rho$ and $\kappa$), Eq. 4 reduces to

$$t = \text{constant} \times W^{2/3} , \qquad (5)$$

reflecting, not surprisingly, an area law. As can be seen from Fig. 4, this agrees rather well with the "data."

This formal type of analysis could also, of course, have been carried out for the metabolic rate and birds' eggs problems. The advantage of such an analysis is that it delineates the assumptions made in reaching conclusions like $B \propto W^{2/3}$ since, in principle, it focuses upon all the relevant variables. Naturally this is crucial in the discussion of any physics problem. For complicated systems, such as birds' eggs, with a very large number of variables, some prior insight or intuition must be used to decide what the important variables are. The dimensions of these variables are determined by the fundamental laws that they obey (such as the diffusion equation). Once the dimensions are known, the structure of the relationship between the variables is determined by Fourier's principle. There is therefore no magic in dimensional analysis, only the art of choosing the "right" variables, ignoring the irrelevant, and knowing the physical laws they obey.

As a simple example, consider the classic problem of the drag force $F$ on a ship moving through a viscous fluid of density $\rho$. We shall choose $F$, $\rho$, the velocity $v$, the viscosity of the fluid $\mu$, some length parameter of the ship $l$, and the acceleration due to gravity $g$ as our

variables. Notice that we exclude other variables, such as the wind velocity and the amplitude of the sea waves because, under calm conditions, these are of secondary importance. Our conclusions may therefore not be valid for sailing ships!

The physics of the problem is governed by the Navier-Stokes equation (which incorporates Newton's law of viscous drag, telling us the dimensions of $\mu$) and the gravitational force law (telling us the dimensions of $g$). Using these dimensions automatically incorporates the appropriate physics. Since we have limited the variables to a set of six, which must be expressible in terms of three basic units (mass $M$, length $L$, and time $T$), there will only be three independent dimensionless combinations. These are

chosen to be $P \equiv F/\rho v^2 l^2$ (the pressure coefficient), $R \equiv vl\rho/\mu$ (Reynold's number), and $N_F \equiv v^2/lg$ (Froude's number). *Although any three similar combinations could have been chosen, these three are special because they delineate the physics.* For example, Reynold's number $R$ relates to the viscous drag on a body moving through a fluid, whereas Froude's number $N_F$ relates to the forces involved with waves and eddies generated on the surface of the fluid by the movement. Thus the rationale for the combinations $R$ and $N_F$ is to separate the role of the viscous forces from that of the gravitational: $R$ does not depend on $g$, and $F$ does not depend on $\mu$. Furthermore, $P$ does not depend on either!

Dimensional analysis now requires that the solution for the pressure coefficient,

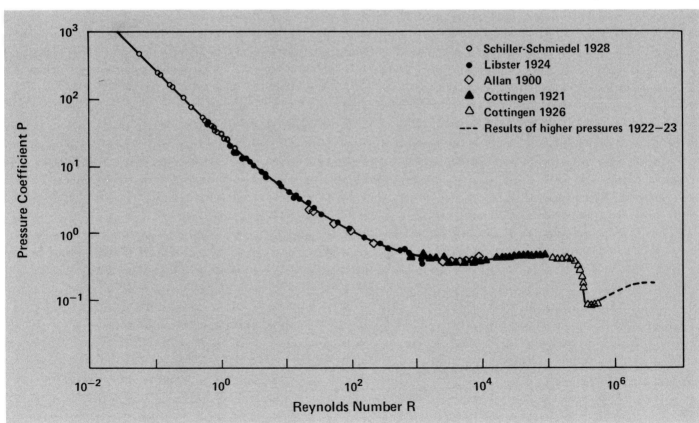

*Fig. 5. The scaling curve for the motion of a sphere through a fluid that results when data from a variety of experiments are plotted in terms of two dimensionless variables: the* *pressure or drag coefficent* **P** *versus Reynolds number* **R.** *(Figure adapted from* **AIP Handbook of Physics,** *2nd edition (1963):section II, p. 253.)*

whatever it is, must be expressible in the dimensionless form

$$P = f(R, N_F) . \tag{6}$$

The actual drag force $F$ can easily be obtained from this equation by re-expressing it in terms of the dimensional variables (see Eq. 8 below).

First, however, consider a situation where surface waves generated by the moving object are *un*important (an extreme example is a submarine). In this case $g$ will not enter the solution since it is manifested as the restoring force for surface waves. $N_F$ can then be dropped from the solution, reducing Eq. 6 to the simple form

$$P = f(R) . \tag{7}$$

In terms of the original dimensional variables, this is equivalent to

$$F = \rho v^2 l^2 f(v l \rho / \mu) . \tag{8}$$

Historically, these last equations have been well tested by measuring the speed of different sizes and types of balls moving through different liquids. If the data are plotted using the dimensionless variables, that is, $P$ versus $R$, then *all* the data should lie on just *one* curve regardless of the size of the ball or the nature of the liquid. Such a curve is called a *scaling curve*, a wonderful example of which is shown in Fig. 5 where one sees a scaling phenomenon that varies over seven orders of magnitude! It is important to recognize that if one had used dimensional variables and plotted $F$ versus $l$, for example, then, instead of a single curve, there would have been *many* different and apparently unrelated curves for the different liquids. Using carefully chosen dimensionless variables (such as Reynold's number) is not only physically more sound but usually greatly simplifies the task of representing the data.

A remarkable consequence of this analysis is that, for similar bodies, the ratio of drag

force to weight *decreases* as the size of the structure increases. From Archimedes' principle the volume of water displaced by a ship is proportional to its weight, that is, $W \propto l^3$ (this, incidentally, is why there is no need to include $W$ as an independent variable in deriving these equations). Combined with Eq. 8 this leads to the conclusion that

$$\frac{F}{W} \propto \frac{1}{l} . \tag{9}$$

This scaling law was extremely important in the 19th century because it showed that *it was cost effective to build bigger ships*, thereby justifying the use of large iron steamboats!

The great usefulness of scaling laws is also illustrated by the observation that the behavior of $P$ for large ships ($l \to \infty$) can be derived from the behavior of small ships moving very fast ($v \to \infty$). This is so because both limits are controlled by the same asymptotic behavior of $f(R) = f(v l \rho / \mu)$. Such observations form the basis of *modeling* theory so crucial in the design of aircraft, ships, buildings, and so forth.

Thomas McMahon, in an article in *Science*, has pointed out another, somewhat more amusing, consequence to the drag force equation. He was interested in how the speed of a rowing boat scales with the number of oarsmen $n$ and argued that, at a steady velocity, the power expended by the oarsmen $E$ to overcome the drag force is given by $Fv$. Thus

$$E = Fv = \rho v^3 l^2 f(R) . \tag{10}$$

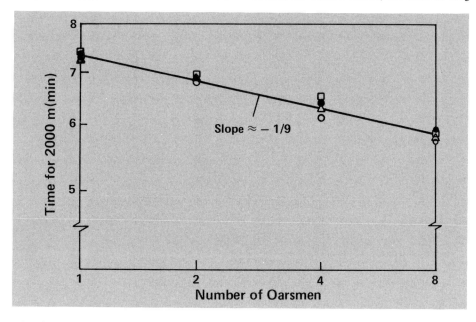

**Fig. 6. The time needed for a rowing boat to complete a 2000-meter course in calm conditions as a function of the number of oarsmen. Data were taken from several international rowing championship events and illustrate the surprisingly slow dropoff predicted by modeling theory. (Adapted from T. A. McMahon, Science 173(1971):349-351.)**

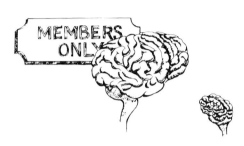

Using Archimedes' principle again and the fact that both $E$ and $W$ should be directly proportional to $n$ leads to the remarkable scaling law

$$v \propto n^{1/9} , \qquad (11)$$

which shows a *very* slow growth with $n$. Figure 6 exhibits data collected by McMahon from various rowing events for the time $t$ ($\propto 1/v$) taken to cover a fixed 2000-meter course under calm conditions. One can see quite plainly the verification of his predicted law—a most satisfying result!

There are many other fascinating and exotic examples of the power of dimensional analysis. However, rather than belaboring the point, I would like to mention a slightly different application of scaling before I turn to the mathematical formulation. All the examples considered so far are of a quantitative nature based on well-known laws of physics. There are, however, situations where the qualitative observation of scaling can be used to scientific advantage to reveal phenomenological "laws."

A nice example (Fig. 7), taken from an article by David Pilbeam and Stephen Jay Gould, shows how the endocranial volume $V$ (loosely speaking, the brain size) scales with body weight $W$ for various hominids and pongids. The behavior for modern pongids is typical of most species in that the exponent $a$, defined by the phenomenological relationship $V \propto W^a$, is approximately 1/3 (for mammals $a$ varies from 0.2 to 0.4). It is very satisfying that a similar behavior is exhibited by australopithecines, extinct cousins of our lineage that died out over a million years ago. However, as Pilbean and Gould point out,

our homo sapiens lineage shows a strikingly different behavior, namely: $a \approx 5/3$. Notice that neither this relationship nor the "standard" behavior ($a \approx 1/3$) is close to the naive geometrical scaling prediction of $a = 1$.

These data illustrate dramatically the qualitative evolutionary advance in the brain development of man. Even though the reasons for $a \approx 1/3$ may not be understood, this value can serve as the "standard" for revealing deviations and provoking speculation concerning evolutionary progress: for example, what is the deep significance of a brain size that grows linearly with height versus a brain size that grows like its fifth power? I shall not enter into such questions here, tempting though they be.

Such phenomenological scaling laws (whether for brain volume, tooth area, or some other measurable parameter of the fossil) can also be used as corroborative evidence for assigning a newly found fossil of some large primate to a particular lineage. The fossil's location on such curves can, in principle, be used to distinguish an australopithecine from a homo. Notice, however, that implicit in all this discussion is knowledge of body weight; presumably, anthropologists have developed verifiable techniques for estimating this quantity. Since they necessarily work with fragments only, some further scaling assumptions *must* be involved in their estimates!

**Relevant Variables.** As already emphasized, the most important and artful aspect of the method of dimensions is the choice of variables relevant to the problem and their grouping into dimensionless combinations that delineate the physics. In spite of the

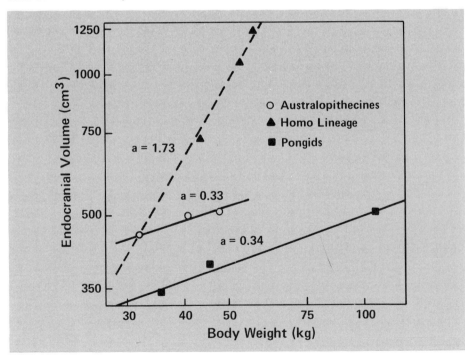

*Fig. 7. Scaling curves for endocranial volume (or brain size) as a function of body weight. The slope of the curve for our homo sapiens lineage (dashed line) is markedly different from those for australopithecines, extinct cousins of the homo lineage, and for modern pongids, which include the chimpanzee, gorilla and orangutan. (Adapted from D. Pilbeam and S. J. Gould,* Science *186(1974):892-901.)*

relative simplicity of the method there are inevitably paradoxes and pitfalls, a famous case of which occurs in Rayleigh's 1915 paper mentioned earlier. His last example concerns the rate of heat lost $H$ by a conductor immersed in a stream of inviscid fluid moving past it with velocity $v$ ("Boussinesq's problem"). Rayleigh showed that, if $K$ is the heat conductivity, $C$ the specific heat of the fluid, $\theta$ the temperature difference, and $l$ some linear dimension of the conductor, then, in dimensionless form,

$$\frac{H}{kl\theta} = f\left( \frac{lvC}{K} \right). \tag{12}$$

Approximately four months after Rayleigh's paper appeared, *Nature* published an eight line comment (half column, yet!) by a D. Riabouchinsky pointing out that Rayleigh's result assumed that temperature was a dimension independent from mass, length, and time. However, from the kinetic theory of gases we know that this is not so: temperature can be defined as the mean kinetic energy of the molecules and so is *not* an independent unit! Thus, according to Riabouchinsky, Rayleigh's expression must be replaced by an expression with an additional dimensionless variable:

$$\frac{H}{kl\theta} = f\left( \frac{lvC}{K}, Cl^{\beta} \right), \tag{13}$$

a much *less* restrictive result.

Two weeks later, Rayleigh responded to Riabouchinsky saying that "it would indeed be a paradox if the *further* knowledge of the nature of heat afforded by molecular theory put us in a *worse* position than before in dealing with a particular problem. . . . It would be well worthy of discussion." Indeed it would; its resolution, which no doubt the reader has already discerned, is left as an exercise (for the time being)! Like all paradoxes, this one cautions us that we occasionally make casual assumptions without quite realizing that we have done so (see "Fundamental Constants and the Rayleigh-Riabouchinsky Paradox").

## Scale Invariance

Let us now turn our attention to a slightly more abstract mathematical formulation that clarifies the relationship of dimensional analysis to *scale invariance*. By scale invariance we simply mean that the structure of physical laws cannot depend on the choice of units. As already intimated, this is automatically accomplished simply by employing dimensionless variables since these clearly do not change when the system of units changes. However, it may not be immediately obvious that this is equivalent to the *form invariance* of physical equations. Since physical laws are usually expressed in terms of dimensional variables, this is an important point to consider: namely, what are the general constraints that follow from the requirement that the laws of physics look the same regardless of the chosen units. The crucial observation here is that implicit in any equation written in terms of dimensional variables are the "hidden" fundamental scales of mass $M$, length $L$, time $T$, and so forth that are relevant to the problem. Of course, one never actually makes these scale parameters explicit precisely because of form invariance.

Our motivation for investigating this question is to develop a language that can be generalized in a natural way to include the subtleties of quantum field theory. Hopefully classical dimensional analysis and scaling will be sufficiently familiar that its generalization to the more complicated case will be relatively smooth! This generalization has been named the *renormalization group* since its origins lie in the renormalization program used to make sense out of the infinities inherent in quantum field theory. It turns out that renormalization requires the introduction of a new arbitrary "hidden" scale that plays a role similar to the role of the scale parameters implicit in any dimensional equation. Thus any equation derived in quantum field theory that represents a physical quantity must not depend upon this choice of hidden scale. The resulting con-

straint will simply represent a generalization of ordinary dimensional analysis; the only reason that it is different is that variables in quantum field theory, such as fields, change in a much more complicated fashion with scale than do their classical counterparts.

Nevertheless, just as dimensional analysis allows one to learn much about the behavior of a system without actually solving the dynamical equations, so the analogous constraints of the renormalization group lead to powerful conclusions about the behavior of a quantum field theory without actually being able to solve it. It is for this reason that the renormalization group has played such an important part in the renaissance of quantum field theory during the past decade or so. Before describing how this comes about, I shall discuss the simpler and more familiar case of scale change in ordinary classical systems.

To begin, consider some physical quantity $F$ that has dimensions; it will, of course, be a function of various dimensional variables $x_i$: $F(x_1, x_2, \ldots, x_n)$. An explicit example is given by Eq. 2 describing the temperature distribution in a cooked turkey or goose.

# Fundamental Constants and the

Let us examine Riabouchinsky's paradox a little more carefully and show how its resolution is related to choosing a system of units where the "fundamental constants" (such as Planck's constant $h$ and the speed of light $c$) can be set equal to unity.

The paradox had to do with whether temperature could be used as an independent dimensional unit even though it can be defined as the mean kinetic energy of the molecular motion. Rayleigh had chosen five physical variables (length $l$, temperature difference $\theta$, velocity $v$, specific heat $C$, and heat conductivity $K$) to describe Boussinesq's problem and had assumed that there were four independent dimensions (energy $E$, length $L$, time $T$, and temperature $\Theta$). Thus the solution for $T/T_0$ necessarily is an arbitrary function of *one* dimensionless combination. To see this explicitly, let us examine the dimensions of the five physical variables:

$$[l] = L, [\theta] = \Theta, [v] = LT^{-1}, [C] = EL^{-3}\Theta^{-1},$$

$$\text{and } [K] = EL^{-1}T^{-1}\Theta^{-1}.$$

Clearly the combination chosen by Rayleigh, $lvC/K$, is dimensionless. Although other dimensionless combinations can be formed, they are not independent of the two combinations ($lvC/K$ and $T/T_o$) selected by Rayleigh.

Now suppose, along with Riabouchinsky, we use our knowledge of the kinetic theory to define temperature "as the mean kinetic energy of the molecules" so that $\Theta$ is no longer an independent dimension. This means there are now only *three* independent dimensions and the solution will depend on an arbitrary function of *two* dimensionless combinations. With $\Theta \propto E$, the dimensions of the physical variables become:

$$[l] = L, [\theta] = E, [v] = LT^{-1}, [C] = L^{-3}, \text{ and } [K] = L^{-1}T^{-1}.$$

It is clear that, in addition to Rayleigh's dimensionless variable, there is now a new *independent* combination, $Cl^3$ for example, that is dimensionless. To reiterate Rayleigh: "it would indeed be a paradox if the *further* knowledge of the nature of heat . . . put us in a *worse* position than before . . . it would be well worthy of discussion."

Like almost all paradoxes, there is a bogus aspect to the argument. It is certainly true that the kinetic theory allows one to express an energy as a temperature. However, this is only useful and *appropriate* for situations where the physics is dominated by molecular considerations. For macroscopic situations such as Boussinesq's problem, the molecular nature of the system is irrelevant; the microscopic variables have been replaced by macroscopic averages embodied in phenomenological properties such as the specific heat and conductivity. To make Riabouchinsky's identification of energy with temperature is to introduce *irrelevant* physics into the problem.

Exploring this further, we recall that such an energy-temperature identification implicitly involves the introduction of Boltzmann's factor $k$. By its very nature, $k$ will only play an explicit role in a physical problem that directly involves the molecular nature of the system; otherwise it will not enter. Thus one could describe the system from the molecular viewpoint (so that $k$ is involved) and then take a macroscopic limit. Taking the limit is equivalent to setting $k = 0$; the presence of a finite $k$ indicates that explicit effects due to the kinetic theory are important.

With this in mind, we can return to Boussinesq's problem and derive Riabouchinsky's result in a somewhat more illuminating fashion. Let us follow Rayleigh and keep $E$, $L$, $T$, and $\Theta$ as the

Each of these variables, including $F$ itself, is always expressible in terms of some standard set of independent units, which can be chosen to be mass $M$, length $L$, and time $T$. These are the hidden scale parameters. Obviously, other combinations could be used. There could even be other independent units, such as temperature (but remember Riabouchinsky!), or more than one independent length (say, transverse and longitudinal). In this discussion, we shall simply use the conventional $M$, $L$, and $T$. Any generalization is straightforward.

In terms of this standard set of units, the magnitude of each $x_i$ is given by

$$x_i = M^{\alpha_i} L^{\beta_i} T^{\gamma_i} \tag{15}$$

The numbers $\alpha_i$, $\beta_i$, and $\gamma_i$ will be recognized

as "the dimensions" of $x_i$. Now suppose we change the system of units by some scale transformation of the form

$$M \to M' = \lambda_M M,$$

$$L \to L' = \lambda_L L,$$

and

$$T \to T' = \lambda_T T. \tag{16}$$

Each variable then responds as follows:

$$x_i \to x_i' = Z_i(\lambda)x_i, \tag{17}$$

where

$$Z_i(\lambda) = \lambda_M^{\alpha_i} \lambda_L^{\beta_i} \lambda_T^{\gamma_i}, \tag{18}$$

and $\lambda$ is shorthand for $\lambda_M$, $\lambda_L$, and $\lambda_T$. Since $F$ is itself a dimensional physical quantity, it transforms in an identical fashion under this scale change:

$$F \to F' = Z(\lambda) F(x_1, x_2, \ldots, x_n), \tag{19}$$

where

$$Z(\lambda) = \lambda_M^{\alpha} \lambda_L^{\beta} \lambda_T^{\gamma}. \tag{20}$$

Here $\alpha$, $\beta$, and $\gamma$ are the dimensions of $F$.

There is, however, an alternate but equivalent way to transform from $F$ to $F'$, namely, by transforming each of the variables $x_i$ separately. Explicitly we therefore also have

# Rayleigh-Riabouchinsky Paradox

independent dimensions but add $k$ (with dimensions $E\Theta^{-1}$) as a new physical variable. The solution will now be an arbitrary function of *two* independent dimensionless variables: $lvC/K$ and $kCl^3$. When Riabouchinsky chose to make $Cl^3$ his other dimensionless variable, he, in effect, chose a system of units where $k = 1$. But that was a terrible thing to do here since the physics dictates that $k = 0$! Indeed, if $k = 0$ we regain Rayleigh's original result, that is, we have only *one* dimensionless variable. It is somewhat ironic that Rayleigh's remarks miss the point: "further knowledge of the nature of heat afforded by molecular theory" does not put one in a better position for solving the problem—rather, it leads to a microscopic description of $K$ and $C$. The important point pertinent to the problem set up by Rayleigh is that knowledge of the molecular theory is irrelevant and $k$ must not enter.

The lesson here is an important one because it illustrates the role played by the fundamental constants. Consider Planck's constant $\hbar \equiv h/2\pi$: it would be completely inappropriate to introduce it into a problem of classical dynamics. For example, any solution of the scattering of two billiard balls will depend on macroscopic variables such as the masses, velocities, friction coefficients, and so on. Since billiard balls are made of protons, it might be tempting to the purist to include as a dependent variable the proton-proton total cross section, which, of course, involves $\hbar$. This would clearly be totally inappropriate but is analogous to what Riabouchinsky did in Boussinesq's problem.

Obviously, if the scattering is between two microscopic "atomic billiard balls" then $\hbar$ *must be* included. In this case it is not only quite legitimate but often convenient to choose a system of units where $\hbar = 1$. However, having done so one cannot directly recover the classical limit corresponding to $\hbar = 0$. With $\hbar = 1$, one is stuck in quantum mechanics just as, with $k = 1$, one is stuck in kinetic theory.

A similar situation obviously occurs in relativity: the velocity of light $c$ must not occur in the classical Newtonian limit. However, in a relativistic situation one is quite at liberty to choose units where $c = 1$. Making that choice, though, presumes the physics involves relativity.

The core of particle physics, relativistic quantum field theory, is a synthesis of quantum mechanics and relativity. For this reason, particle physicists find that a system of units in which $\hbar = c = 1$ is not only convenient but is a manifesto that quantum mechanics and relativity are the basic physical laws governing their area of physics. In quantum mechanics, momentum $p$ and wavelength $\lambda$ are related by the de Broglie relation: $p = 2\pi\hbar/\lambda$; similarly, energy $E$ and frequency $\omega$ are related by Planck's formula: $E = \hbar\omega$. In relativity we have the famous Einstein relation: $E = mc^2$. Obviously if we choose $\hbar = c = 1$, all energies, masses, and momenta have the same units (for example, electron volts ($eV$)), and these are the same as *inverse* lengths and times. Thus larger energies and momenta *inevitably* correspond to shorter times and lengths.

Using this choice of units automatically incorporates the profound physics of the uncertainty principle: to probe short space-time intervals one needs large energies. A useful number to remember is that $10^{-13}$ centimeter, or 1 fermi (fm), equals the reciprocal of 200 MeV. We then find that the electron mass ($\approx 1/2$ MeV) corresponds to a length of $\approx 400$ fm—its Compton wavelength. Or the 20 TeV ($2 \times 10^7$ MeV) typically proposed for a possible future facility corresponds to a length of $10^{-18}$ centimeter. This is the scale distance that such a machine will probe! ∎

$$F \to F' = F(Z_1(\lambda)x_1, Z_2(\lambda)x_2, \ldots, Z_n(\lambda)x_n) . \tag{21}$$

Equating these two different ways of effecting a scale change leads to the identity

$$F(Z_1(\lambda)x_1, Z_2(\lambda)x_2, \ldots, Z_n(\lambda)x_n) = Z(\lambda) F(x_1, x_2, \ldots, x_n) . \tag{22}$$

As a concrete example, consider the equation $E = mc^2$. To change scale one can either transform $E$ directly or transform $m$ and $c$ separately and multiply the results appropriately—obviously the final result must be the same.

We now want to ensure that the resulting form of the equation does not depend on $\lambda$. This is best accomplished using Euler's trick

of taking $\partial/\partial\lambda$ and then setting $\lambda = 1$. For example, if we were to consider changes in the mass scale, we would use $\partial/\partial\lambda_M$ and the chain rule for partial differentiation to arrive at

$$\sum_{i=1}^{n} x_i \frac{\partial Z_i}{\partial \lambda_M} \frac{\partial F}{\partial x_i'} = \frac{\partial Z}{\partial \lambda_M} F . \tag{23}$$

When we set $\lambda_M = 1$, differentiation of Eqs. 18 and 20 yields

$$\left( \frac{\partial Z_i}{\partial \lambda_M} \right)_{\lambda_M = 1} = \alpha_i ,$$

$$\left( \frac{\partial Z}{\partial \lambda_M} \right)_{\lambda_M = 1} = \alpha , \tag{24}$$

and $x_i' = x_i$, so that Eq. 23 reduces to

$$\alpha_1 x_1 \frac{\partial F}{\partial x_1} + \alpha_2 x_2 \frac{\partial F}{\partial x_2} + \alpha_3 x_3 \frac{\partial F}{\partial x_3} + \ldots$$

$$+ \alpha_n x_n \frac{\partial F}{\partial x_n} = \alpha F . \tag{25}$$

Obviously this can be repeated with $\lambda_L$ and $\lambda_T$ to obtain a set of three coupled partial differential equations expressing the fundamental *scale invariance of physical laws (that is, the invariance of the physics to the choice of units)* implicit in Fourier's original work. These equations can be solved without too much difficulty; their solution is, in fact, a special case of the solution to the renormalization group equation (given explicitly as Eq. 35 below). Not too surprisingly, one finds that the solution is precisely equivalent to the constraints of dimensional analysis. Thus there is never any explicit need to use these rather cumbersome equations: ordinary dimensional analysis takes care of it for you!

## Quantum Field Theory

We have gone through this little mathematical exercise to illustrate the well-known relationship of dimensional analysis to scale and form invariance. I now want to discuss how the formalism must be amended when applied to quantum field theory and give a sense of the profound consequences that follow. Using the above chain of reasoning as a guide, I shall examine the response of a quantum field theoretic system to a change in scale and derive a partial differential equation analogous to Eq. 25. This equation is known as the *renormalization group equation* since its origins lay in the somewhat arcane area of the renormalization procedure used to tame the infinities of quantum field theory. I shall therefore have to digress momentarily to give a brief résumé of this subject before returning to the question of scale change.

**Renormalization.** Perhaps the most unnerving characteristic of quantum field theory for the beginning student (and possibly also for the wise old men) is that almost all calculations of its physical consequences naively lead to infinite answers. These infinities stem from divergences at high momenta associated with virtual processes that are always present in any transition amplitude. The renormalization scheme, developed by Richard P. Feynman, Julian S. Schwinger, Sin-Itiro Tomonaga, and Freeman Dyson, was invented to make sense out of this for quantum electrodynamics (QED).

To get a feel for how this works I shall focus on the photon, which carries the force associated with the electromagnetic field. At the classical limit the *propagator** for the

photon represents the usual static $1/r$ Coulomb potential. The corresponding Fourier transform (that is, the propagator's representation in momentum space) in this limit is $1/q^2$, where $q$ is the momentum carried by the photon. Now consider the "classical" scattering of two charged particles (represented by the Feynman diagram in Fig. 8 (a)). For this event the exchange of a single photon gives a transition amplitude proportional to $e_0^2/q^2$, where $e_0$ is the charge (or coupling constant) occurring in the Lagrangian. A standard calculation results in the classical Rutherford formula, which can be extended relativistically to the spin-1/2 case embodied in the diagram.

A typical quantum-mechanical correction to the scattering formula is illustrated in Fig. 8 (b). The exchanged photon can, by virtue of the uncertainty principle, create for a very short time a virtual electron-positron pair, which is represented in the diagram by the loop. We shall use $k$ to denote the momentum carried around the loop by the two particles.

There are, of course, many such corrections that serve to modify the $1/q^2$ single-

photon behavior, and this is represented schematically in part (c). It is convenient to include all these corrections in a single multiplicative factor $D_0$ that represents deviations from the single-photon term. The "full" photon propagator including all possible radiative corrections is therefore $D_0/q^2$. The reason for doing this is that $D_0$ is a *dimensionless* function that gives a measure of the polarization of the vacuum caused by the production of virtual particles. (The origin of the Lamb shift is vacuum polarization.)

We now come to the central problem: upon evaluation it is found that contributions from diagrams like (b) are infinite because there is no restriction on the magnitude of the momentum $k$ flowing *in* the loop! Thus, typical calculations lead to integrals of the form

$$\int_0^\infty \frac{dk^2}{k^2 + aq^2}, \qquad (26)$$

which diverge logarithmically. Several prescriptions have been invented for making such integrals finite; they all involve "reg-

---

*Roughly speaking, the photon propagator can be thought of as the Green's function for the electromagnetic field. In the relativistically covariant Lorentz gauge, the classical Maxwell's equations read*

$$\Box^2 \, \mathbf{A(x)} = \mathbf{j(x)} \,,$$

*where $\mathbf{A(x)}$ is the vector potential and $\mathbf{j(x)}$ is the current source term derived in QED from the motion of the electrons. (To keep things simple I am suppressing all space-time indices, thereby ignoring spin.) This equation can be solved in the standard way using a Green's function:*

$$\mathbf{A(x)} = \int \mathbf{d^4 x'} \, G(x' - x) \, \mathbf{j(x')} \,,$$

*with*

$$\Box^2 \, G(x) = \delta(x) \,.$$

*Now a transition amplitude is proportional to the interaction energy, and this is given by*

$$H_I = \int \mathbf{d^4} \, x \, \mathbf{j}(x) \, \mathbf{A}(x) =$$

$$\int \mathbf{d^4} \, x \int \mathbf{d^4} x' \, \mathbf{j}(x) \, G(x-x') \, \mathbf{j}(x') \,,$$

*illustrating how G "mediates" the force between two currents separated by a space-time interval (x-x'). It is usually more convenient to work with Fourier transforms of these quantities (that is, in momentum space). For example, the momentum space solution for G is $\tilde{G}(\mathbf{q}) = 1/\mathbf{q}^2$, and this is usually called the free photon propagator since it is essentially classical. The corresponding "classical" transition amplitude in momentum space is just $\tilde{\mathbf{j}}(\mathbf{q})(1/\mathbf{q}^2)\tilde{\mathbf{j}}(\mathbf{q})$, which is represented by the Feynman graph in Fig. 8 (a).*

*In quantum field theory, life gets much more complicated because of radiative corrections as discussed in the text and illustrated in (b) and (c) of Fig. 8. The definition of the propagator is generally in terms of a correlation function in which a photon is created at point x out of the vacuum for a period x-x' and then returns to the vacuum at point x'. Symbolically, this is represented by*

$$G(x-x') \sim \langle vac|\mathbf{A(x')} \, \mathbf{A(x)}|vac\rangle \,.$$

*During propagation, anything allowed by the uncertainty principle can happen—these are the radiative corrections that make an exact calculation of G almost impossible.*

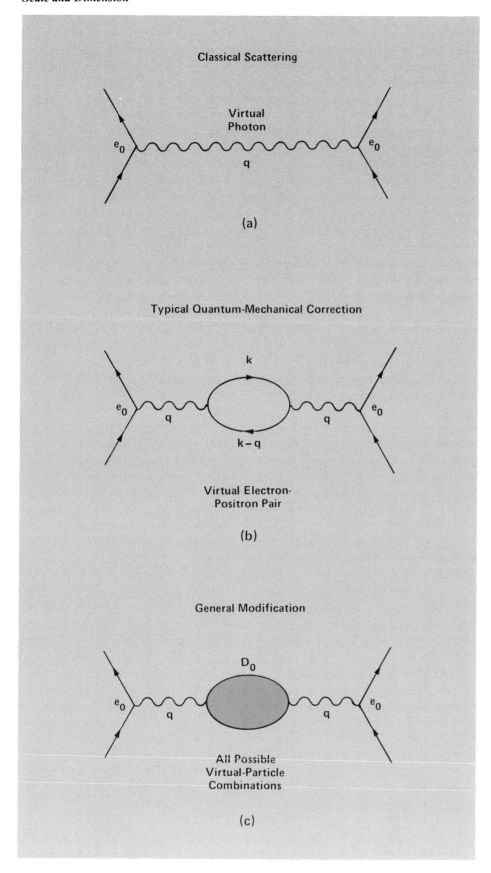

**Classical Scattering**

**Virtual Photon**

$e_0$  $q$  $e_0$

(a)

**Typical Quantum-Mechanical Correction**

$k$

$e_0$  $q$  $q$  $e_0$

$k-q$

**Virtual Electron-Positron Pair**

(b)

**General Modification**

$D_0$

$e_0$  $q$  $q$  $e_0$

**All Possible Virtual-Particle Combinations**

(c)

**Fig. 8. Feynman diagrams for (a) the classical scattering of two particles of charge $e_0$, (b) a typical correction that must be made to that scattering—here because of the creation of a virtual electron-positron pair—and (c) a diagram representing all such possible corrections. The matrix element is proportional for (a) to $e_0^2/q^2$ and for (c) to $D_0/q^2$ where $D_0$ includes all corrections.**

ularizing" the integrals by introducing some large mass parameter $\Lambda$. A standard technique is the so-called Pauli-Villars scheme in which a factor $\Lambda^2/(k^2 + \Lambda^2)$ is introduced into the integrand with the understanding that $\Lambda$ is to be taken to infinity at the end of the calculation (notice that in this limit the regulating factor approaches one). With this prescription, the above integral is therefore replaced by

$$\lim_{\Lambda^2 \to \infty} \int_0^\infty \frac{dk^2 \Lambda^2}{(k^2 + aq^2)(k^2 + \Lambda^2)}$$
$$= \ln \frac{\Lambda^2}{aq^2}. \tag{27}$$

The integral can now be evaluated and its divergence expressed in terms of the (infinite) mass parameter $\Lambda$. All the infinities arising from quantum fluctuations can be dealt with in a similar fashion with the result that the following series is generated:

$$D_0(q,e_0) \approx 1 + a_1 e_0^2 \left( \ln \frac{\Lambda^2}{q^2} + \ldots \right) +$$
$$e_0^4 \left[ a_2 \left( \ln \frac{\Lambda^2}{q^2} \right)^2 + b_2 \ln \frac{\Lambda^2}{q^2} + \ldots \right] + \ldots. \tag{28}$$

In this way the structure of the infinite divergences in the theory are parameterized in terms of $\Lambda$, which can serve as a finite *cutoff* in the integrals over virtual momenta.*

The remarkable triumph of the renormalization program is that, rather than imposing such an arbitrary cutoff, all these divergences can be swallowed up by an *infinite rescaling* of the fields and coupling con-

*In this discussion I assumed, for simplicity, that the original Lagrangian was massless; that is, it contained no explicit mass parameter. The addition of such a mass term would only complicate the discussion unnecessarily without giving any new insights.*

stants. Thus, a *finite* propagator $D$, that does not depend on $\Lambda$, can be derived from $D_0$ by rescaling if, at the same time, one rescales the charge similarly. These rescalings take the form

$$D = Z_D D_0 \text{ and } e = Z_e e_0 . \qquad (29)$$

The crucial property of these scaling factors is that they are independent of the physical momenta (such as $q$) but depend on $\Lambda$ in such a way that when the cutoff is removed, $D$ and $e$ remain finite. In other words, when $\Lambda \to \infty$, $Z_D$ and $Z_e$ must develop infinities of their own that precisely compensate for the infinities of $D_0$ and $e_0$. The original so-called *bare* parameters in the theory calculated from the Lagrangian ($D_0$ and $e_0$) therefore have no physical meaning—only the renormalized parameters ($D$ and $e$) do.

Now let us apply some ordinary dimensional analysis to these remarks. Because they are simply scale factors, the $Z$'s must be dimensionless. However, the $Z$'s are functions of $\Lambda$ but not of $q$. But that is very peculiar: a dimensionless function cannot depend on a *single* mass parameter! Thus, in order to express the $Z$'s in dimensionless form, *a new finite mass scale $\mu$ must be introduced* so that one can write $Z = Z(\Lambda^2/\mu^2, e_0)$. *An immediate consequence of renormalization is therefore to induce a mass scale not manifest in the Lagrangian.* This is extremely interesting because it provides a possible mechanism for generating mass even though no mass parameter appears in the Lagrangian. We therefore have the exciting possibility of being able to calculate the masses of *all* the elementary particles in terms of just *one* of them. Similar considerations for the dimensionless $D$'s clearly require that they be expressible as $D_0 = D_0(q^2/\Lambda^2, e_0)$, as in Eq. 28, and $D = D(q^2/\mu^2, e)$. (The dream of particle theorists is to write down a Lagrangian with *no* mass parameter that describes all the interations in terms of just *one* coupling constant. The mass spectrum and scattering amplitudes for all the elementary particles

would then be calculable in terms of the value of this single coupling at some given scale! A wonderful fantasy.)

To recapitulate, the physical finite renormalized propagator $D$ is related to its bare and divergent counterpart $D_0$ (calculated from the Lagrangian using a cutoff mass) by an infinite rescaling:

$$D\left(\frac{q^2}{\mu^2}, e\right) = \lim_{\Lambda \to \infty} Z_D\left(\frac{\mu^2}{\Lambda^2}, e_0\right) D_0\left(\frac{q^2}{\Lambda^2}, e_0\right) . \qquad (30)$$

Similarly, the physical finite charge $e$ is given by an infinite rescaling of the bare charge $e_0$ that occurs in the Lagrangian

$$e = \lim_{\Lambda \to \infty} Z_e\left(\frac{\mu^2}{\Lambda^2}, e_0\right) e_0 . \qquad (31)$$

Notice that the physical coupling $e$ now depends implicitly on the renormalization scale parameter $\mu$. Thus, in QED, for example, it is not strictly sufficient to state that the fine structure constant $\alpha \approx 1/137$; rather, one must also specify the corresponding scale. From this point of view there is nothing magic about the particular number 137 since a change of scale would produce a different value.

At this stage, some words of consolation to a possibly bewildered reader are in order. It is not intended to be obvious how such infinite rescalings of infinite complex objects lead to consistent finite results! An obvious question is what happens with more complicated processes such as scattering amplitudes and particle production? These are surely even more divergent than the relatively simple photon propagator. How does one know that a similar rescaling procedure can be carried through in the general case?

The proof that such a procedure does indeed work consistently for *any* transition amplitude in the theory was a real tour de force. A crucial aspect of this proof was the remarkable discovery that in QED only a *finite* number (three) of such rescalings was

necessary to render the theory finite. This is terribly important because it means that once we have renormalized a few basic entities, such as $e_0$, all further rescalings of more complicated quantities are completely determined. Thus, the theory retains predictive power—in marked contrast to the highly unsuitable scenario in which each transition amplitude would require its own infinite rescaling to render it finite. Such theories, termed nonrenormalizable, would apparently have no predictive power. High energy physicists have, by and large, restricted their attention to renormalizable theories just because all their consequences can, in principle, be calculated and predicted in terms of just a few parameters (such as the physical charge and some masses).

I should emphasize the phrase "in principle" since in practice there are very few techniques available for actually carrying out honest calculations. The most prominent of these is perturbation theory in the guise of Feynman graphs. Most recently a great deal of effort, spurred by the work of K. G. Wilson, has gone into trying to adapt quantum field theory to the computer using lattice gauge theories.* In spite of this it remains sadly true that perturbation theory is our only "global" calculational technique. Certainly its success in QED has been nothing less than phenomenal.

Actually only a very small class of renormalizable theories exist and these are characterized by dimensionless coupling constants. Within this class are gauge theories like QED and its non-Abelian extension in which the photon interacts with itself. All modern particle physics is based upon such theories. One of the main reasons for their popularity, besides the fact they are renormalizable, is that they possess the property of being *asymptotically free*. In such theories one finds that the renormalization group constraint, to be discussed shortly, requires that the large momentum behavior

---

*In recent years there has been some effort to come to grips analytically with the nonperturbative aspects of gauge theories.*

be equivalent to the small coupling limit; thus for large momenta the renormalized coupling effectively vanishes thereby allowing the use of perturbation theory to calculate physical processes.

This idea was of paramount importance in substantiating the existence of quarks from deep inelastic electron scattering experiments. In these experiments quarks behaved as if they were quasi-free even though they must be bound with very strong forces (since they are never observed as free particles). Asymptotic freedom gives a perfect explanation for this: the effective coupling, though strong at low energies, gets vanishingly small as $q^2$ becomes large (or equivalently, as distance becomes small).

In seeing how this comes about we will be led back to the question of *how the field theory responds to scale change*. We shall follow the exact same procedure used in the classical case: first we scale the hidden parameter ($\mu$, in this case) and see how a typical transition amplitude, such as a propagator, responds. A partial differential equation, analogous to Eq. 25, is then derived using

Euler's trick. This is solved to yield the general constraints due to renormalization analogous to the constraints of dimensional analysis. I will then show how these constraints can be exploited, using asymptotic freedom as an example.

**The Renormalization Group Equation.** As already mentioned, renormalization makes the bare parameters occurring in the Lagrangian effectively irrelevant; the theory has been transformed into one that is now specified by the value of its physical coupling constants at some mass scale $\mu$. In this sense $\mu$ plays the role of the hidden scale parameter $M$ in ordinary dimensional analysis by setting the scale of units by which all quantities are measured.

This analogy can be made almost exact by considering a scale change for the arbitrary parameter $\mu$ in which $\mu \rightarrow \lambda^{1/2}\mu$. This change allows us to rewrite Eq. 30 in a form that expresses the response of $D$ to a scale change:

$$D\left(\frac{q^2}{\lambda\mu^2},g(\lambda\mu^2)\right) = Z(\lambda)\, D\left(\frac{q^2}{\mu^2},g(\mu^2)\right). \tag{32}$$

(From now on I will use $g$ to denote the coupling rather than $e$ because $e$ is usually reserved for the electric charge in QED.)

The scale factor $Z(\lambda)$, which is independent of $q^2$ and $g$, must, unlike the $Z$'s of Eqs. 30 and 31, be *finite* since it relates two finite quantities. Notice that all explicit reference to the bare quantities has now been eliminated. The structure of this equation is *identical* to Eq. 22, the scaling equation derived for the classical case; *the crucial difference is that $Z(\lambda)$ no longer has the simple power law behavior expressed in Eq. 18.* In fact, the general structure of $Z(\lambda)$ and $g(\mu)$ are not known in field theories of interest. Nevertheless we can still learn much by converting this equation to the differential form analogous to Eq. 25 that expresses scale invariance. As before we simply take $\partial/\partial\lambda$ and set $\lambda = 1$, thereby deriving the so-called *renormalization group equation:*

$$-q^2\frac{\partial D}{\partial q^2} + \beta(g)\frac{\partial D}{\partial g} = \gamma(g)\,D\,, \tag{33}$$

where

$$\beta(g) = \mu^2\frac{\partial g}{\partial\mu^2} \tag{34}$$

and

$$\gamma(g) = \frac{\partial\ln Z(\lambda)}{\partial\lambda}\Big|_{\lambda=1}\,. \tag{35}$$

Comparing Eq. 33 with the scaling equation of classical dimensional analysis (Eq. 25), we see that the role of the dimension is played by $\gamma$. For this reason, and to distinguish it from ordinary dimensions, $\gamma$ is usually called the *anomalous dimension* of $D$, a phrase originally coined by Wilson. (We say anomalous because, in terms of ordinary dimensions and again by analogy with Eq. 25, $D$ is actually dimensionless!) It would similarly have been natural to call $\beta(g)/g$ the anomalous dimension of $g$; however, conventionally, one simply refers to $\beta(g)$ as the $\beta$-function. Notice that $\beta(g)$ characterizes the theory as a whole (as does $g$ itself since it represents the coupling) whereas $\gamma(g)$ is a property of the particular object or field one is examining.

The general solution of the renormalization group equation (Eq. 33) is given by

$$D\left(\frac{q^2}{\mu^2},g\right) = e^{A(g)}f\left(\frac{q^2}{\mu^2}\,e^{K(g)}\right), \tag{36}$$

where

$$A(g) = \int^g dg\,\frac{\gamma(g)}{\beta(g)} \tag{37}$$

and

$$K(g) = \int^g \frac{dg}{\beta(g)}\,. \tag{38}$$

The arbitrary function $f$ is, in principle, fixed by imposing suitable boundary conditions. (Equation 25 can be viewed as a special and rather simple case of Eq. 33. If this is done, the analogues of $\gamma(g)$ and $\beta(g)/g$ are constants, resulting in trivial integrals for $A$ and $K$. One can then straightforwardly use this general solution (Eq. 36) to verify the claim that the scaling equation (Eq. 22) is indeed exactly equivalent to using ordinary dimensional analysis.) The general solution reveals what is perhaps the most profound consequence of the renormalization group, namely, that in quantum field theory the momentum variables and the coupling constant are inextricably linked. The photon propagator $(D/q^2)$, for instance, appears at first sight to depend separately on the momentum $q^2$ and the coupling constant $g$. Actually, however, the renormalizability of the theory constrains it to depend effectively, as shown in Eq. 36, on only *one* variable $(q^2 e^{K(g)}/\mu^2)$. This, of course, is exactly what happens in ordinary dimensional analysis. For example, recall the turkey cooking problem. The temperature distribution at first sight depended on several different variables: however, scale invariance, in the guise of dimensional analysis, quickly showed that there was in fact only a single relevant variable.

The observation that renormalization introduces an arbitrary mass scale upon which no physical consequences must depend was first made in 1953 by E. Stueckelberg and A. Peterman. Shortly thereafter Murray Gell-Mann and F. Low attempted to exploit this idea to understand the high-energy structure of QED and, in so doing, exposed the intimate connection between $g$ and $q^2$. Not much use was made of these general ideas until the pioneering work of Wilson in the late 1960s. I shall not review here his seminal work on phase transitions but simply remark that the scaling constraint implicit in the renormalization group can be applied to correlation functions to learn about critical exponents.* Instead I shall concentrate on the

*Since the photon propagator is defined as the correlation function of two electromagnetic fields in the vacuum it is not difficult to imagine that the formalism discussed here can be directly applied to the correlation functions of statistical physics.*

particle physics successes, including Wilson's, that led to the discovery that non-Abelian gauge theories were asymptotically free. Although the foci of particle and condensed matter physics are quite different, they become unified in a spectacular way through the language of field theory and the renormalization group. The analogy with dimensional analysis is a good one, for, as we saw in the first part of this article, its constraints can be applied to completely diverse problems to give powerful and insightful results. In a similar fashion, the renormalization group can be applied to *any* problem that can be expressed as a field theory (such as particle physics or statistical physics).

Often in physics, progress is made by examining the system in some asymptotic regime where the underlying dynamics simplifies sufficiently for the general structure to become transparent. With luck, having understood the system in some extreme region, one can work backwards into the murky regions of the problem to understand its more complex structures. This is essentially the philosophy behind bigger and bigger accelerators: keep pushing to higher energies in the hope that the problem will crack, revealing itself in all its beauty and simplicity. 'Tis indeed a faithful quest for the holy grail. As I shall now demonstrate, the paradigm of looking first for simplicity in

asymptotic regimes is strongly supported by the methodology of the renormalization group.

In essence, we use the same modeling-theory scaling technique used by ship designers. Going back to Eq. 36, one can see immediately that the high-energy or short-distance limit ($q^2 \to \infty$ with $g$ fixed) is identical to keeping $q^2$ fixed while taking $K \to \infty$. However, from its definition (Eq. 38), $K$ diverges whenever $\beta(g)$ has a zero. Similarly, the low-energy or long-distance limit ($q^2 \to 0$ while $g$ is fixed) is equivalent to $K \to -\infty$, which also occurs when $\beta \to 0$. Thus *knowledge of the zeros of $\beta$, the so-called fixed points of the equation, determines the high- and low-energy behaviors of the theory.*

If one assumes that for small coupling quantum field theory is governed by ordinary perturbation theory, then the $\beta$-function has a zero at zero coupling ($g \to 0$). In this limit one typically finds $\beta(g) \approx -bg^3$ where $b$ is a calculable coefficient. Of course, $\beta$ might have other zeroes, but, in general, this is unknown. In any case, for small $g$ we find (using Eq. 38) that $K(g) \approx (2bg^2)^{-1}$, which diverges to either $+\infty$ or $-\infty$ depending on the sign of $b$. In QED, the case originally studied by Gell-Mann and Low, $b < 0$ so that $K \to -\infty$, which is equivalent to the low-energy limit. One can think of this as an explanation of why perturbation theory works so well in the low-energy regime of QED: the smaller the energy, the smaller the effective coupling constant.

**Quantum Chromodynamics.** It appears that some non-Abelian gauge theories and, in particular, QCD (see "Particle Physics and the Standard Model") possess the unique property of having a *positive* $b$. This marvelous observation was first made by H. D. Politzer and independently by D. J. Gross and F. A. Wilczek in 1973 and was crucial in understanding the behavior of quarks in the famous deep inelastic scattering experiments at the Stanford Linear Accelerator Center. As a result, it promoted QCD to the star position of being a member of "the standard model." With $b > 0$ the high-energy limit is related to perturbation theory and is therefore calculable and understandable. I shall now give an explicit example of how this comes about.

First we note that no boundary conditions have yet been imposed on the general solution (Eq. 36). The one boundary condition that *must* be imposed is the known free field theory limit ($g = 0$). For the photon in QED, or the gluon in QCD, the propagator $G$ ($= D/q^2$) in this limit is just $1/q^2$. Thus $D(q^2/\mu^2, 0) = 1$. Imposing this on Eq. 36 gives

$$D\left(\frac{q^2}{\mu^2}, 0\right) = \lim_{g \to 0} e^{A(g)} f\left(\frac{q^2}{\mu^2} e^{K(g)}\right)$$
$$= 1 . \tag{39}$$

Now when $g \to 0$, $\gamma(g) \approx -ag^2$, where $a$ is a calculable coefficient. Combining this with the fact that $\beta(g) \approx -bg^3$ leads, by way of Eq. 37, to $A(g) \approx (a/b) \ln g$. Since $K(g) \approx (2bg^2)^{-1}$, the boundary condition (Eq. 39) gives

$$\lim_{q \to 0} f\left(\frac{q^2}{\mu^2} e^{1/(2bg^2)}\right) = g^{-a/b} . \tag{40}$$

Defining the dimensionless variable in the function $f$ as

$$x \equiv \left(\frac{q^2}{\mu^2}\right) e^{1/(2bg^2)} , \tag{41}$$

it can be shown that with $b > 0$ Eq. 40 is equivalent to

$$\lim_{x \to \infty} f(x) = (2b \ln x)^{a/2b} . \tag{42}$$

An important point here is that the $x \to \infty$ limit can be reached either by letting $g \to 0$ or by taking $q^2 \to \infty$. Since the $g \to 0$ limit is calculable, so is the $q^2 \to \infty$ limit. The free field ($g \to 0$) boundary condition therefore

determines the large $x$ behavior of $f(x)$, and, once again, the "modeling technique" can be used—here to determine the large $q^2$ behavior of the propagator $G$.

In fact, combining Eq. 36 with Eq. 42 leads to the conclusion that

$$\lim_{q^2 \to \infty} D\left(\frac{q^2}{\mu^2}, g\right) = e^{A(g)}\left(2b \ln \frac{q^2}{\mu^2}\right)^{a/2b} . \tag{43}$$

This is the generic structure that finally emerges: the high-energy or large-$q^2$ behavior of the propagator $G = D/q^2$ is given by free field theory ($1/q^2$) modulated by calculable powers of logarithms. The wonderful miracle that has happened is that all the powers of $\ln(\Lambda^2/q^2)$ originally generated from the divergences in the "bare" theory (as illustrated by the series in Eq. 28) have been summed by the renormalization group to give the simple expression of Eq. 43. The amazing thing about this "exact" result is that is far easier to calculate than having to sum an infinite number of individual terms in a series. Not only does the methodology do the summing, but, more important, it justifies it!

I have already mentioned that asymptotic freedom (that is, the equivalence of vanishingly small coupling with increasing momentum) provides a natural explanation of the apparent paradox that quarks could appear free in high-energy experiments even though they could not be isolated in the laboratory. Furthermore, with lepton probes, where the theoretical analysis is least ambiguous, the predicted logarithmic modulation of free-field theory expressed in Eq. 43 has, in fact, been brilliantly verified. Indeed, this was the main reason that QCD was accepted as the standard model for the strong interactions.

There is, however, an even more profound consequence of the application of the renormalization group to the standard model that leads to interesting speculations con-

cerning unified field theories. As discussed in "Particle Physics and the Standard Model," QED and the weak interactions are partially unified into the electroweak theory. Both of these have a negative $b$ and so are *not* asymptotically free; their effective couplings grow with energy rather than decrease. By the same token, the QCD coupling should grow as the energy *decreases*, ultimately leading to the confinement of quarks. Thus as energy increases, the two small electroweak couplings grow and the relatively large QCD coupling decreases. In 1974, Georgi, Quinn, and Weinberg made the remarkable observation that all *three* couplings eventually became equal at an energy scale of about $10^{14}$ GeV! The reason that this energy turns out to be so large is simply due to the very slow logarithmic variation of the couplings. This is a very suggestive result because it is extremely tempting to conjecture that beyond $10^{14}$ GeV (that is, at distances below $10^{-27}$ cm) all three interactions become unified and are governed by the same *single* coupling. Thus, the strong, weak, and electromagnetic forces, which at low energies appear quite disparate, may actually be manifestations of the same field theory. The search for such a unified field theory (and its possible extension to gravity) is certainly one of the central themes of present-day particle physics. It has proven to be a very exciting but frustrating quest that has sparked the imagination of many physicists. Such ideas are, of course, the legacy of Einstein, who devoted the last twenty years of his life to the search for a unified field theory. May his dreams become reality! On this note of fantasy and hope we end our brief discourse about the role of scale and dimension in understanding the world—or even the universe—around us. The seemingly innocuous investigations into the size and scale of animals, ships, and buildings that started with Galileo have led us, via some minor diversions, into baked turkey, incubating eggs, old bones, and the obscure infinities of Feynman diagrams to the ultimate question of unified field theories. Indeed, similitudes have been used and visions multiplied. ∎

**Geoffrey B. West** was born in the county town of Taunton in Somerset, England. He received his B.A. from Cambridge University in 1961 and his Ph.D. from Stanford in 1966. His thesis, under the aegis of Leonard Schiff, dealt mostly with the electromagnetic interaction, an interest he has sustained throughout his career. He was a postdoctoral fellow at Cornell and Harvard before returning to Stanford in 1970 as a faculty member. He came to Los Alamos in 1974 as Leader of what was then called the High-Energy Physics Group in the Theoretical Division, a position he held until 1981 when he was made a Laboratory Fellow. His present interests revolve around the structure and consistency of quantum field theory and, in particular, its relevance to quantum chromodynamics and unified field theories. He has served on several advisory panels and as a member of the executive committee of the Division of Particles and Fields of the American Physical Society.

## Further Reading

The following are books on the classical application of dimensional analysis:

Percy Williams Bridgman. *Dimensional Analysis.* New Haven: Yale University Press, 1963.

Leonid Ivanovich Sedov. *Similarity and Dimensional Methods in Mechanics.* New York: Academic Press, 1959.

Garrett Birkhoff. *Hydrodynamics: A Study in Logic, Fact and Similitude.* Princeton: Princeton University Press, 1960.

D'Arcy Wentworth Thompson. *On Growth and Form.* Cambridge: Cambridge University Press, 1917. This book is, in some respects, comparable to Galileo's and should be required reading for all budding young scientists.

Benoit B. Mandelbrot. *The Fractal Geometry of Nature.* New York: W. H. Freeman, 1983. This recent, very interesting book represents a modern evolution of the subject into the area of fractals; in principle, the book deals with related problems, though I find it somewhat obscure in spite of its very appealing format.

Examples of classical scaling were drawn from the following:

Thomas McMahon. "Size and Shape in Biology." *Science* 179(1973):1201-1204.

Hermann Rahn, Amos Ar, and Charles V. Paganelli. "How Bird Eggs Breathe." *Scientific American* 240(February 1979):46-55.

Thomas A. McMahon. "Rowing: a Similarity Analysis." *Science* 173(1971):349-351.

David Pilbeam and Stephen Jay Gould. "Size and Scaling in Human Evolution." *Science* 186(1974): 892-901.

The Rayleigh-Riabouchinsky exchange is to be found in:

Rayleigh. "The Principle of Similitude." *Nature* 95(1915):66-68.

D. Riabouchinsky. "Letters to Editor." *Nature* 95(1915):591.

Rayleigh. "Letters to Editor." *Nature* 95(1915):644.

Books on quantum electrodynamics (QED) include:

Julian Schwinger, editor. *Selected Papers on Quantum Electrodynamics*. New York: Dover, 1958. This book gives a historical perspective and general review.

James D. Bjorken and Sidney D. Drell. *Relativistic Quantum Mechanics*. New York: McGraw-Hill, 1964.

N. N. Bogoliubov and D. V. Shirkov. *Introduction to the Theory of Quantized Fields*. New York: Interscience, 1959.

H. David Politzer. "Asymptotic Freedom: An Approach to Strong Interactions." *Physics Reports* 14(1974):129-180. This and the previous reference include a technical review of the renormalization group.

Claudio Rebbi. "The Lattice Theory of Quark Confinement." *Scientific American* 248(February 1983):54-65. This reference is also a nontechnical review of lattice gauge theories.

For a review of the deep inelastic electron scattering experiments see:

Henry W. Kendall and Wolfgang K. H. Panofsky. "The Structure of the Proton and the Neutron." *Scientific American* 224(June 1971):60-76.

Geoffrey B. West. "Electron Scattering from Atoms, Nuclei and Nucleons." *Physics Reports* 18(1975):263-323.

References dealing with detailed aspects of renormalization and its consequences are:

Kenneth G. Wilson. "Non-Lagrangian Models of Current Algebra." *Physical Review* 179(1969):1499-1512.

Geoffrey B. West. "Asymptotic Freedom and the Infrared Problem: A Novel Solution to the Renormalization-Group Equations." *Physical Review D* 27(1983):1402-1405.

E. C. G. Stueckelberg and A. Petermann. "La Normalisation des Constantes dans la Theorie des Quanta." *Helvetica Physica Acta* 26(1953):499-520.

M. Gell-Mann and F. E. Low. "Quantum Electrodynamics at Small Distances." *Physical Review* 95(1954):1300-1312.

H. David Politzer. "Reliable Perturbative Results for Strong Interactions?" *Physical Review Letters* 30(1973):1346-1349.

David J. Gross and Frank Wilczck. "Ultraviolet Behavior of Non-Abelian Gauge Theories." *Physical Review Letters* 30(1973):1343-1346.

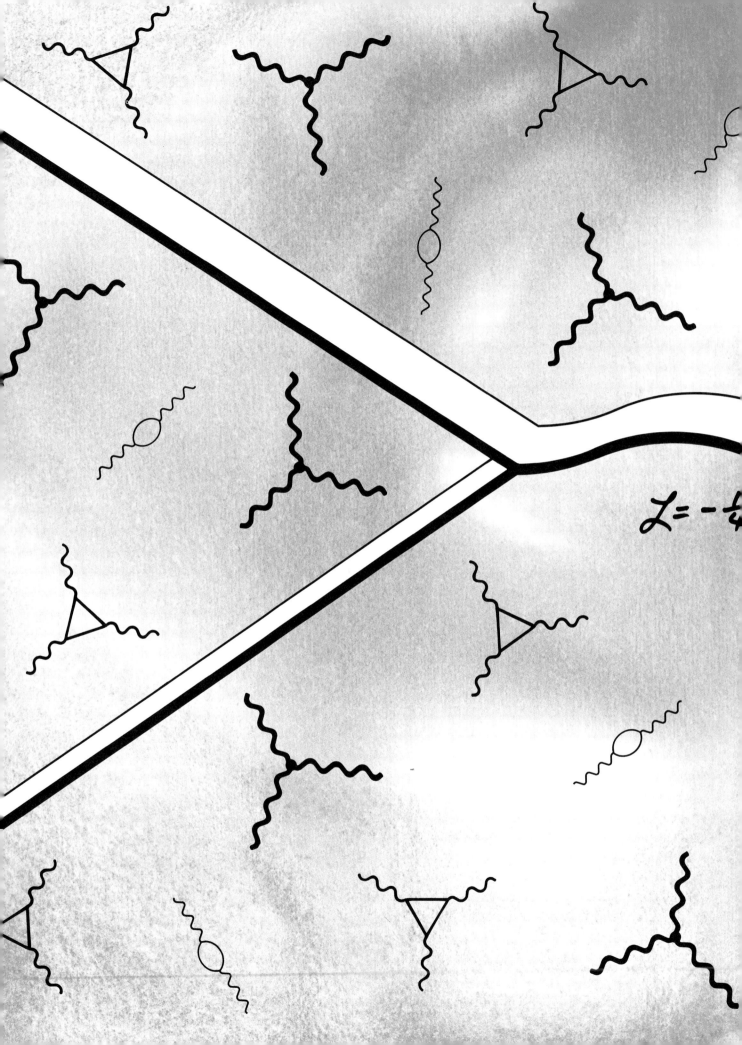

# Particle Physics and the Standard Model

$$F_{\mu\nu}F^{\mu\nu} + i\bar{\Psi}\gamma^{\mu}(\partial_{\mu} - ieA_{\mu})\Psi$$

by Stuart A. Raby, Richard C. Slansky, and Geoffrey B. West

Until the 1930s all natural phenomena were presumed to have their origin in just two basic forces—gravitation and electromagnetism. Both were described by classical fields that permeated all space. These fields extended out to infinity from well-defined sources, mass in the one case and electric charge in the other. Their benign rule over the physical universe seemed securely established.

As atomic and subatomic phenomena were explored, it became apparent that two completely novel forces had to be added to the list; they were dubbed the weak and the strong. The strong force was necessary in order to understand how the nucleus is held together: protons bound together in a tight nuclear ball ($10^{-12}$ centimeter across) must be subject to a force much stronger than electromagentism to prevent their flying apart. The weak force was invoked to understand the transmutation of a neutron in the nucleus into a proton during the particularly slow form of radioactive decay known as beta decay.

Since neither the weak force nor the strong force is directly observed in the macroscopic world, both must be very short-range relative to the more familiar gravitational and electromagnetic forces. Furthermore, the relative strengths of the forces associated with all four interactions are quite different, as can be seen in Table 1. It is therefore not too surprising that for a very long period these interactions were thought to be quite separate. In spite of this, there has always been a lingering suspicion (and hope) that in some miraculous fashion all four were simply manifestations of one source or principle and could therefore be described by a single unified field theory.

*The color force among quarks and gluons is described by a generalization of the Lagrangian $\mathscr{L}$ of quantum electrodynamics shown above. The large interaction vertex dominating these pages is a common feature of the strong, the weak, and the electromagnetic forces. A feature unique to the strong force, the self-interaction of colored gluons, is suggested by the spiral in the background.*

## Table 1

The four basic forces. Differences in strengths among the basic interactions are observed by comparing characteristic cross sections and particle lifetimes. (Cross sections are often expressed in barns because the cross-sectional areas of nuclei are of this order of magnitude; one barn equals $10^{-24}$ square centimeter.) The stronger the force, the larger is the effective scattering area, or cross section, and the shorter the lifetime of the particle state. At 1 GeV strong processes take place $10^2$ times faster than electromagnetic processes and $10^5$ times faster than weak processes.

| Name | Relative Strength between Two Protons at $10^{-13}$cm | Typical Scattering Cross Sections at 1 GeV in millibarns | Typical Lifetimes of Particle States | Representative Effects |
|---|---|---|---|---|
| **Strong** $\alpha_s \sim 1$ $\alpha_s = \dfrac{g_s^2}{4\pi\hbar c}$ | Proton-Neutron Elastic Scattering $n + p \rightarrow n + p$ $\sigma \sim 10^1$ mb | Strong Decay of the Delta $\Delta^+ \rightarrow p + \pi^0$ $\tau \sim 10^{-23}$ s | Stable Proton (bound states of quarks) Stable Nuclei Fission Fusion |
| **Electromagnetic** $\alpha \sim \dfrac{1}{137}$ $\alpha = \dfrac{e^2}{4\pi\hbar c}$ | Scattering of Light $\gamma + p \rightarrow \gamma + p$ $\sigma \sim 10^{-1}$ mb | Electromagnetic Decay of the Rho $\rho \rightarrow \pi + \gamma$ $\tau \sim 10^{-21}$ s | Light Radiowaves Stable Atoms and Molecules Chemical Reactions |
| **Weak** $G_F m_p^2 \sim 10^{-5}$ $G_F = \dfrac{\sqrt{2}\,g^2}{8M_W^2}$ | Neutrino-Nucleon Scattering $\nu + p \rightarrow \nu + p$ $\sigma \sim 10^{-11}$ mb | Weak Decay of the Muon $\mu^- \rightarrow \nu_\mu + e^- + \bar{\nu}_e$ $\tau \sim 10^{-6}$ s | Beta Decay Unstable Neutron Pion Decay Muon Decay |
| **Gravitational** $G_N m_p^2 \sim 10^{-38}$ | Graviton-Proton Scattering $g + p \rightarrow g + p$ $\sigma \sim 10^{-77}$ mb | | Galaxies Solar Systems Curved Space-Time and Cosmology |

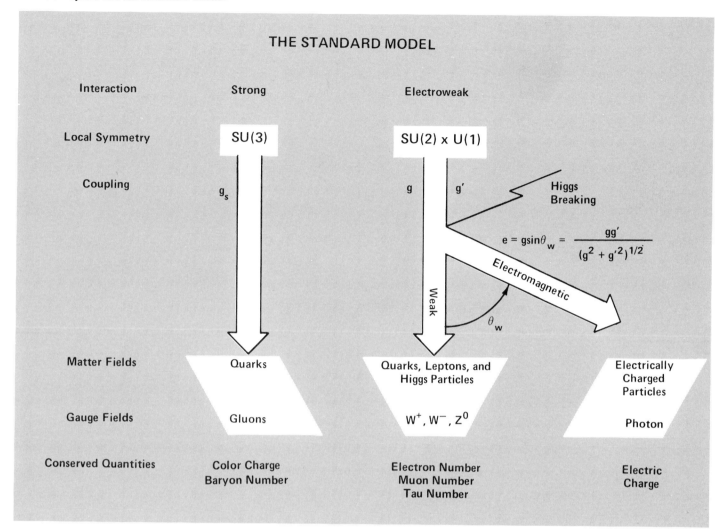

*Fig. 1. The main features of the standard model. The strong force and the electroweak force are each induced by a local symmetry group, SU(3) and SU(2) × U(1), respectively. These two symmetries are entirely independent of each other. SU(3) symmetry (called the color symmetry) is exact and therefore predicts conservation of color charge. The SU(2) × U(1) symmetry of the electroweak theory is an exact sym-metry of the Lagrangian of the theory but not of the solu-tions to the theory. The standard model ascribes this sym-metry breaking to the Higgs particles, particles that create a nonzero weak charge in the vacuum (the lowest energy state of the system). The only conserved quantity that remains after the symmetry breaking is electric charge.*

The spectacular progress in particle phys-ics over the past ten years or so has renewed this dream; many physicists today believe that we are on the verge of uncovering the structure of this unified theory. The theoreti-cal description of the strong, weak, and elec-tromagnetic interactions is now considered well established, and, amazingly enough, the theory shows these forces to be quite similar despite their experimental differences. The weak and strong forces have sources analogous to, but more complicated than, electric charge, and, like the electromagnetic force, both can be described by a special type of field theory called a local gauge theory. This formulation has been so successful at explaining all known phenomenology up to energies of 100 GeV (1 GeV = $10^9$ electron volts) that it has been coined "the standard model" and serves as the point of departure for discussing a grand unification of all forces, including that of gravitation.

The elements of the standard model are summarized in Fig. 1. In this description the basic constituents of matter are quarks and leptons, and these constituents interact with each other through the exchange of gauge particles (vector bosons), the modern analogue of force fields. These so-called local gauge interactions are inscribed in the lan-guage of Lagrangian quantum field theory, whose rich formalism contains mysteries that escape even its most faithful practi-tioners. Here we will introduce the central themes and concepts that have led to the standard model, emphasizing how its for-malism enables us to describe all phenomenology of the strong, weak, and electromagnetic interactions as different manifestations of a single symmetry prin-ciple, the principle of local symmetry. As we shall see, the standard model has many arbitrary parameters and leaves unanswered a number of important questions. It can hardly be regarded as a thing of great beauty—unless one keeps in mind that it embodies a single unifying principle and therefore seems to point the way toward a grander unification.

For those readers who are more mathematically inclined, the arguments here are complemented by a series of lecture notes immediately following the main text and entitled "From Simple Field Theories to the Standard Model." The lecture notes in-troduce Lagrangian formalism and stress the symmetry principles underlying construction of the standard model. The main emphasis is on the classical limit of the model, but indications of its quantum generalizations are also included.

## Unification and Extension

Two central themes of physics that have led to the present synthesis are "unification" and "extension." By "unification" we mean the coherent description of phenomena that are at first sight totally unrelated. This takes the form of a mathematical description with specific rules of application. A theory must not only describe the known phenomena but also make predictions of new ones. Almost all theories are incomplete in that they provide a description of phenomena only within a specific range of parameters. Typically, a theory changes as it is extended to explain phenomena over a larger range of parameters, and sometimes it even simplifies. Hence, the second theme is called extension—and refers in particular to the extension of theories to new length or energy scales. It is usually extension and the resulting simplification that enable unification.

Perhaps the best-known example of extension and unification is Newton's theory of gravity (1666), which unifies the description of ordinary-sized objects falling to earth with that of the planets revolving around the sun. It describes phenomena over distance scales ranging from a few centimeters up to $10^{25}$ centimeters (galactic scales). Newton's theory is superceded by Einstein's theory of relativity only when one tries to describe phenomena at extremely high densities and/or velocities or relate events over cosmological distance and time scales.

The other outstanding example of unification in classical physics is Maxwell's theory of electrodynamics, which unifies electricity with magnetism. Coulomb (1785) had established the famous inverse square law for the force between electrically charged bodies, and Biot and Savart (1820) and Ampère (1820-1825) had established the law relating the magnetic field $\mathbf{B}$ to the electric current as well as the law for the force between two

electric currents. Thus it was known that static charges give rise to an electric field $\mathbf{E}$ and that moving charges give rise to a magnetic field $\mathbf{B}$. Then in 1831 Faraday discovered that the field itself has a life of its own, independent of the sources. A time-dependent magnetic field induces an electric field. This was the first clear hint that electric and magnetic phenomena were manifestations of the same force field.

Until the time of Maxwell, the basic laws of electricity and magnetism were expressed in a variety of different mathematical forms, all of which left the central role of the fields obscure. One of Maxwell's great achievements was to rewrite these laws in a single formalism using the fields $\mathbf{E}$ and $\mathbf{B}$ as the fundamental physical entities, whose sources are the charge density $\rho$ and the current density $\mathbf{J}$, respectively. In this formalism the laws of electricity and magnetism are expressed as differential equations that manifest a clear interrelationship between the two fields. Nowadays they are usually written in standard vector notation as follows.

Coulomb's law: $\qquad \nabla \cdot \mathbf{E} = 4\pi\rho/\varepsilon_0;$

Ampère's law: $\qquad \nabla \times \mathbf{B} = 4\pi\mu_0\mathbf{J};$

Faraday's law: $\qquad \nabla \times \mathbf{E} + \partial\mathbf{B}/\partial t = 0;$

and the absence of
magnetic monopoles: $\qquad \nabla \cdot \mathbf{B} = 0 .$

The parameters $\varepsilon_0$ and $\mu_0$ are determined by measuring Coulomb's force between two static charges and Ampère's force between two current-carrying wires, respectively.

Although these equations clearly "unite" $\mathbf{E}$ with $\mathbf{B}$, they are incomplete. In 1865 Maxwell realized that the above equations were not consistent with the conservation of electric charge, which requires that

$$\nabla \cdot \mathbf{J} + \partial\rho/\partial t = 0 .$$

This inconsistency can be seen from Ampère's law, which in its primitive form requires that

$$\nabla \cdot \mathbf{J} = (4\pi\mu_0)^{-1}\nabla \cdot (\nabla \times \mathbf{B}) \equiv 0 .$$

Maxwell obtained a consistent solution by amending Ampère's law to read

$$\nabla \times \mathbf{B} = 4\pi\mu_0\mathbf{J} + \varepsilon_0\mu_0 \frac{\partial\mathbf{E}}{\partial t} .$$

With this new equation, Maxwell showed that both $\mathbf{E}$ and $\mathbf{B}$ satisfy the wave equation. For example,

$$\left(\nabla^2 - \varepsilon_0\mu_0 \frac{\partial^2}{\partial t^2}\right)\mathbf{E} = 0 .$$

This fact led him to propose the electromagnetic theory of light. Thus, from Maxwell's unification of electric and magnetic phenomena emerged the concept of electromagnetic waves. Moreover, the speed $c$ of the electromagnetic waves, or light, is given by $(\varepsilon_0\mu_0)^{-1/2}$ and is thus determined uniquely in terms of purely static electric and magnetic measurements alone!

It is worth emphasizing that apart from the crucial change in Ampère's law, Maxwell's equations were well known to natural philosophers before the advent of Maxwell! The unification, however, became manifest only through his masterstroke of expressing them in terms of the "right" set of variables, namely, the fields $\mathbf{E}$ and $\mathbf{B}$.

## Extension to Small Distance Scales

Maxwell's unification provides an accurate description of large-scale electromagnetic phenomena such as radio waves, current flow, and electromagnets. This theory can also account for the effects of a medium, provided macroscopic concepts such as conductivity and permeability are introduced. However, if we try to extend it to very short distance scales, we run into trouble; the granularity, or quantum nature, of matter and of the field itself becomes important, and Maxwell's theory must be altered.

Determining the physics appropriate to each length scale is a crucial issue and has been known to cause confusion (see "Fundamental Constants and the Rayleigh-

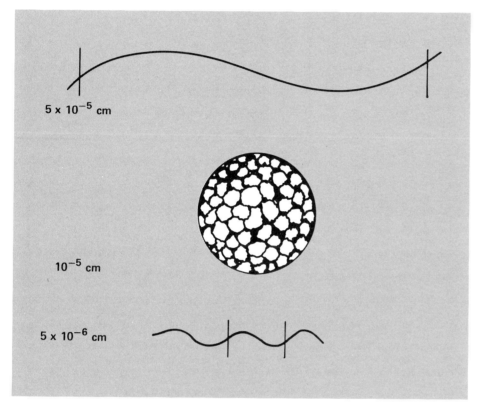

*Fig. 2. The wavelength of the probe must be smaller than the scale of the structure one wants to resolve. Viruses, which are approximately $10^{-5}$ centimeter in extent, cannot be resolved with visible light, the average wavelength of which is $5 \times 10^{-5}$ centimeter. However, electrons with momentum p of about 20 eV/c have de Broglie wavelengths short enough to resolve them.*

the basis of the often-stated fact that resolving smaller distances requires particles of greater momentum or energy. Notice, incidentally, that for sufficiently short wavelengths, one is *forced* to incorporate special relativity since the corresponding particle momentum becomes so large that Newtonian mechanics fails.

The marriage of quantum mechanics and special relativity gave birth to quantum field theory, the mathematical and physical language used to construct theories of the elementary particles. Below we will give a brief review of its salient features. Here we simply want to remind the reader that quantum field theory automatically incorporates quantum ideas such as Heisenberg's uncertainty principle and the dual wave-particle properties of all of matter, as well as the equivalence of mass and energy.

Since the wavelength of our probe determines the size of the object that can be studied (Fig. 2), we need extremely short wavelength (high energy) probes to investigate particle phenomena. To gain some perspective, consider the fact that with visible light we can see without aid objects as small as an amoeba (about $10^{-2}$ centimeter) and with an optical microscope we can open up the world of bacteria at about $10^{-4}$ centimeter. This is the limiting scale of light probes because wavelengths in the visible spectrum are on the order of $5 \times 10^{-5}$ centimeter.

To resolve even smaller objects we can exploit the wave-like aspects of energetic particles as is done in an electron microscope. For example, with "high-energy" electrons ($E \approx 20$ eV) we can view the world of viruses at a length scale of about $10^{-5}$ centimeter. With even higher energy electrons we can see individual molecules (about $10^{-7}$ centimeter) and atoms ($10^{-8}$ centimeter). To probe down to nuclear ($10^{-12}$ centimeter) and subnuclear scales, we need the particles available from high-energy accelerators. Today's highest energy accelerators produce 100-GeV particles, which probe distance scales as small as $10^{-16}$ centimeter.

This brings us to the second type of change

Riabouchinsky Paradox"). For example, the structure of the nucleus is completely irrelevant when dealing with macroscopic distances of, say, 1 centimeter, so it would be absurd to try to describe the conductivity of iron over this distance in terms of its quark and lepton structure. On the other hand, it would be equally absurd to extrapolate Ohm's law to distance intervals of $10^{-13}$ centimeter to determine the flow of electric current. Relevant physics changes with scale!

The thrust of particle physics has been to study the behavior of matter at shorter and shorter distance scales in hopes of understanding nature at its most fundamental level. As we probe shorter distance scales, we encounter two types of changes in the phys-

ics. First there is the fundamental change resulting from having to use quantum mechanics and special relativity to describe phenomena at very short distances. According to quantum mechanics, particles have both wave and particle properties. Electrons can produce interference patterns as waves and can deposit all their energy at a point as a particle. The wavelength $\lambda$ associated with the particle of momentum $p$ is given by the de Broglie relation

$$\lambda = \frac{h}{p},$$

where $h$ is Planck's constant ($h/2\pi \equiv \hbar = 1.0546 \times 10^{-27}$ erg · second). This relation is

in appropriate physics with change in scale, namely, changes in the forces themselves. Down to distances of approximately $10^{-12}$ centimeter, electromagnetism is the dominant force among the elementary particles. However, at this distance the strong force, heretofore absent, suddenly comes into play and completely dominates the interparticle dynamics. The weak force, on the other hand, is present at all scales but only as a small effect. At the shortest distances being probed by present-day accelerators, the weak and electromagnetic forces become comparable in strength but remain several orders of magnitude weaker than the strong force. It is at this scale however, that the fundamental similarity of all three forces begins to emerge. Thus, as the scale changes, not only does each force itself change, but its relationship to the other forces undergoes a remarkable evolution. In our modern way of thinking, which has come from an understanding of the renormalization, or scaling, properties of quantum field theory, these changes in physics are in some ways analogous to the paradigm of phase transitions. To a young and naive child, ice, water, and steam appear to be quite different entities, yet rudimentary observations quickly teach that they are different manifestations of the same stuff, each associated with a different temperature scale. The modern lesson from renormalization group analysis, as discussed in "Scale and Dimension—From Animals to Quarks," is that the physics of the weak, electromagnetic, and strong forces may well represent different aspects of the same unified interaction. This is the philosophy behind grand unified theories of all the interactions.

## Quantum Electrodynamics and Field Theory

Let us now return to the subject of electromagnetism at small distances and describe quantum electrodynamics (QED), the relativistic quantum field theory, developed in the 1930s and 1940s, that extends Maxwell's theory to atomic scales. We emphasize that the standard model is a generalization of

# Antiproton

By 1931 the negative energy states yielded for the electron by Dirac's relativistic quantum theory of 1928 had been interpreted, not as states of the proton (Dirac's initial thought), but as states of a particle with the same mass as the electron but opposite electric charge. Such a particle was found by chance in 1932 among the products of cosmic-ray collisions with nuclei. Searches for the antiproton (or negative proton) in the same environment proved unsuccessful, and physicists began considering its production by bombarding nuclei with energetic protons from an accelerator. Since electric charge and baryon number must be conserved in strong interactions, the production process involves creation of a proton-antiproton pair:

$$p + p \,(\text{or } n) \rightarrow (p + \bar{p}) + p + p \,(\text{or } n).$$

This reaction has a threshold of approximately 5.7 GeV for the kinetic energy of the incident proton.

The Berkeley Bevatron was designed with antiproton production in mind, and this 6-GeV synchrotron enabled O. Chamberlain, E. Segrè, C. Wiegand, and T. Ypsilantis to make the first laboratory observation of the antiproton in 1955. Their identification method involved sorting out, from among the many products of the proton-nucleon collisions, negatively charged particles of a certain momentum with a bending magnet and further sorting out particles of the appropriate velocity, and hence mass, with two scintillation detectors spaced a known distance apart. Discovery of the antiproton strongly supported the idea that for every particle there exists an antiparticle with the same mass but opposite values of electric charge or other quantum properties. ∎

this first and most successful quantum field theory.

In quantum field theory every particle has associated with it a mathematical operator, called a quantum field, that carries the particle's characteristic quantum numbers. Probably the most familiar quantum number is spin, which corresponds to an intrinsic angular momentum. In classical mechanics angular momentum is a continuous variable, whereas in quantum mechanics it is restricted to multiples of ½ when measured in units of $\hbar$. Particles with ½-integral spin (1/2, 3/2, 5/2, ...) are called fermions; particles with integral spin (0, 1, 2, 3, ... ) are called bosons. Since no two identical fermions can occupy the same position at the same time (the

famous Pauli exclusion principle), a collection of identical fermions must necessarily take up some space. This special property of fermions makes it natural to associate them with matter. Bosons, on the other hand, can crowd together at a point in space-time to form a classical field and are naturally regarded as the mediators of forces.

In the quantized version of Maxwell's theory, the electromagnetic field (usually in the guise of the vector potential $A_\mu$) is a boson field that carries the quantum numbers of the photon, namely, mass $m = 0$, spin $s = 1$, and electric charge $Q = 0$. This quantized field, by the very nature of the mathematics, automatically manifests dual wave-particle properties. Electrically charged particles,

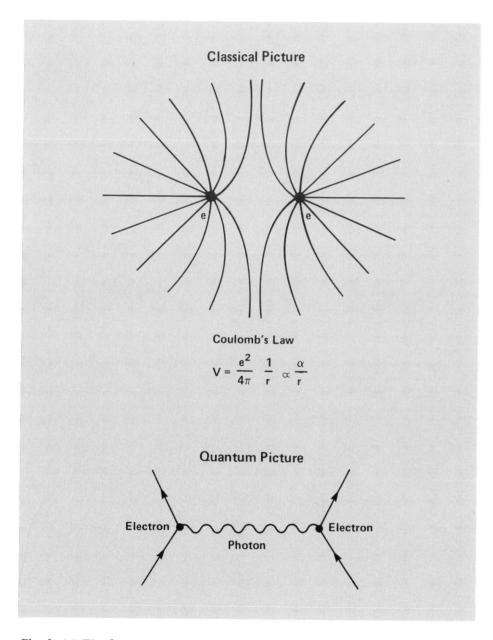

**Classical Picture**

**Coulomb's Law**

$$V = \frac{e^2}{4\pi} \; \frac{1}{r} \; \propto \; \frac{\alpha}{r}$$

**Quantum Picture**

Electron

Photon

Electron

such as electrons and positrons, are also represented by fields, and, as in the classical theory, they interact with each other through the electromagnetic field. In QED, however, the interaction takes place via an exchange of photons. Two electrons "feel" each other's presence by passing photons back and forth between them. Figure 3 pictures the interaction with a "Feynman diagram": the straight lines represent charged particles and the wavy line represents a photon. (In QED such diagrams correspond to terms in a perturbative expansion for the scattering between charged particles (see Fig. 5). Similarly, most Feynman diagrams in this issue represent lowest order contributions to the particle reactions shown.)

These exchanged photons are rather special. A real photon, say in the light by which you see, must be massless since only a massless particle can move at the speed of light. On the other hand, consider the left-hand vertex of Fig. 3, where a photon is emitted by an electron; it is not difficult to convince oneself that if the photon is massless, energy and momentum are not conserved! This is no sin in quantum mechanics, however, as Heisenberg's uncertainty principle permits such violations provided they occur over sufficiently small space-time intervals. Such is the case here: the violating photon is absorbed at the right-hand vertex by another electron in such a way that, overall, energy and momentum are conserved. The exchanged photon is "alive" only for a period concomitant with the constraints of the uncertainty principle. Such photons are referred to as virtual photons to distinguish them from real ones, which can, of course, live forever.

The uncertainty principle permits all sorts of virtual processes that momentarily violate energy-momentum conservation. As illustrated in Fig. 4, a virtual photon being exchanged between two electrons can, for a very short time, turn into a virtual electron-positron pair. This conversion of energy into mass is allowed by the famous equation of special relativity, $E = mc^2$. In a similar fashion almost anything that *can* happen *will*

*Fig. 3. (a) The force between two electrons is described classically by Coulomb's law. Each electron creates a force field (shown as lines emanating from the charge (e) that is felt by the other electron. The potential energy V is the energy needed to bring the two electrons to within a distance r of each other. (b) In quantum field theory two electrons feel each other's presence by exchanging virtual photons, or virtual particles of light. Photons are the quanta of the electromagnetic field. The Feynman diagram above represents the (lowest order, see Fig. 5) interaction between two electrons (straight lines) through the exchange of a virtual photon (wavy line).*

happen, given a sufficiently small space-time interval. It is the countless multitude of such virtual processes that makes quantum field theory so rich and so difficult.

Given the immense complexity of the theory, one wonders how any reliable calculation can ever be made. The saving grace of quantum electrodynamics, which has made its predictions the most accurate in all of physics, is the smallness of the coupling between the electrons and the photons. The coupling strength at each vertex where an electron spews out a virtual photon is just the electronic charge $e$, and, since the virtual photon must be absorbed by some other electron, which also has charge $e$, the probability for this virtual process is of magnitude $e^2$. The corresponding *dimensionless* parameter that occurs naturally in this theory is denoted by $\alpha$ and defined as $e^2/4\pi\hbar c$. It is approximately equal to 1/137. The probabilities of more complicated virtual processes involving many virtual particles are proportional to higher powers of $\alpha$ and are therefore very much smaller relative to the probabilities for simpler ones. Put slightly differently, the smallness of $\alpha$ implies that perturbation theory is applicable, and we can control the level of accuracy of our calculations by including higher and higher order virtual processes (Fig. 5). In fact, quantum electrodynamic calculations of certain atomic and electronic properties agree with experiment to within one part in a billion.

As we will elaborate on below, the quantum field theories of the electroweak and the strong interactions that compose the standard model bear many resemblances to quantum electrodynamics. Not too surprisingly, the coupling strength of the weak interaction is also small (and in fact remains small at all energy or distance scales), so perturbation theory is always valid. However, the analogue of $\alpha$ for the strong interaction is not always small, and in many calculations perturbation theory is inadequate. Only at the high energies above 1 GeV, where the theory is said to be asymptotically free, is the analogue of $\alpha$ so small that perturbation theory is valid. At low and moderate energies

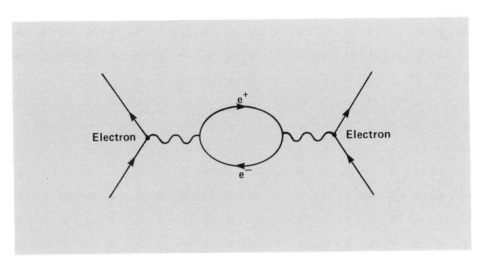

Fig. 4. A virtual photon being exchanged between two electrons can, for a very short time, turn into a virtual electron-positron (e⁺-e⁻) pair. This virtual process is one of many that contribute to the electromagnetic interaction between electrically charged particles (see Fig. 5).

(for example, those that determine the properties of protons and neutrons) the strong-interaction coupling strength is large, and analytic techniques beyond perturbation theory are necessary. So far such techniques have not been very successful, and one has had to resort to the nasty business of numerical simulations!

As discussed at the end of the previous section, these changes in coupling strengths with changes in scale are the origin of the changes in the forces that might lead to a unified theory. For an example see Fig. 3 in "Toward a Unified Theory."

## Symmetries

One cannot discuss the standard model without introducing the concept of symmetry. It has played a central role in classifying the known particle states (the ground states of 200 or so particles plus excited states) and in predicting new ones. Just as the chemical elements fall into groups in the periodic table, the particles fall into multiplets characterized by similar quantum numbers. However, the use of symmetry in particle physics goes well beyond mere classification. In the construction of the standard model, the special kind of symmetry known as local symmetry has become *the* guiding dynamical principle; its aesthetic influence in the search for unification is reminiscent of the quest for beauty among the ancient Greeks. Before we can discuss this dynamical principle, we must first review the general concept of symmetry in particle physics.

In addition to electric charge and mass, particles are characterized by other quantum numbers such as spin, isospin, strangeness, color, and so forth. These quantum numbers reflect the symmetries of physical laws and are used as a basis for classification and, ultimately, unification.

Although quantum numbers such as spin and isospin are typically the distinguishing features of a particle, it is probably less well known that the mass of a particle is sometimes its only distinguishing feature. For example, a muon ($\mu$) is distinguished from an electron ($e$) only because its mass is 200 times greater that that of the electron. Indeed, when the muon was discovered in 1938, Rabi was reputed to have made the remark, "Who ordered *that*?" And the tau

**Electron Scattering**

(Interaction)$^2 \propto \alpha$

(Interaction)$^4 \propto \alpha^2$

(Interaction)$^6 \propto \alpha^3$

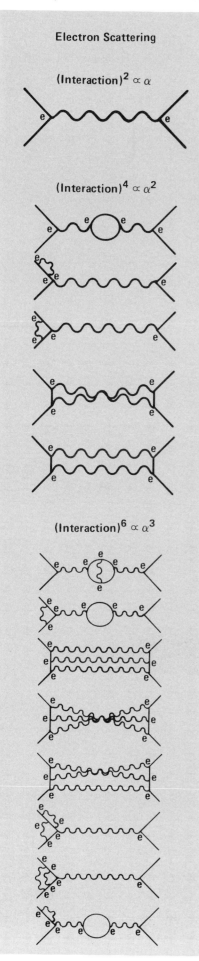

**Electromagnetic Interaction** $= eJ^\mu A_\mu$

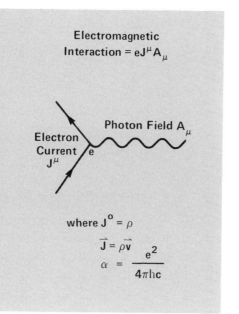

Photon Field A$_\mu$

Electron Current J$^\mu$

where $J^0 = \rho$

$$\vec{J} = \rho\vec{v}$$

$$\alpha = \frac{e^2}{4\pi\hbar c}$$

*Fig. 5. As shown above, the basic interaction vertex of quantum electrodynamics is an electron current $J^\mu$ interacting with the electromagnetic field $A_\mu$. Because the coupling strength $\alpha$ is small, the amplitude for processes involving such interactions can be approximated by a perturbation expansion on a free field theory. The terms in such an expansion, shown at left for electron scattering, are proportional to various powers of $\alpha$. The largest contribution to the electron-scattering amplitude is proportional to $\alpha$ and is represented by a Feynmann diagram in which the interaction vertex appears twice. Successively smaller contributions arise from terms proportional to $\alpha^2$ with four interaction vertices, from terms proportional to $\alpha^3$ with six interaction vertices, and so on.*

($\tau$), discovered in 1973, is 3500 times heavier than an electron yet again identical to the electron in other respects. One of the great unsolved mysteries of particle physics is the origin of this apparent hierarchy of mass among these leptons. (A lepton is a fundamental fermion that has no strong interactions.) Are there even more such particles? Is there a reason why the mass hierarchy among the leptons is paralleled (as we will describe below) by a similar hierarchy among the quarks? It is believed that when we understand the origin of fermion masses, we will also understand the origin of CP violation in nature (see box). These questions are frequently called the family problem and are discussed in the article by Goldman and Nieto.

**Groups and Group Multiplets**. Whether or not the similarity among $e$, $\mu$, and $\tau$ reflects a fundamental symmetry of nature is not known. However, we will present several possibilities for this family symmetry to introduce the language of groups and the significance of internal symmetries.

Consider a world in which the three leptons have the same mass. In this world atoms with muons or taus replacing electrons would be indistinguishable: they would have identical electromagnetic absorption or emission bands and would form identical elements. We would say that this world is *invariant* under the interchange of electrons, muons, and taus, and we would call this invariance a *symmetry of nature*. In the real world these particles don't have the same mass; therefore our hypothetical symmetry, if it exists, is broken and we can distinguish a muonic atom from, say, its electronic counterpart.

We can describe our hypothetical invariance or family symmetry among the three leptons by a set of symmetry operations that form a mathematical construct called a *group*. One property of a group is that any two symmetry operations performed in succession also corresponds to a symmetry operation in that group. For example, replacing an electron with a muon, and then replacing a muon with a tau can be defined as two *discrete* symmetry operations that when performed in succession are equivalent to the discrete symmetry operation of replacing an electron with a tau. Another group property is that every operation must have an inverse. The inverse of replacing an electron with a muon is replacing a muon with an electron. This set of discrete operations on $e$, $\mu$, and $\tau$ forms the discrete six-element group $\pi_3$ (with $\pi$ standing for permutation). In this language $e$, $\mu$, and $\tau$ are called a *multiplet* or *representation* of $\pi_3$ and are said to transform as a triplet under $\pi_3$.

Another possibility is that the particles $e$, $\mu$, and $\tau$ transform as a triplet under a group of *continuous* symmetry operations. Consider Fig. 6, where $e$, $\mu$, and $\tau$ are represented as three orthogonal vectors in an abstract

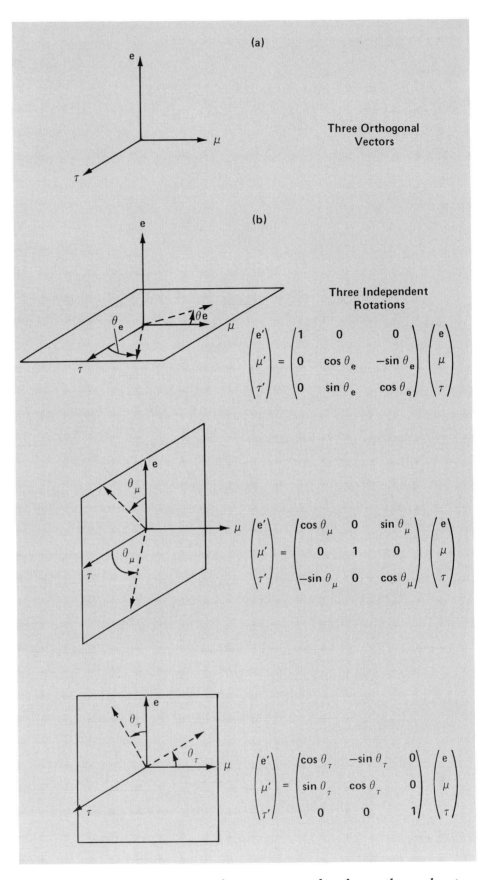

three-dimensional space. The set of continuous rotations of the three vectors about three independent axes composes the group known as the three-dimensional rotation group and denoted by SO(3). As shown in Fig. 6, SO(3) has three independent transformations, which are represented by orthogonal $3 \times 3$ matrices. (Note that $\pi_3$ is a subset of SO(3).)

Suppose that SO(3) were an unbroken family symmetry of nature and $e$, $\mu$, and $\tau$ transformed as a triplet under this symmetry. How would it be revealed experimentally? The SO(3) symmetry would add an extra degree of freedom to the states that could be formed by $e$, $\mu$, and $\tau$. For example, the spatially symmetric ground state of helium, which ordinarily must be antisymmetric under the interchange of the two electron spins, could now be antisymmetric under the interchange of either the spin or the family quantum number of the two leptons. In particular, the ground state would have three different antisymmetric configurations and the threefold degeneracy might be split by spin-spin interactions among the leptons and by any SO(3) symmetric interaction. Thus the ground state of known helium would probably be replaced by sets of degenerate levels with small hyperfine energy splittings.

In particle physics we are always interested in the largest group of operations that leaves all properties of a system unchanged. Since $e$, $\mu$, and $\tau$ are described by complex fields, the largest group of operations that could act on this triplet is U(3) (the group of all unitary 3 $\times$ 3 matrices $U$ satisfying $U^\dagger U = 1$). Another possibility is SU(3), a subgroup of U(3) satisfying the additional constraint that det $U = 1$.

This list of symmetries that may be reflected in the similarity of $e$, $\mu$, and $\tau$ is not exhaustive. We could invoke a group of symmetry operations that acts on any subset of the three particles, such as SU(2) (the group of $2 \times 2$ unitary matrices with det $U = 1$) acting, say, on $e$ and $\mu$ as a doublet and on $\tau$ as a singlet. Any one of these possibilities may be realized in nature, and each possibility has different experimentally observable

**Fig. 6. (a)** *The three leptons* e, $\mu$, *and* $\tau$ *are represented as three orthogonal vectors in an abstract three-dimensional space.* **(b)** *The set of rotations about the three orthogonal axes defines SO(3), the three-dimensional rotation group. SO(3) has three charges (or generators) associated with the infinitesimal transformations about the three independent axes. These generators have the same Lie algebra as the generators of the group SU(2), as discussed in Lecture Note 4 following this article.*

consequences. However, the known differences in the masses of $e$, $\mu$, and $\tau$ imply that *any* symmetry used to describe the similarity among them is a broken symmetry. Still, a broken symmetry will retain traces of its consequences (if the symmetry is broken by a small amount) and thus also provides useful predictions.

Our hypothetical broken symmetry among $e$, $\mu$, and $\tau$ is but one example of an approximate *internal* global symmetry. Another is the symmetry between, say, the neutron and the proton in strong interactions, which is described by the group known as strong-isospin SU(2). The neutron and proton transform as a doublet under this symmetry and the three pions transform as a triplet. We will discuss below the classifica-

# CP Violation

The faith of physicists in symmetries of nature, so shaken by the observation of parity violation in 1956, was soon restored by invocation of a new symmetry principle—CP conservation—to interpret parity-violating processes. This principle states that a process is indistinguishable from its mirror image provided all particles in the mirror image are replaced by their antiparticles. Alas, in 1964 this principle also was shattered with the results of an experiment on the decay of neutral kaons.

According to the classic analysis of M. Gell-Mann and A. Pais, neutral kaons exist in two forms: $K_S^0$, with an even CP eigenvalue and decaying with a relatively short lifetime of $10^{-10}$ second into two pions, and $K_L^0$, with an odd CP eigenvalue and decaying with a lifetime of about $5 \times 10^{-8}$ second into three pions. CP conservation prohibited the decay of the longer lived $K_L^0$ into two pions. But in an experiment at Brookhaven, J. Christenson, J. Cronin, V. Fitch, and R. Turlay found that about 1 in 500 $K_L^0$ mesons decays into two pions. This first observation of CP violation has been confirmed in many other experiments on the neutral kaon system, but to date no other CP-violating effects have been found. The underlying mechanism of CP violation remains to be understood, and an implication of the phenomenon, the breakdown of time-reversal invariance (which is necessary to maintain CPT conservation), remains to be observed. ∎

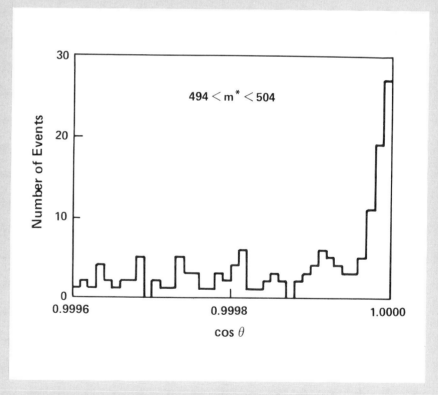

*Evidence for the CP-violating decay of* $K_L^0$ *into two pions. Here the number of events in which the invariant mass (m\*) of the decay products was in close proximity to the mass of the neutral kaon is plotted versus the cosine of the angle $\theta$ between the* $K_L^0$ *beam and the vector sum of the momenta of the decay products. The peak in the number of events at* $\cos \theta \cong 1$ *(indicative of two-body decays) could only be explained as the decay of* $K_L^0$ *into two pions with a branching ratio of about* $2 \times 10^{-3}$. *(Adapted from "Evidence for the $2\pi$ Decay of the $K_L^0$ Meson" by J. H. Christenson, J. W. Cronin, V. L. Fitch, and R. Turlay,* **Physical Review Letters** *13(1964):138.)*

tion of strongly interacting particles into multiplets of SU(3), a scheme that combines strong isospin with the quantum number called strangeness, or strong hypercharge. (For a more complete discussion of continuous symmetries and internal global symmetries such as SU(2), see Lecture Notes 2 and 4.)

Exact, or unbroken, symmetries also play a fundamental role in the construction of theories: exact rotational invariance leads to the exact conservation of angular momentum, and exact translational invariance in space-time leads to the exact conservation of energy and momentum. We will now discuss how the exact phase invariance of electrodynamics leads to the exact conservation of electric charge.

**Global U(1) Invariance and Conservation Laws.** In quantum field theory the dynamics of a system are encoded in a function of the fields called a Lagrangian, which is related to the energy of the system. The Lagrangian is the most convenient means for studying the symmetries of the theory because it is usually a simple task to check if the Lagrangian remains unchanged under particular symmetry operations.

An electron is described in quantum field theory by a complex field,

$$\psi_{\text{electron}} = (\psi_1 + i\psi_2)/\sqrt{2} ,$$

and a positron is described by the complex conjugate of that field,

$$\psi_{\text{positron}} = (\psi_1 - i\psi_2)/\sqrt{2} .$$

Although the real fields $\psi_1$ and $\psi_2$ are separately each able to describe a spin-½ particle, the two together are necessary to describe a particle with electric charge.*

The Lagrangian of quantum electrodynamics is unchanged by the continuous operation of multiplying the electron field by

*The real fields $\psi_1$ and $\psi_2$ are four-component Majorana fields that together make up the standard four-component complex Dirac spinor field.*

an arbitrary phase, that is, by the transformation

$$\psi \rightarrow e^{i\Lambda Q}\psi ,$$

where $\Lambda$ is an arbitary real number and $Q$ is the electric charge operator associated with the field. The eigenvalue of $Q$ is $-1$ for an electron and $+1$ for a positron. This set of phase transformations forms the global symmetry group U(1) (the set of unitary $1 \times 1$ matrices). In QED this symmetry is unbroken, and electric charge is a conserved quantum number of the system.

There are other global U(1) symmetries relevant in particle physics, and each one implies a conserved quantum number. For example, baryon number conservation is associated with a U(1) phase rotation of all baryon fields by an amount $e^{i\Lambda B}$, where $B = 1$ for protons and neutrons, $B = \frac{1}{3}$ for quarks, and $B = 0$ for leptons. Analogously, electron number is conserved if the field of the electron neutrino is assigned the same electron number as the field of the electron and all other fields are assigned an electron number of zero. The same holds true for muon number and tau number. Thus a global U(1) phase symmetry seems to operate on each type of lepton. (Possible violation of muon-number conservation is discussed in "Experiments To Test Unification Schemes.")

## The Principle of Local Symmetry

We are now ready to distinguish a global phase symmetry from a local one and examine the dynamical consequences that emerge from the latter. Figure 7 illustrates what hap-

pens to the electron field under the global phase transformation $\psi \rightarrow e^{i\Lambda Q}\psi$. For convenience, space-time is represented by a set of discrete points labeled by the index $j$. The phase of the electron field at each point is represented by an arrow that rotates about the point, and the kinetic energy of the field is represented by springs connecting the arrows at different space-time points. A *global* U(1) transformation rotates every two-dimensional vector by the same arbitrary angle $\Lambda$: $\theta_j \rightarrow \theta_j + Q\Lambda$, where $Q$ is the electric charge. In order for the Lagrangian to be invariant under this global phase rotation, it is clearly sufficient for it to be a function only of the phase *differences* $(\theta_i - \theta_j)$. Both the free electron terms and the interaction terms in the QED Lagrangian are invariant under this continuous global symmetry.

A local U(1) transformation, in contrast, rotates every two-dimensional vector by a *different* angle $\Lambda_j$. This local transformation, unlike its global counterpart, does *not* leave the Lagrangian of the free electron invariant. As represented in Fig. 7 by the stretching and compressing of the springs, the kinetic energy of the electron changes under local phase transformations. Nevertheless, the full Lagrangian of quantum electrodynamics *is* invariant under these local U(1) transformations. The electromagnetic field $(A_\mu)$ precisely compensates for the local phase rotation and the Lagrangian is left invariant. This is represented in Fig. 7 by restoring the stretched and compressed springs to their initial tension. Thus, the kinetic energy of the electron (the energy stored in the springs) is the same before and after the local phase transformation.

In our discrete notation, the full La-

*Fig. 7. Global versus local phase transformations. The arrows represent the phases of an electron field at four discrete points labeled by $j = 1, 2, 3,$ and 4. The springs represent the kinetic energy of the electrons. A global phase transformation does not change the tension in the springs and therefore costs no energy. A local phase transformation without gauge interactions stretches and compresses the springs and thus does cost energy. However introduction of the gauge field (represented by the white haze) exactly compensates for the local phase transformation of the electron field and the springs return to their original tension so that local phase transformations with gauge interactions do not cost energy.*

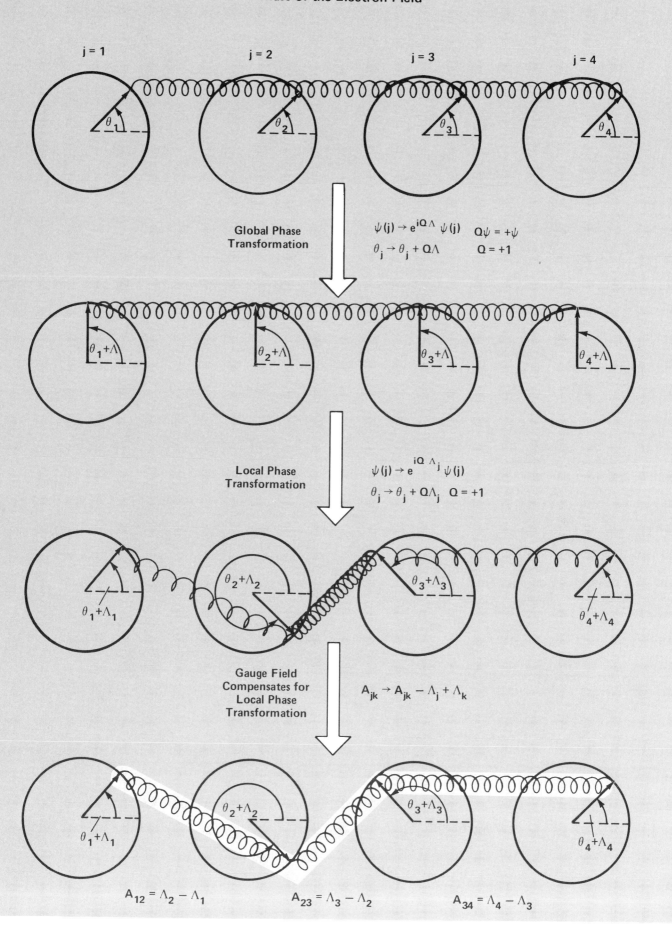

**Phase of the Electron Field**

grangian is a function of $\theta_j - \theta_k + A_{jk}Q$ and is invariant under the simultaneous transformations

$$\theta_j \rightarrow \theta_j + Q\Lambda_j \quad \text{and} \quad A_{jk} \rightarrow A_{jk} - \Lambda_j + \Lambda_k \,.$$

The matrix with elements $A_{jk}$ is the discrete space-time analogue of the electromagnetic potential defined on the links between the points $k$ and $j$. Thus, if one starts with a theory of free electrons with no interactions and demands that the physics remain invariant under local phase transformation of the electron fields, then one induces the standard electromagnetic interactions between the electron current $J^\mu$ and photon field $A_\mu$, as shown in Figs. 5 and 8. From this point of view, Maxwell's equations can be viewed as a *consequence* of the local U(1) phase invariance. Although this local invariance was originally viewed as a curiosity of QED, it is now viewed as the guiding principle for constructing field theories. The invariance is usually termed *gauge invariance*, and the photon is referred to as a gauge particle since it mediates the U(1) gauge interaction. It is worth emphasizing that local U(1) invariance implies that the photon is massless because the term that would describe a massive photon is *not* itself invariant under local U(1) transformations.

The local gauge invariance of QED is the prototype for theories of both the weak and the strong interactions. Obviously, since neither of these is a long-range interaction, some additional features must be at work to account for their different properties. Before turning to a discussion of these features, we stress that in theories based on local gauge invariance, currents always play an important role. In classical electromagnetism the fundamental interaction takes place between the vector potential and the electron current; this is reflected in quantum electrodynamics by Feynman diagrams: the virtual photon (the gauge field) ties into the current produced by the moving electron (see Fig. 8). As will become clear below, a similar situation exists in the strong interaction and, more important, in the weak interaction.

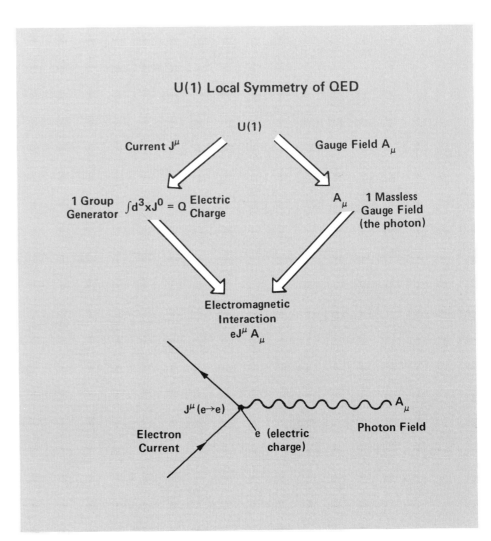

Fig. 8. *The U(1) local symmetry of QED implies the existence of a gauge field to compensate for the local phase transformation of the electrically charged matter fields. The generator of the U(1) local phase transformation is Q, the electric charge operator defined in the figure in terms of the current density J⁰. The gauge field Aμ interacts with the electrically charged matter fields through the current Jμ. The coupling strength is e, the charge of the electron.*

## The Strong Interaction

In an atom electrons are bound to the nucleus by the Coulomb force and occupy a region about $10^{-8}$ centimeter in extent. The nucleus itself is a tightly bound collection of protons and neutrons confined to a region about $10^{-12}$ centimeter across. As already emphasized, the force that binds the protons and neutrons together to form the nucleus is much stronger and considerably shorter in range than the electromagnetic force. Leptons do not feel this strong force; particles that *do* participate in the strong interactions are called hadrons.

# $\Omega^-$

In 1961 M. Gell-Mann and independently Y. Ne'eman proposed a system for classifying the roughly one hundred baryons and mesons known at the time. This "Eightfold Way" was based on the SU(3) group, which has eight independent symmetry operations. According to this system, hadrons with the same baryon number, spin angular momentum, and parity and with electric charge, strangeness (or hypercharge), isotopic spin, and mass related by certain rules were grouped into large multiplets encompassing the already established isospin multiplets, such as the neutron and proton doublet or the negative, neutral, and positive pion triplet. Most of the known hadrons fit quite neatly into octets. However, the decuplet partly filled by the quartet of $\Delta$ baryons and the triplet of $\Sigma(1385)$ baryons lacked three members. Discovery of the $\Xi(1520)$ doublet was announced in 1962, and these baryons satisfied the criteria for membership in the decuplet. This partial confirmation of the Eightfold Way motivated a search at Brookhaven for the remaining member, already named $\Omega^-$ and predicted to be stable against strong and electromagnetic interactions, decaying (relatively slowly) by the weak interaction. Other properties predicted for this particle were a baryon number of 1, a spin angular momentum of 3/2, positive parity, negative electric charge, a strangeness of $-3$, an isotopic spin of 0, and a mass of about 1676 MeV.

A beam of 5-GeV negative kaons produced at the AGS was directed into a liquid-hydrogen bubble chamber, where the $\Omega^-$ was to be produced by reaction of the kaons with protons. The tracks of the decay products of the new particle were then sought in the bubble-chamber photographs. In early 1964 a candidate event was found for decay of an $\Omega^-$ into a $\pi^-$ and a $\Xi^0$, one of three possible decay modes. Within several weeks, by coincidence and good fortune, another $\Omega^-$ was found, this time decaying into a $\Lambda^0$ and a $K^-$, the mode now known to be dominant.

Analysis of the tracks for these two events confirmed the predicted mass and strangeness, and further studies confirmed the predicted spin and parity. Discovery of the $\Omega^-$ established the Eightfold Way as a viable description of hadronic states. ∎

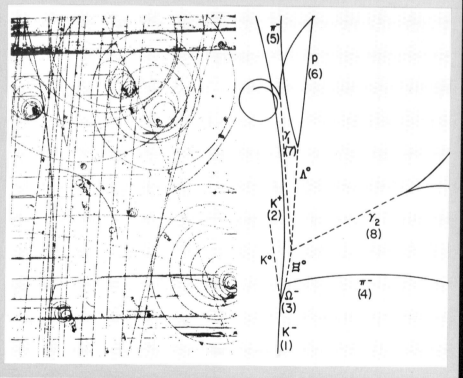

*The $\Omega^-$ was first detected in the bubble-chamber photograph reproduced above. A $K^-$ entered the bubble chamber from the bottom (track 1) and collided with a proton. The collision produced an $\Omega^-$ (track 3), a $K^+$ (track 2), and a $K^0$, which, being neutral, left no track and must have decayed outside the bubble chamber. The $\Omega^-$ decayed into a $\pi^-$ (track 4) and a $\Xi^0$. The $\Xi^0$ in turn decayed into a $\Lambda^0$ and a $\pi^0$. The $\Lambda^0$ decayed into a $\pi^-$ (track 5) and a proton (track 6), and the $\pi^0$ very quickly decayed into two gamma rays, one of which (track 7) created an $e^-$-$e^+$ pair within the bubble chamber. (Photo courtesy of the Niels Bohr Library of the American Institute of Physics and Brookhaven National Laboratory.)*

## Table of "Elementary Particles"

### BARYONS

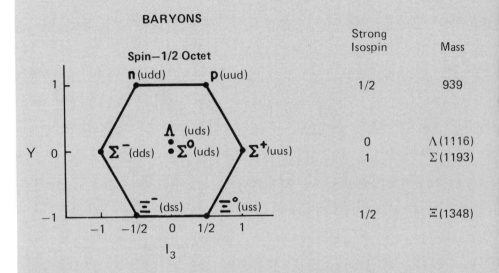

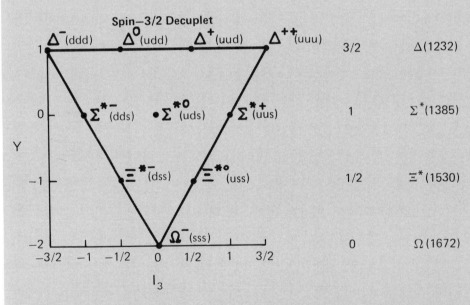

| | Strong Isospin | Mass |
|---|---|---|
| **Spin—1/2 Octet** | | |
| n, p | 1/2 | 939 |
| Λ | 0 | Λ (1116) |
| Σ | 1 | Σ (1193) |
| Ξ | 1/2 | Ξ (1348) |
| **Spin—3/2 Decuplet** | | |
| Δ | 3/2 | Δ (1232) |
| Σ* | 1 | Σ* (1385) |
| Ξ* | 1/2 | Ξ* (1530) |
| Ω | 0 | Ω (1672) |

*Fig. 9. The Eightfold Way classified the hadrons into multiplets of the symmetry group SU(3). Particles of each SU(3) multiplet that lie on a horizontal line form strong-isospin (SU(2)) multiplets. Each particle is plotted according to the quantum numbers $I_3$ (the third component of strong isospin) and strong hypercharge Y (Y = S + B, where S is strangeness and B is baryon number). These quantum numbers correspond to the two diagonal generators of SU(3). The quantum numbers of each particle are easily understood in terms of its fundamental quark constituents. Baryons contain three quarks and mesons contain quark-antiquark pairs. Baryons in the spin-3/2 decuplet are obtained from baryons in the spin-1/2 octet by changing the spin and SU(3) flavor quantum numbers of the three quark wave functions. For example, the three quarks that compose the neutron in the spin-1/2 octet can reorient their spins to form the $\Delta^0$ in the spin-3/2 decuplet. Similar changes in the meson quark-antiquark wave functions change the spin-0 meson octet into the spin-1 meson octet.*

The mystery of the strong force and the structure of nuclei seemed very intractable as little as fifteen years ago. Studying the relevant distance scales requires machines that can accelerate protons or electrons to energies of 1 GeV and beyond. Experiments with less energetic probes during the 1950s revealed two very interesting facts. First, the strong force does not distinguish between protons and neutrons. (In more technical language, the proton and the neutron transform into each other under isospin rotations, and the Lagrangian of the strong interaction is invariant under these rotations.) Second, the structure of protons and neutrons is as rich as that of nuclei. Furthermore, many new hadrons were discovered that were apparently just as "elementary" as protons and neutrons.

The table of "elementary particles" in the mid-1960s displayed much of the same complexity and symmetry as the periodic table of the elements. In 1961 both Gell-Mann and Ne'eman proposed that all hadrons could be classified in multiplets of the symmetry group called SU(3). The great triumph of this proposal was the prediction and subsequent discovery of a new hadron, the omega minus. This hadron was needed to fill a vacant space in one of the SU(3) multiplets (Fig. 9).

In spite of the SU(3) classification scheme, the belief that all of these so-called elementary particles were truly elementary became more and more untenable. The most contradictory evidence was the finite size of hadrons (about $10^{-13}$ centimeter), which drastically contrasted with the point-like nature of the leptons. Just as the periodic table was eventually explained in terms of a few basic building blocks, so the hadronic zoo was eventually tamed by postulating the existence of a small number of "truly elementary point-like particles" called quarks. In 1963 Gell-Mann and, independently, Zweig realized that all hadrons could be constructed from three spin-½ fermions, designated u, d, and s (up, down, and strange). The SU(3) symmetry that manifested itself in the table of "elementary particles" arose from an invariance of the La-

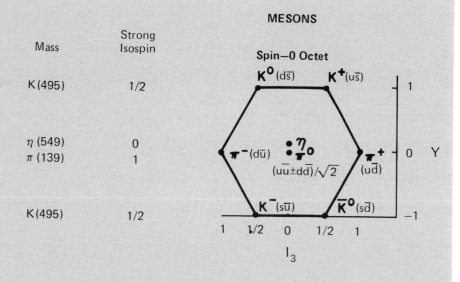

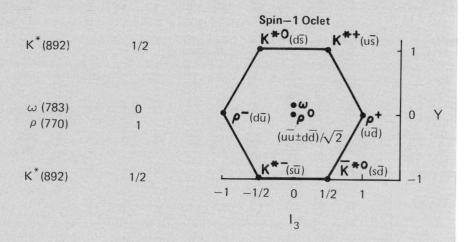

| Mass | Strong Isospin | | | |
|------|---------------|--|--|--|

**MESONS**

**Spin—0 Octet**

K (495)  1/2

η (549)  0
π (139)  1

K (495)  1/2

K*(892)  1/2

ω (783)  0
ρ (770)  1

K*(892)  1/2

**Quarks**

| Name | Symbol | Electric Charge | Y |
|------|--------|-----------------|-----|
| Up | u | 2/3 | 1/3 |
| Down | d | −1/3 | 1/3 |
| Strange | s | −1/3 | −2/3 |

grangian of the strong interaction to rotations among these three objects. This global symmetry is exact only if the $u$, $d$, and $s$ quarks have identical masses, which implies that the particle states populating a given SU(3) multiplet also have the same mass. Since this is certainly not the case, SU(3) is a broken global symmetry. The dominant breaking is presumed to arise, as in the example of $e$, $\mu$, and $\tau$, from the differences in the masses of the $u$, $d$, and $s$ quarks. The origin of these quark masses is one of the great unanswered questions. It is established, however, that SU(3) symmetry among the $u$, $d$, and $s$ quarks is preserved by the strong interaction. Nowadays, one refers to this SU(3) as a *flavor* symmetry, with $u$, $d$, and $s$ representing different quark flavors. This nomenclature is to distinguish it from another and quite different SU(3) symmetry possessed by quarks, a local symmetry that is associated directly with the strong force and has become known as the SU(3) of *color*. The theory resulting from this symmetry is called quantum chromodynamics (QCD), and we now turn our attention to a discussion of its properties and structure.

The fundamental structure of quantum chromodynamics mimics that of quantum electrodynamics in that it, too, is a gauge theory (Fig. 10). The role of electric charge is played by three "colors" with which each quark is endowed—red, green, and blue. The three color varieties of each quark form a triplet under the SU(3) local gauge symmetry. A local phase transformation of the quark field is now considerably extended since it can rotate the color and thereby change a red quark into a blue one. The local gauge transformations of quantum electrodynamics simply change the phase of an electron, whereas the color transformations of QCD actually change the particle. (Note that these two types of phase transformation are totally independent of each other.)

We explained earlier that the freedom to change the local phase of the electron field forces the introduction of the photon field (sometimes called the gauge field) to keep the Lagrangian (and therefore the resulting physics) invariant under these local phase changes. This is the principle of local symmetry. A similar procedure applied to the quark field induces the so-called chromodynamic force. There are eight independent symmetry transformations that change the color of a quark and these must be compensated for by the introduction of *eight* gauge fields, or spin-1 bosons (analogous to the single photon of quantum electrodynamics). Extension of the local U(1) gauge invariance of QED to more complicated symmetries such as SU(2) and SU(3) was first done by Yang and Mills in 1954. These larger symmetry groups involve so-called non-Abelian, or non-commuting algebras (in which $AB \neq BA$), so it has become customary to refer to this class of theories as "non-Abelian gauge theories." An alternative term is simply "Yang-Mills theories."

The eight gauge bosons of QCD are referred to by the bastardized term "gluon," since they represent the glue that holds the physical hadrons, such as the proton, together. The interactions of gluons with quarks are depicted in Fig. 10. Although gluons are the counterpart to photons in that they have unit spin and are massless, they possess one crucial property not shared by photons: they themselves carry color. Thus they not only mediate the color force but also carry it; it is as if photons were charged. This difference (it is the difference between an Abelian and a non-Abelian gauge theory) has many profound physical consequences. For example, because gluons carry color they can (unlike photons) interact with themselves (see Fig. 10) and, in effect, weaken the force of the color charge at *short* distances. The opposite effect occurs in quantum electrodynamics: screening effects weaken the effective electric charge at *long* distances. (As mentioned above, a virtual photon emanating from an electron can create a virtual electron-positron pair. This polarization screens, or effectively decreases, the electron's charge.)

The weakening of color charge at short distances goes by the name of *asymptotic freedom*. Asymptotic freedom was first ob-

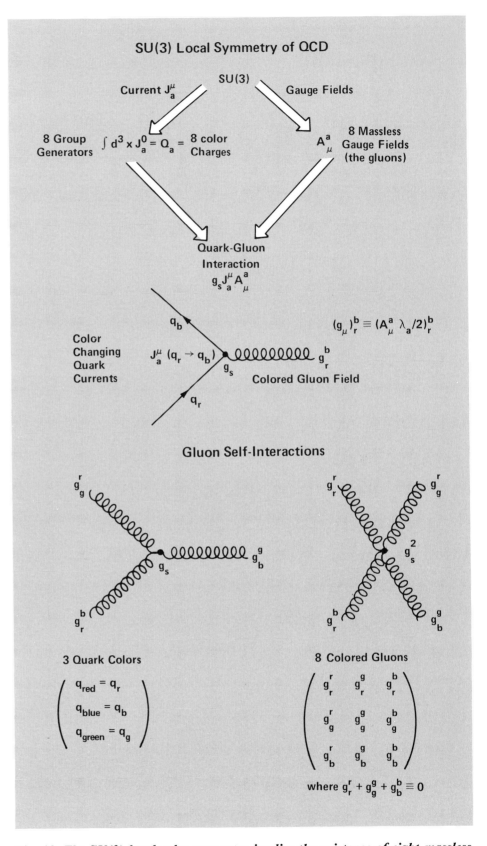

Fig. 10. *The SU(3) local color symmetry implies the existence of eight massless gauge fields (the gluons) to compensate for the eight independent local transformations of the colored quark fields. The subscripts r, g, and b on the gluon and quark fields correspond respectively to red, green, and blue color charges. The eight gluons carry color and obey the non-Abelian algebra of the SU(3) generators (see Lecture Note 4). The interactions induced by the local SU(3) color symmetry include a quark-gluon coupling as well as two types of gluon self-interactions (one proportional to the couping $g_s$ and the other proportional to $g_s^2$).*

# QCD on a Cray:
# the masses of elementary particles

*by Gerald Guralnik, Tony Warnock, and Charles Zemach*

How can we extract answers from QCD at energies below 1 GeV? As noted in the text, the confinement of quarks suggests that weak-coupling perturbative methods are *not* going to be successful at these energies. Nevertheless, if QCD is a valid theory it must explain the multiplicities, masses, and couplings of the experimentally observed strongly interacting particles. These would emerge from the theory as bound states and resonances of quarks and gluons. A valid theory must also account for the apparent absence of isolated quark states and might predict the existence and properties of particles (such as glueballs) that have not yet been seen.

The most promising nonperturbative formulation of QCD exploits the Feynman path integral. Physical quantities are expressed as integrals of the quark and gluon fields over the space-time continuum with the QCD Lagrangian appearing in an exponential as a kind of Gibbs weight factor. This is directly analogous to the partition function formulation of statistical mechanics. The path integral prescription for strong interaction dynamics becomes well defined mathematically when the space-time continuum is approximated by a discrete four-dimensional lattice of finite size and the integrals are evaluated by Monte Carlo sampling.

The original Monte Carlo ideas of Metropolis and Ulam have now been applied to QCD by many researchers. These efforts have given credibility, but not confirmation, to the hope that computer simulations might indeed provide critical tests of QCD and significant numerical results. With considerable patience (on the order of many months of computer time) a VAX 11/780 can be used to study universes of about 3000 space-time points. Such a universe is barely large enough to contain a proton and not really adequate for a quantitative calculation. Consequently, with these methods, any result from a computer of VAX power is, at best, only an indication of what a well-done numerical simulation might produce.

We believe that a successful computer simulation must combine the following: (1) physical and mathematical ingenuity to search out the best formulations of problems still unsolved in principle; (2) sophisticated numerical analysis and computer programming; and (3) a computer with the speed, memory, and input/output rate of the Cray XMP with a solid-state disk (or better). We have done calculations of particle masses on a lattice of 55,296 space-time points using the Cray XMP. Using new methods developed with coworkers R. Gupta, J. Mandula, and A. Patel, we are examining glueball masses, renormalization group behavior, and the behavior of the theory on much larger lattices. The results to date support the belief that QCD describes interactions of the elementary particles and that these numerical methods are currently the most powerful means for extracting the predictive content of QCD.

The calculations, which have two input parameters (the pion mass and the long-range quark-quark force constant in units of the lattice spacing), provide estimates of many measurable quantities. The accompanying table shows some of our results on elementary particle masses and certain meson coupling strengths. These results represent several hundred hours of Cray time. The quoted relative errors derive from the statistical analysis of the Monte Carlo calculation itself rather than from a comparison with experimental data. Significantly more computer time would significantly reduce the errors in the calculated masses and couplings.

Our work would not have been possible without the support of C Division and many of its staff. We have received generous support from Cray Research and are particularly indebted to Bill Dissly and George Spix for contribution of their skills and their time. ∎

Calculated and experimental values for the masses and coupling strengths of some mesons and baryons.

| | Calculated Value (MeV/$c^2$) | Relative Error (%) | Experimental Value (MeV/$c^2$) |
|---|---|---|---|
| **Masses** | | | |
| $\rho$ meson | 767 | 18 | 769 |
| Excited $\rho$ | 1426 | 27 | 1300? |
| $\delta$ meson | 1154 | 15 | 983 |
| $A_1$ meson | 1413 | 17 | 1275 |
| Proton | 989 | 23 | 940 |
| $\Delta$ baryon | 1199 | 17 | 1210 |
| **Couplings** | | | |
| $f_\pi$ | 121 | 21 | 93 |
| $f_\rho$ | 211 | 15 | 144 |

*observable* hadrons are necessarily colorless, whereas quarks and gluons are permanently confined. This is just as well since gluons are massless, and by analogy with the photon, unconfined massless gluons should give rise to a long-range, Coulomb-like, color force in the strong interactions. Such a force is clearly at variance with experiment! Even though color is confined, residual strong color forces can still "leak out" in the form of color-neutral pions or other hadrons and be responsible for the binding of protons and neutrons in nuclei (much as residual electromagnetic forces bind atoms together to form molecules).

The success of QCD in explaining short-distance behavior and its aesthetic appeal as a generalization of QED have given it its place in the standard model. However, confidence in this theory still awaits convincing calculations of phenomena at distance scales of $10^{-13}$ centimeter, where the "strong" nature of the force becomes dominant and perturbation theory is no longer valid. (Lattice gauge theory calculations of the hadronic spectrum are becoming more and more reliable. See "QCD on a Cray: The Masses of Elementary Particles.")

## The Weak Interaction

Many nuclei are known to be unstable and to emit several kinds of particles when they decay; historically these particles were called alpha particles, beta rays, and gamma rays. These three are now associated with three quite different modes of decay. An alpha particle, itself a helium nucleus, is emitted during the strong-interaction decay mode known as fission. Large nuclei that are only loosely bound by the strong force (such as uranium-238) can split into two stable pieces, one of which is an alpha particle. A gamma ray is simply a photon with "high" energy (above a few MeV) and is emitted during the decay of an excited nucleus. A beta ray is an electron emitted when a neutron in a nucleus decays into a proton, an electron, and an electron antineutrino ($n \rightarrow p + e^- + \bar{\nu}_e$, see Fig. 11). The proton remains in

served in deep inelastic scattering experiments (see "Scaling in Deep Inelastic Scattering"). This phenomenon explains why hadrons at high energies behave as if they were made of almost free quarks even though one knows that quarks must be tightly bound together since they have never been experimentally observed in their free state. The weakening of the force at high energies means that we can use perturbation theory to calculate hadronic processes at these energies.

The self-interaction of the gluons also explains the apparently permanent confinement of quarks. At long distances it leads to such a proliferation of virtual gluons that the color charge effectively grows without limit, forbidding the propagation of *all* colored particles. Only bleached, or color-neutral, states (such as baryons, which have equal proportions of red, blue, and green, or mesons which have equal proportions of red-antired, green-antigreen, and blue-antiblue) are immune from this confinement. Thus all

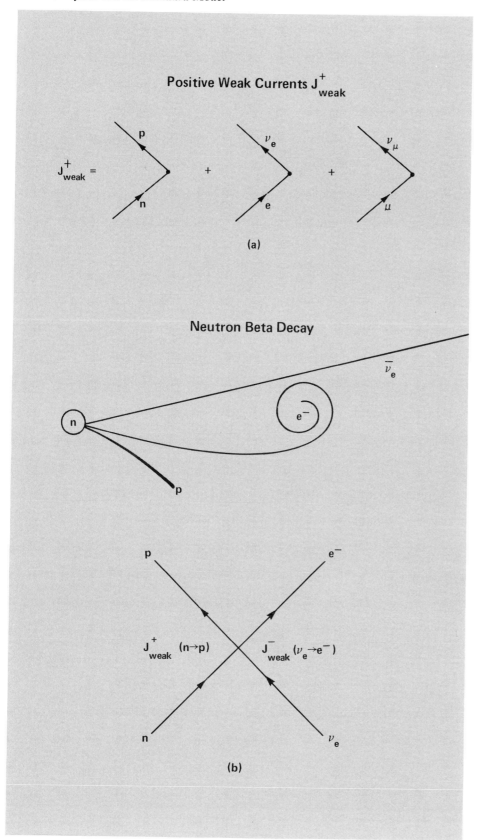

**Fig. 11. (a)** *Components of the charge-raising weak current* $J^+_{weak}$ *are represented in the figure by Feynman diagrams in which a neutron changes into a proton, an electron into an electron neutrino, and a muon into a muon neutrino. The charge-lowering current* $J^-_{weak}$ *is represented by reversing the arrows.* **(b)** *Beta decay (shown in the figure) and other low-energy weak processes are well described by the Fermi interaction* $J^+_{weak} \times J^-_{weak}$. *The figure shows the Feynman diagram of the Fermi interaction for beta decay.*

the nucleus, and the electron and its antineutrino escape. This decay mode is characterized as weak because it proceeds much more slowly than most electromagnetic decays (see Table 1). Other baryons may also undergo beta decay.

Beta decay remained very mysterious for a long time because it seemed to violate energy-momentum conservation. The free neutron was observed to decay into two particles, a proton and an electron, each with a spectrum of energies, whereas energy-momentum conservation dictates that each should have a unique energy. To solve this dilemma, Pauli invoked the neutrino, a massless, neutral fermion that participates only in weak interactions.

**The Fermi Theory.** Beta decay is just one of many manifestations of the weak interaction. By the 1950s it was known that all weak processes could be concisely described in terms of the current-current interaction first proposed in 1934 by Fermi. The charged weak currents $J^+_{weak}$ and $J^-_{weak}$ change the electric charge of a fermion by one unit and can be represented by the sum of the Feynman diagrams of Fig. 11a. In order to describe the maximal parity violation, (that is, the maximal right-left asymmetry) observed in weak interactions, the charged weak current includes only left-handed fermion fields. (These are defined in Fig. 12 and Lecture Note 8.)

Fermi's current-current interaction is then given by all the processes included in the product $(G_F/\sqrt{2})(J^+_{weak} \times J^-_{weak})$ where $J^-_{weak}$ means all arrows in Fig. 11a are reversed. This interaction is in marked contrast to quantum electrodynamics in which two currents interact through the exchange of a virtual photon (see Fig. 3). In weak processes two charge-changing currents appear to interact locally (that is, at a single point) without the help of such an intermediary. The coupling constant for this local interaction, denoted by $G_F$ and called the Fermi constant, is not dimensionless like the coupling parameter $\alpha$ in QED, but has the

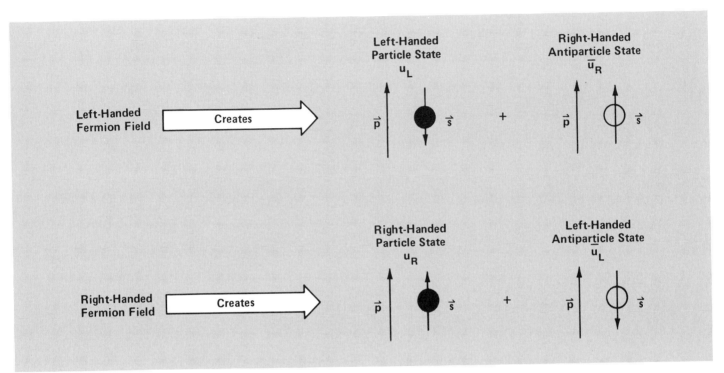

*Fig. 12. A Dirac spinor field can be decomposed into left-and right-handed pieces. A left-handed field creates two types of particle states at ultrarelativistic energies—$u_L$, a particle with spin opposite to the direction of motion, and $\bar{u}_R$, an antiparticle with spin along the direction of motion. Only left-handed fields contribute to the weak charged currents shown in Fig. 11. The left- and right-handedness (or chirality) of a field describes a Lorentz covariant decomposition of Dirac spinor fields.*

dimension of mass$^{-2}$ or energy$^{-2}$. In units of energy, the measured value of $G_F^{-1/2}$ equals 293 GeV. Thus the strength of the weak processes seems to be determined by a specific energy scale. But why?

**Predictions of the W boson.** An explanation emerges if we postulate the existence of an intermediary for the weak interactions. Recall from Fig. 3 that the exchanged, or virtual, photons in QED basically correspond to the Coulomb potential $\alpha/r$, whose Fourier transform is $\alpha/q^2$, where $q$ is the momentum of the virtual photon. It is tempting to suggest that the nearly zero range of the weak interaction is only apparent in that the two charged currents interact through a potential of the form $\alpha'[\exp(-M_W r)]/r$ (a form originally proposed by Yukawa for the short-

range force between nucleons), where $\alpha'$ is the analogue of $\alpha$ and the mass $M_W$ is so large that this potential has essentially no range. The Fourier transform of this potential, $\alpha'/(q^2 + M_W^2)$, suggests that, if this idea is correct, the interaction between the weak currents is mediated by a "heavy photon" of mass $M_W$. Nowadays this particle is called the W boson; its existence explains the short range of the weak interactions.

Notice that at low energies, or, equivalently, when $M_W^2 \gg q^2$, the Fourier transform, or so-called propagator of the W boson, reduces to $\alpha'/(M_W^2)$, and since this factor multiplies the two currents, it must be proportional to Fermi's constant. Thus the existence of the W boson gives a natural explanation of why $G_F$ is not dimensionless.

Now, since both the weak and electro-

magnetic interactions involve electric charge, these two might be manifestations of the same basic force. If they were, then $\alpha'$ might be the same as $\alpha$ and $G_F$ would be proportional to $\alpha/M_W^2$. Thus the existence of a very massive W boson can explain not only the short range but also the weakness of weak interactions relative to electromagnetic interactions! This argument not only predicts the existence of a W boson but also yields a rough estimate of its mass:

$$M_W \approx \sqrt{\alpha/G_F} = 25 \text{ GeV}/c^2 .$$

This prediction of a new particle was made in the 1950s, when such energies were well beyond reach of the existing accelerators.

Arguments like the one above convinced physicists that a theoretical unification of

## Table 2

### Multiplets and quantum numbers in the SU(2) × U(1) electroweak theory.

| | | Weak Isotopic Charge $I_3$ | Weak Hypercharge $Y$ | Electric Charge $Q$ $(=I_3 + \frac{1}{2}Y)$ |
|---|---|---|---|---|
| **Quarks** | | | | |
| SU(2) Doublet | $u_L$ | $\frac{1}{2}$ | $\frac{1}{3}$ | $\frac{2}{3}$ |
| | $d_L$ | $-\frac{1}{2}$ | $\frac{1}{3}$ | $-\frac{1}{3}$ |
| SU(2) Singlets | $u_R$ | $0$ | $4/3$ | $\frac{2}{3}$ |
| | $d_R$ | $0$ | $-\frac{2}{3}$ | $-\frac{1}{3}$ |
| **Leptons** | | | | |
| SU(2) Doublet | $(\nu_e)_L$ | $\frac{1}{2}$ | $-1$ | $0$ |
| | $e_L$ | $-\frac{1}{2}$ | $-1$ | $-1$ |
| SU(2) Singlet | $e_R$ | $0$ | $-2$ | $-1$ |
| **Gauge Bosons** | | | | |
| SU(2) Triplet | $W^+$ | $1$ | $0$ | $1$ |
| | $W_3$ | $0$ | $0$ | $0$ |
| | $W^-$ | $-1$ | $0$ | $-1$ |
| SU(2) Singlet | $B$ | $0$ | $0$ | $0$ |
| **Higgs Boson** | | | | |
| SU(2) Doublet | $\varphi^+$ | $\frac{1}{2}$ | $1$ | $1$ |
| | $\varphi^0$ | $-\frac{1}{2}$ | $1$ | $0$ |

energies, where $q^2 \gg M_W^2$, the weak interaction becomes comparable in strength to the electromagnetic. Thus we see explicitly how the apparent strength of the interaction depends on the wavelength of the probe.

**The SU(2) × U(1) Electroweak Theory.** Since quantum electrodynamics is a gauge theory based on local U(1) invariance, it is not too surprising that the theory unifying the electromagnetic and weak forces is also a gauge theory. Construction of such a theory required overcoming both technical and phenomenological problems.

The technical problem concerned the fact that an electroweak gauge theory is necessarily a Yang-Mills theory (that is, a theory in which the gauge fields interact with each other); the gauge fields, namely the $W$ bosons, must be charged to mediate the charge-changing weak interactions and therefore by definition must interact with each other electromagnetically through the photon. Moreover, the local gauge symmetry of the theory must be broken because an unbroken symmetry would require all the gauge particles to be massless like the photon and the gluons, whereas the $W$ boson must be massive. A major theoretical difficulty was understanding how to break a Yang-Mills gauge symmetry in a consistent way. (The solution is presented below.)

In addition to the technical issue, there was the phenomenological problem of choosing the correct local symmetry group. The most natural choice was SU(2) because the low-lying states (that is, the observed quarks and leptons) seemed to form doublets under the weak interaction. For example, a $W^-$ changes $\nu_e$ into $e$, $\nu_\mu$ into $\mu$, or $u$ into $d$ (where all are left-handed fields), and the $W^+$ effects the reverse operation. Moreover, the three gauge bosons required to compensate for the three independent phase rotations of a local SU(2) symmetry could be identified with the $W^+$, the $W^-$, and the photon. Unfortunately, this simplistic scenario does not work: it gives the wrong electric charge assignments for the quarks and leptons in the SU(2) doublets. Specifically, electric charge

electromagnetic and weak interactions must be possible. Several attempts were made in the 1950s and 1960s, notably by Schwinger and his student Glashow and by Ward and Salam, to construct an "electroweak theory" in terms of a local gauge (Yang-Mills) theory that generalizes QED. Ultimately, Weinberg set forth the modern solution to giving masses to the weak bosons in 1967, although it was not accepted as such until 't Hooft and Veltman showed in 1971 that it constituted a consistent quantum field theory. The success of the electroweak theory culminated in 1982 with the discovery at CERN of the $W$ boson at almost exactly the prediced mass. Notice, incidentally, that at sufficiently high

$Q$ would be equal to the SU(2) charge $I_3$, and the values of $I_3$ for a doublet are $\pm\frac{1}{2}$. This is clearly the wrong charge. In addition, SU(2) would not distinguish the charges of a quark doublet ($\frac{2}{3}$ and $-\frac{1}{3}$) from those of a lepton doublet (0 and $-1$).

To get the correct charge assignments, we can either put quarks and leptons into SU(2) triplets (or larger multiplets) instead of doublets, or we can enlarge the local symmetry group. The first possibility requires the introduction of new heavy fermions to fill the multiplets. The second possibility requires the introduction of at least one new U(1) symmetry (let's call it weak hypercharge $Y$), which yields the correct electric charge assignments if we define

$$Q = I_3 + \frac{1}{2}\,Y\,.$$

This is exactly the possibility that has been confirmed experimentally. Indeed, the electroweak theory of Glashow, Salam, and Weinberg is a local gauge theory with the symmetry group SU(2) $\times$ U(1). Table 2 gives the quark and lepton multiplets and their associated quantum numbers under SU(2) $\times$ U(1), and Fig. 13 displays the interactions defined by this local symmetry. There is one coupling associated with each factor of SU(2) $\times$ U(1), a coupling $g$ for SU(2) and a coupling $g'/2$ for U(1).

The addition of the local U(1) symmetry introduces a new uncharged gauge particle into the theory that gives rise to the so-called neutral-current interactions. This new type of weak interaction, which allows a neutrino to interact with matter without changing its identity, had not been observed when the neutral weak boson was first proposed in 1961 by Glashow. Not until 1973, after all the technical problems with the SU(2) $\times$ U(1) theory had been worked out, were these interactions observed in data taken at CERN in 1969 (see Fig. 14).

The physical particle that mediates the weak interaction between neutral currents is the massive $Z^0$. The electromagnetic interaction between neutral currents is mediated by the familiar massless photon. These two

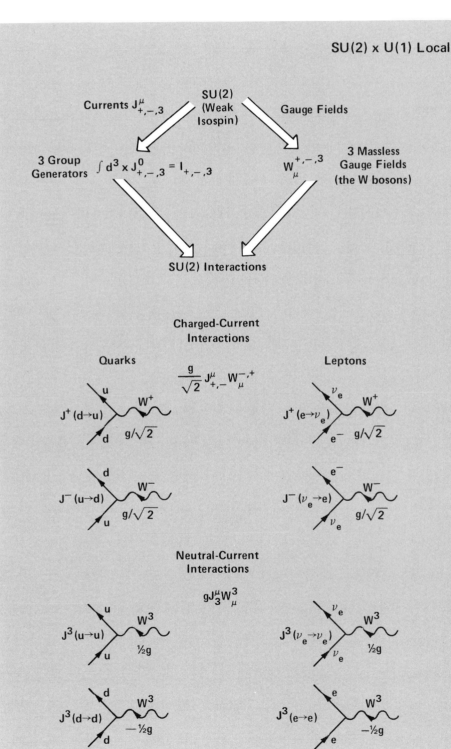

46

## Symmetry of Electroweak Interactions

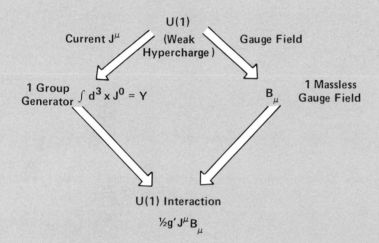

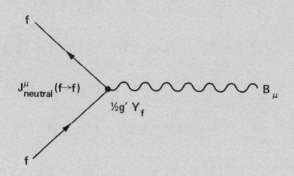

*Fig. 13. The unbroken SU(2) × U(1) local symmetry of the electroweak theory has associated with it gauge fields, currents, and interactions analogous to those of QED and QCD (see Figs. 5 and 8). The figure shows the lowest order interactions between the fermion fields and the gauge fields. The SU(2) interaction involves left-handed quark and lepton fields only. The f in the U(1) interaction stands for both left- and right-handed fermion fields with charge $Y_f$. ($Y_f$ differs for the left- and right-handed components.) Although the gauge fields are self-interacting as in the case of QCD, the SU(2) × U(1) symmetry is broken and the gauge fields are massive so that their self-interactions contribute only very small corrections to the lowest order diagrams and are not shown.*

physical particles are different from the two neutral gauge particles ($B$ and $W_3$) associated with the unbroken SU(2) × U(1) symmetry shown in Fig. 13. In fact, the photon and the $Z^0$ are linear combinations of the neutral gauge particles $W_3$ and $B$:

$$A = B \cos \theta_W + W_3 \sin \theta_W$$

and

$$Z^0 = B \sin \theta_W - W_3 \cos \theta_W .$$

The mixing of SU(2) and U(1) gauge particles to give the physical particles is one result of the fact that the SU(2) × U(1) symmetry must be a broken symmetry.

**Spontaneous Symmetry Breaking.** The astute reader may well be wondering how a local gauge theory, which in QED required the photon to be massless, can allow the mediator of the weak interactions to be massive, especially since the two forces are to be unified. The solution to this paradox lies in the curious way in which the SU(2) × U(1) symmetry is broken.

As Nambu described so well, this breaking is very much analogous to the symmetry breaking that occurs in a superconductor. A superconductor has a local U(1) symmetry, namely, electromagnetism. The ground state, however, is not invariant under this symmetry since it is an ordered state of bound electron-electron pairs (the so-called Cooper pairs) and therefore has a nonzero electric charge distribution. As a result of this asymmetry, photons inside the superconductor acquire an effective mass, which is responsible for the Meissner effect. (A magnetic field cannot penetrate into a superconductor; at the surface it decreases exponentially at a rate proportional to the effective mass of the photon.)

In the weak interactions the symmetry is also assumed to be broken by an asymmetry of the ground state, which in this case is the "vacuum." The asymmetry is due to an ordered state of electrically neutral bosons that carry the weak charge, the so-called Higgs bosons. They break the SU(2) × U(1) symmetry to give the U(1) of electromagnetism in such a way that the $W^\pm$ and the $Z^0$ obtain masses and the photon remains massless. As a result the charges $I_3$ and $Y$ associated with SU(2) $\times$ U(1) are not conserved in weak processes because the vacuum can absorb these quantum numbers. The electric charge $Q$ associated with U(1) of electromagnetism remains conserved.

The asymmetry of the ground state is frequently referred to as spontaneous symmetry breaking; it does not destroy the symmetry of the Lagrangian but destroys only the symmetry of the states. This symmetry breaking mechanism allows the electroweak Lagrangian to remain invariant under the local symmetry transformations while the gauge particles become massive (see Lecture Notes 3, 6, and 8 for details).

In the spontaneously broken theory the electromagnetic coupling $e$ is given by the expression $e = g\sin\theta_W$, where

$$\sin^2\theta_W \equiv g'^2/(g^2 + g'^2) .$$

Thus, $e$ and $\theta_W$ are an alternative way of expressing the couplings $g$ and $g'$, and just as $e$ is not determined in QED, the equally important mixing angle $\theta_W$ is not determined by the electroweak theory. It is, however, measured in the neutral-current interactions. The experimental value is $\sin^2\theta_W = 0.224 \pm 0.015$. The theory predicts that

$$M_W/M_Z = \cos\theta_W$$

and

$$M_W = \left(\frac{\pi\alpha}{\sqrt{2}\,G_F}\right)^{1/2} \frac{1}{\sin\theta_W} .$$

These relations (which are changed only slightly by small quantum corrections) and the experimental value for the weak angle $\theta_W$ predict masses for the $W^\pm$ and $Z^0$ that are in very good agreement with the 1983 observations of the $W^\pm$ and $Z^0$ at CERN.

In the electroweak theory quarks and leptons also obtain mass by interacting with the ordered vacuum state. However, the values of their masses are not predicted by the

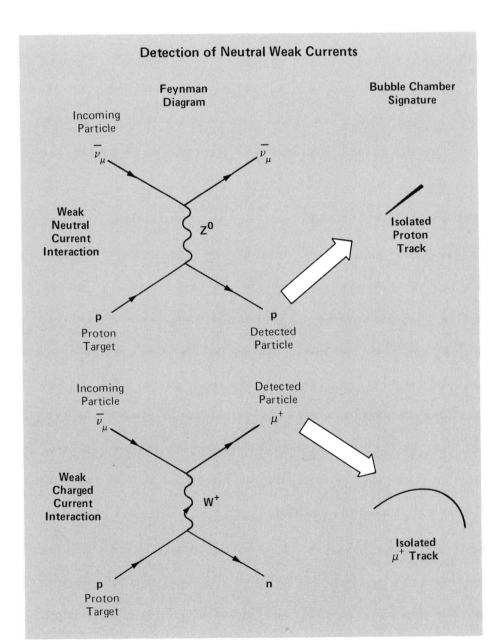

**Detection of Neutral Weak Currents**

*Fig. 14. Neutral-current interactions were first identified in 1973 in photographs taken with the CERN Gargamelle bubble chamber. The figure illustrates the difference between neutral-current and charged-current interactions and shows the bubble-chamber signature of each. The bubble tracks are created by charged particles moving through superheated liquid freon. The incoming antineutrinos interact with protons in the liquid. A neutral-current interaction leaves no track from a lepton, only a track from the positivley charged proton and perhaps some tracks from pions. A charged-current interaction leaves a track from a positively charged muon only.*

# Electronic Weak Neutral Current

Nonconservation of parity was first proposed by C. N. Yang and T. D. Lee in 1956 as a solution to the so-called $\tau$-$\theta$ puzzle: the decay products of the $\tau$ meson (three $\pi$ mesons) differed in parity from the decay products of the $\theta$ meson (two $\pi$ mesons), yet in all other respects the two mesons (now known as $K_L^0$ and $K_S^0$) appeared identical. Yang and Lee's heretical suggestion was proved correct only months later by the cobalt-60 experiment of C. S. Wu and E. Ambler. This experiment, which revealed a decided asymmetry in the direction of emission of beta particles from spin-aligned cobalt-60 nuclei, established parity violation as a feature of charged-current weak interactions and thus of the $\tau$ and $\theta$ decays.

According to the Glashow-Weinberg-Salam theory unifying electromagnetic and weak interactions, parity violation should be a feature also of neutral-current weak interactions but at a low level because of competing electromagnetic interactions. In 1978 a group of twenty physicists headed by C. Prescott observed a parity violation of almost exactly the predicted magnitude in a beautifully executed experiment at the Stanford Linear Accelerator Center. The experiment clearly revealed a small difference (of order 1 part in 10,000) between the cross sections for scattering of right- and left-handed longitudinally polarized electrons by deuterons or protons. ■

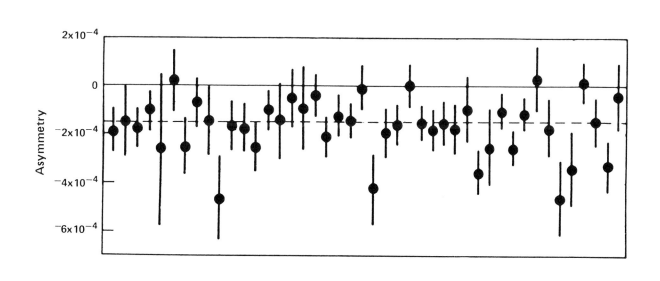

*Data from the SLAC experiment demonstrating parity violation in neutral-current weak interactions. The asymmetry plotted here is defined as the ratio of the difference between the scattering cross sections for right- and left-handed longitudinally polarized electrons to the sum of the cross sections. The dashed line is the mean value of the forty-four asymmetry measurements. (Adapted from* SLAC Beam Line, *Report No. 8, October, 1978.)*

theory but are proportional to arbitrary parameters related to the strength of the coupling of the quarks and leptons to the Higgs boson.

**The Higgs Boson.** In the simplest version of the spontaneously broken electroweak model, the Higgs boson is a complex SU(2) doublet consisting of four real fields (see Table 2). These four fields are needed to transform massless gauge fields into massive ones. A massless gauge boson such as the photon has only two orthogonal spin components (both transverse to the direction of motion), whereas a massive gauge boson has three (two transverse and one longitudinal, that is, in the direction of motion). In the electroweak theory the $W^+$, the $W^-$, and the $Z^0$ absorb three of the four real Higgs fields to form their longitudinal spin components and in so doing become massive. In more picturesque language, the gauge bosons "eat" the Higgs boson and become massive from the feast. The remaining neutral Higgs field is not used up in this magic transformation from massless to massive gauge bosons and therefore should be observable as a particle in its own right. Unfortunately, its mass is not fixed by the theory. However, it can decay into quarks and leptons with a definite signature. It is certainly a necessary component of the theory and is presently being looked for in high-energy experiments at CERN. Its absence is a crucial missing link in the confirmation of the standard model.

**Open Problems.** Our review of the standard model would not be complete without mention of some questions that it leaves unanswered. We discussed above how the three charged leptons ($e$, $\mu$, and $\tau$) may form a triplet under some broken symmetry. This is only part of the story. There are, in fact, three quark-lepton families (Table 3), and these three families may form a triplet under such a broken symmetry. (There is a missing state in this picture: conclusive evidence for the top quark $t$ has yet to be presented. The bottom quark $b$ has been observed in $e^+e^-$ annihilation experiments at SLAC and

# $W^-, W^+, Z^0$

In January 1983 two groups announced the results of separate searches at the CERN proton-antiproton collider for the $W^-$ and $W^+$ vector bosons of the electroweak model. One group, headed by C. Rubbia and A. Astbury, reported definite identification, from among about a billion proton-antiproton collisions, of four $W^-$ decays and one $W^+$ decay. The mass reported by this group ($81 \pm 5$ GeV/$c^2$) agrees well with that predicted by the electroweak model ($82 \pm 2.4$ GeV/$c^2$). The other group, headed by P. Darriulat and using a different detector, reported identification of four possible $W^\pm$ decays, again from among a billion events. The charged vector bosons were produced by annihilation of a quark inside a proton ($uud$) with an antiquark inside an antiproton ($\bar{u}\bar{u}\bar{d}$):

$$d + \bar{u} \to W^-$$

and

$$u + \bar{d} \to W^+ .$$

Since these reactions have a threshold energy equal to the mass of the charged bosons, the colliding proton and antiproton beams were each accelerated to about 270 GeV to provide the quarks with an average center-of-mass energy slightly above the threshold energy. (Only one-half of the energy of a proton or antiproton is carried by its three quark constituents; the other half is carried by the gluons.) Rubbia's group distinguished the two-body decay of the bosons (into a charged and neutral lepton pair such as $e^+\nu_e$) by two methods: selection of events in which the charged lepton possessed a large momentum transverse to the axis of the colliding beams, and selection of events in which a large amount of energy appeared to be missing, presumably carried off by the (undetected) neutrino. Both methods converged on the same events.

By mid 1983 each of the two groups had succeeded also in finding $Z^0$, the neutral vector boson of the electroweak model. They reported slightly different mass values ($96.5 \pm 1.5$ and $91.2 \pm 1.7$ GeV/$c^2$), both in agreement with the predicted value of $94.0 \pm 2.5$ GeV/$c^2$. For $Z^0$ the production and decay processes are given by

$$u + \bar{u} \,(\text{or } d + \bar{d}) \to Z^0 \to e^- + e^+ \,(\text{or } \mu^- + \mu^+) .$$

In addition, both groups reported an asymmetry in the angular distribution of charged leptons from the many more decays of $W^-$ and $W^+$ that had been seen since their discovery. This parity violation confirmed that the particles observed are truly electroweak vector bosons. ∎

## Table 3

**The three families of quarks and leptons and their masses. The subscripts R and L denote right- and left-handed particles as defined in Fig. 12.**

| Quark Mass (MeV/$c^2$) | Quarks | | | Leptons | | | Lepton Mass (MeV/$c^2$) |
|---|---|---|---|---|---|---|---|
| | | | | **First Family** | | | |
| 5 | up | $u_L$ | $u_R$ | $(\nu_e)_L$ | | electron neutrino | 0 |
| 8 | down | $d_L$ | $d_R$ | $e_L$ | $e_R$ | electron | 0.511 |
| | | | | **Second Family** | | | |
| 1270 | charm | $c_L$ | $c_R$ | $(\nu_\mu)_L$ | | muon neutrino | 0 |
| 175 | strange | $s_L$ | $s_R$ | $\mu_L$ | $\mu_R$ | muon | 105.7 |
| | | | | **Third Family** | | | |
| 45000 (?) | top | $t_L$ | $t_R$ | $(\nu_\tau)_L$ | | tau neutrino | 0 |
| 4250 | bottom | $b_L$ | $b_R$ | $\tau_L$ | $\tau_R$ | tau | 1784 |

Cornell.) The standard model says nothing about why three identical families of quarks and leptons should exist, nor does it give any clue about the hierarchical pattern of their masses (the $\tau$ family is heavier than the $\mu$ family, which is heavier than the $e$ family). This hierarchy is both puzzling and intriguing. Perhaps there are even more undiscovered families connected to the broken family symmetry. The symmetry could be global or local, and either case would predict new, weaker interactions among quarks and leptons.

Table 3 brings up two other open questions. First, we have listed the neutrinos as being massless. Experimentally, however, there exist only upper limits on their possible masses. The most restrictive limit comes from cosmology, which requires the sum of neutrino masses to be less then 100 eV. It is known from astrophysical observations that most of the energy in the universe is in a form that does not radiate electromagnetically. If neutrinos have mass, they could, in fact, be the dominant form of energy in the universe today.

Second, we have listed $u$ and $d$, $c$ and $s$, and $t$ and $b$ as doublets under weak SU(2). This is, however, only approximately true. As a result of the *broken* family symmetry, states with the same electric charge (the $d$, $s$, and $b$ quarks or the $u$, $c$, and $t$ quarks) can mix, and the weak doublets that couple to the $W^\pm$ bosons are actually given by $u$ and $d'$, $c$ and $s'$, and $t$ and $b'$. A $3 \times 3$ unitary matrix known as the Kobayashi-Maskawa (K-M) matrix rotates the mass eigenstates (states of definite mass) $d$, $s$, and $b$ into the weak doublet states $d'$, $s'$, and $b'$. The K-M matrix is conventionally written in terms of three mixing angles and an arbitrary phase. The largest mixing is between the $d$ and $s$ quarks and is characterized by the Cabibbo angle $\theta_C$ (see Lecture Note 9), which is named for the man who studied strangeness-changing weak decays such as $\Sigma^0 \rightarrow p + e^- + \bar{\nu}_e$. The observed value of sin $\theta_C$ is about 0.22. The other mixing angles are all at least an order of magnitude smaller. The structure of the K-M matrix, like the masses of the quarks and leptons, is a complete mystery.

## Conclusions

Although many mysteries remain, the standard model represents an intriguing and compelling theoretical framework for our present-day knowledge of the elementary particles. Its great virtue is that all of the known forces can be described as local gauge theories in which the interactions are generated from the single unifying principle of local gauge invariance. The fact that in quantum field theory interactions can drastically change their character with scale is crucial to

# $J/\psi$

In 1974 two experimental groups pursuing completely different lines of research at different laboratories simultaneously discovered the same particle. (In deference to the different names adopted by the two groups, the particle is now referred to as $J/\psi$.) At SPEAR, the electron-positron storage ring at the Stanford Linear Accelerator Center, a group led by B. Richter was investigating, as a function of incident energy, the process of electron-positron annihilation to hadrons. They found an enormous and very narrow resonance at a collision energy of about 3.1 GeV and attributed it to the formation of a new particle $\psi$. Meanwhile, at the Brookhaven AGS, a group led by S. Ting was investigating essentially the inverse process, the formation of electron-positron pairs in collisions of protons with nucleons. They determined the number of pair-producing events as a function of the mass of the parent particle (as deduced from the energy and angular separation of each electron-positron pair) and found a very large, well-defined increase at a mass of about $3.1$ GeV/$c^2$. This resonance also was attributed to the formation of a new particle $J$.

The surprisingly long lifetime of $J/\psi$, as indicated by the narrowness of the resonance, implied that its decay to lighter hadrons (all, according to the original quark model, composed of the up, down, and strange quarks) was somehow inhibited. This inhibition was given two possible interpretations: $J/\psi$ was perhaps a form of matter exhibiting a net "color" (a quantum property of quarks), or it was perhaps a meson containing the postulated charmed and anti-charmed quarks. The latter interpretation was soon adopted, and in

those terms the production of $J/\psi$ in the two experiments can be written

$$e^+ + e^- \leftrightarrow c + \bar{c}.$$

For further elucidation of the $J/\psi$ system, electron-positron annihilation proved more fruitful than the hadronic production process.

This discovery of a fourth quark (which had been postulated by S. Glashow and J. Bjorken in 1964 to achieve a symmetry between the number of quarks and the known number of leptons and again by Glashow, J. Iliopoulos, and L. Maiani in 1970 to reconcile the weak-interaction selection rules and the electroweak model) convinced theorists that renormalizable gauge field theories, in conjunction with spontaneous symmetry breaking, were the right tool for understanding the fundamental interactions of nature. ■

*The group from M.I.T. and Brookhaven that discovered $J/\psi$ in proton-nucleon collisions at the AGS, together with a graph of their evidence. (Photo courtesy of the Niels Bohr Library of the American Institute of Physics and Brookhaven National Laboratory.)*

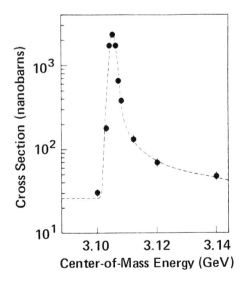

*Graph of the evidence for formation of $J/\psi$ in electron-positron annihilations at SPEAR. (Adapted from* SLAC Beam Line, *Volume 7, Number 11, November 1976.)*

Ϫ

In 1977 a group led by L. Lederman provided evidence for a fifth, or bottom, quark with the discovery of Ϫ, a long-lived particle three times more massive than $J/\psi$. In an experiment similar to that of Ting and coworkers and performed at the Fermilab proton accelerator, the group determined the number of events giving rise to muon-antimuon pairs as a function of the mass of the parent particle and found a sharp increase at about 9.5 GeV/$c^2$. Like the $J/\psi$ system, the Ϫ system has been elucidated in detail from experiments involving electron-positron collisions rather than proton collisions, in this case at Cornell's electron storage ring, CESR.

The existence of the bottom quark, and of a sixth, or top, quark, was expected on the basis of the discovery of the tau lepton at SPEAR in 1975 and Glashow and Bjorken's 1964 argument of quark-lepton symmetry. Recent results from high-energy proton-antiproton collision experiments at CERN have been interpreted as possible evidence for the top quark with a mass somewhere between 30 and 50 GeV/$c^2$. ∎

this approach. The essence of the standard model is to put the physics of the apparently separate strong, weak, and electromagnetic interactions in the single language of local gauge field theories, much as Maxwell put the apparently separate physics of Coulomb's, Ampère's, and Faraday's laws into the single language of classical field theory.

It is very tempting to speculate that, because of the chameleon-like behavior of quantum field theory, all the interactions are simply manifestations of a *single* field theory. Just as the "undetermined parameters" $\varepsilon_0$ and $\mu_0$ were related to the velocity of light through Maxwell's unification of electricity and magnetism, so the undetermined parameters of the standard model (such as quark and lepton masses and mixing angles) might be fixed by embedding the standard model in some grand unified theory.

A great deal of effort has been focused on this question during the past few years, and some of the problems and successes are discussed in "Toward a Unified Theory" and "Supersymmetry at 100 GeV." Although hints of a solution have emerged, it is fair to say that we are still a long way from for-mulating an ultimate synthesis of all physical laws. Perhaps one of the reasons for this is that the role of gravitation still remains mysterious. This weakest of all the forces, whose effects are so dramatic in the macroscopic world, may well hold the key to a truly deep understanding of the physical world. Many particle physicists are therefore turning their attention to the Einsteinian view in which geometry becomes the language of expression. This has led to many weird and wonderful speculations concerning higher dimensions, complex manifolds, and other arcane subjects.

An alternative approach to these questions has been to peel yet another skin off the onion and suggest that the quarks and leptons are themselves composite objects made of still more elementary objects called preons. After all, the proliferation of quarks, leptons, gauge bosons, and Higgs particles is beginning to resemble the situation in the early 1960s when the proliferation of the observed hadronic states made way for the introduction of quarks. Maybe introducing preons can account for the mystery of flavor: $e$, $\mu$, and $\tau$, for example, may simply be bound states of such objects.

Regardless of whether the ultimate understanding of the structure of matter, should there be one, lies in the realm of preons, some single primitive group, higher dimensions, or whatever, the standard model represents the first great step in that direction. The situation appears ripe for some kind of grand unification. Where are you, Maxwell? ∎

# Further Reading

Gerard 't Hooft. "Gauge Theories of the Forces Between Elementary Particles." *Scientific American*, June 1980, pp. 104-137.

Howard Georgi. "A Unified Theory of Elementary Particles and Forces." *Scientific American*, April 1981, pp. 48-63.

# Lecture Notes
## from simple field theories to the standard model

*by Richard C. Slansky*

The standard model of electroweak and strong interactions consists of two relativistic quantum field theories, one to describe the strong interactions and one to describe the electromagnetic and weak interactions. This model, which incorporates all the known phenomenology of these fundamental interactions, describes spinless, spin-½, and spin-1 fields interacting with one another in a manner determined by its Lagrangian. The theory is relativistically invariant, so the mathematical form of the Lagrangian is unchanged by Lorentz transformations.

Although rather complicated in detail, the standard model Lagrangian is based on just two basic ideas beyond those necessary for a quantum field theory. One is the concept of local symmetry, which is encountered in its simplest form in electrodynamics. Local symmetry determines the form of the interaction between particles, or fields, that carry the charge associated with the symmetry (not necessarily the electric charge). The interaction is mediated by a spin-1 particle, the vector boson, or gauge particle. The second concept is spontaneous symmetry breaking, where the vacuum (the state with no particles) has a nonzero charge distribution. In the standard model the nonzero weak-interaction charge distribution of the vacuum is the source of most masses of the particles in the theory. These two basic ideas, local symmetry and spontaneous symmetry breaking, are exhibited by simple field theories. We begin these lecture notes with a Lagrangian for scalar fields and then, through the extensions and generalizations indicated by the arrows in the diagram below, build up the formalism needed to construct the standard model.

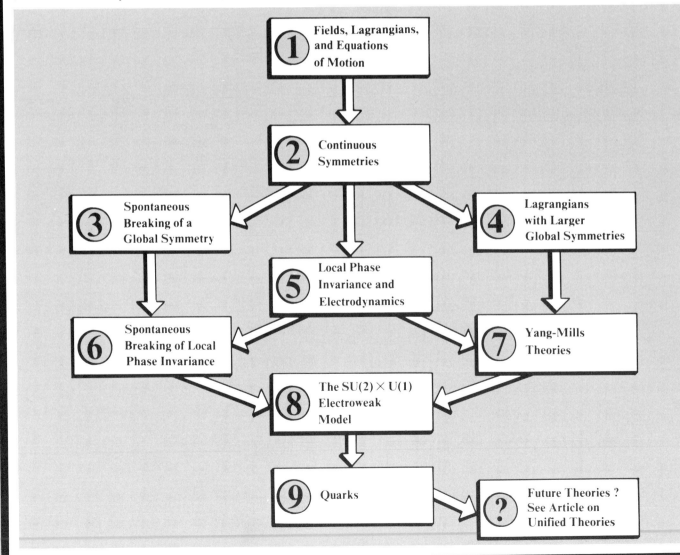

# Fields, Lagrangians, and Equations of Motion

We begin this introduction to field theory with one of the simplest theories, a complex scalar field theory with independent fields $\varphi(x)$ and $\varphi^\dagger(x)$. ($\varphi^\dagger(x)$ is the complex conjugate of $\varphi(x)$ if $\varphi(x)$ is a classical field, and, if $\varphi(x)$ is generalized to a column vector or to a quantum field, $\varphi^\dagger(x)$ is the Hermitian conjugate of $\varphi(x)$.) Since $\varphi(x)$ is a complex function in classical field theory, it assigns a complex number to each four-dimensional point $x = (ct, \mathbf{x})$ of time and space. The symbol $x$ denotes all four components. In quantum field theory $\varphi(x)$ is an operator that acts on a state vector in quantum-mechanical Hilbert space by adding or removing elementary particles localized around the space-time point $x$.

In this note we present the case in which $\varphi(x)$ and $\varphi^\dagger(x)$ correspond respectively to a spinless charged particle and its antiparticle of equal mass but opposite charge. The charge in this field theory is like electric charge, except it is not yet coupled to the electromagnetic field. (The word "charge" has a broader definition than just electric charge.) In Note 3 we show how this complex scalar field theory can describe a quite different particle spectrum: instead of a particle and its antiparticle of equal mass, it can describe a particle of zero mass and one of nonzero mass, each of which is its own antiparticle. Then the scalar theory exhibits the phenomenon called spontaneous symmetry breaking, which is important for the standard model.

A complex scalar theory can be defined by the Lagrangian density,

$$\mathscr{L}(\varphi, \partial_\mu\varphi, \varphi^\dagger, \partial_\mu\varphi^\dagger) = \partial^\mu\varphi^\dagger\partial_\mu\varphi - m^2\varphi^\dagger\varphi - \lambda(\varphi^\dagger\varphi)^2 , \tag{1a}$$

where $\partial_\mu\varphi \equiv \partial\varphi/\partial x^\mu$. (Upper and lower indices are related by the metric tensor, a technical point not central to this discussion.) The Lagrangian itself is

$$\mathrm{L}(t_1,t_2) \equiv \int_{t_2}^{t_1} dt \int d^3\mathbf{x}\, \mathscr{L} . \tag{1b}$$

The first term in Eq. 1a is the kinetic energy of the fields $\varphi(x)$ and $\varphi^\dagger(x)$, and the last two terms are the negative of the potential energy. Terms quadratic in the fields, such as the $-m^2\varphi^\dagger\varphi$ term in Eq. 1a, are called mass terms. If $m^2 > 0$, then $\varphi(x)$ describes a spinless particle and $\varphi^\dagger(x)$ its antiparticle of identical mass. If $m^2 < 0$, the theory has spontaneous symmetry breaking.

The equations of motion are derived from Eq. 1 by a variational method. Thus, let us change the fields and their derivatives by a small amount $\delta\varphi(x)$ and $\delta\partial_\mu\varphi(x) = \partial_\mu\delta\varphi(x)$. Then,

$$\delta\mathrm{L}(t_1,t_2) = \int_{t_2}^{t_2} \left[ \frac{\partial\mathscr{L}}{\partial\varphi}\delta\varphi + \frac{\partial\mathscr{L}}{\partial\varphi^\dagger}\delta\varphi^\dagger + \frac{\partial\mathscr{L}}{\partial(\partial_\mu\varphi)}\partial_\mu\delta\varphi \right.$$
$$\left. + \frac{\partial\mathscr{L}}{\partial(\partial_\mu\varphi^\dagger)}\partial_\mu\delta\varphi^\dagger \right] d^4x , \tag{2}$$

where the variation is defined with the restrictions $\delta\varphi(\mathbf{x},t_1) = \delta\varphi(\mathbf{x},t_2) = \delta\varphi^\dagger(\mathbf{x},t_1) = \delta\varphi^\dagger(\mathbf{x},t_2) = 0$, and $\delta\varphi(x)$ and $\delta\varphi^\dagger(x)$ are independent. The last two terms are integrated by parts, and the surface term is dropped since the integrand vanishes on the boundary. This procedure yields the Euler-Lagrange equations for $\varphi^\dagger(x)$,

$$\partial_\mu\left( \frac{\partial\mathscr{L}}{\partial(\partial_\mu\varphi)} \right) - \frac{\partial\mathscr{L}}{\partial\varphi} = 0 , \tag{3}$$

and for $\varphi(x)$. (The Euler-Lagrange equation for $\varphi(x)$ is like Eq. 3 except that $\varphi^\dagger$ replaces $\varphi$. There are two equations because $\delta\varphi(x)$ and $\delta\varphi^\dagger(x)$ are independent.) Substituting the Lagrangian density, Eq. 1a, into the Euler-Lagrange equations, we obtain the equations of motion,

$$\partial^\mu\partial_\mu\varphi + m^2\varphi + 2\lambda(\varphi^\dagger\varphi)\varphi = 0 , \tag{4}$$

plus another equation of exactly the same form with $\varphi(x)$ and $\varphi^\dagger(x)$ exchanged.

This method for finding the equations of motion can be easily generalized to more fields and to fields with spin. For example, a field theory that is incorporated into the standard model is electrodynamics. Its list of fields includes particles that carry spin. The electromagnetic vector potential $A_\mu(x)$ describes a "vector" particle with a spin of 1 (in units of the quantum of action $\hbar = 1.0546 \times 10^{-27}$ erg second), and its four spin components are enumerated by the space-time vector index $\mu$ ( $= 0, 1, 2, 3$, where 0 is the index for the time component and 1, 2, and 3 are the indices for the three space components). In electrodynamics only two of the four components of $A_\mu(x)$ are independent. The electron has a spin of $\frac{1}{2}$, as does its antiparticle, the positron. Electrons and positrons of both spin projections, $\pm\frac{1}{2}$, are described by a field $\psi(x)$, which is a column vector with four entries. Many calculations in electrodynamics are complicated by the spins of the fields.

There is a much more difficult generalization of the Lagrangian formalism: if there are constraints among the fields, the procedure yielding the Euler-Lagrange equations must be modified, since the field variations are not all independent. This technical problem complicates the formulation of electrodynamics and the standard model, especially when computing quantum corrections. Our examination of the theory is not so detailed as to require a solution of the constraint problem.

# ②Continuous Symmetries

It is often possible to find sets of fields in the Lagrangian that can be rearranged or transformed in ways described below without changing the Lagrangian. The transformations that leave the Lagrangian unchanged (or invariant) are called symmetries. First, we will look at the form of such transformations, and then we will discuss implications of a symmetrical Lagrangian. In some cases symmetries imply the existence of conserved currents (such as the electromagnetic current) and conserved charges (such as the electric charge), which remain constant during elementary-particle collisions. The conservation of energy, momentum, angular momentum, and electric charge are all derived from the existence of symmetries.

Let us consider a continuous linear transformation on three real spinless fields $\varphi_i(x)$ (where $i = 1, 2, 3$) with $\varphi_i(x) = \varphi_i^\dagger(x)$. These three fields might correspond to the three pion states. As a matter of notation, $\varphi(x)$ is a column vector, where the top entry is $\varphi_1(x)$, the second entry is $\varphi_2(x)$, and the bottom entry is $\varphi_3(x)$. We write the linear transformation of the three fields in terms of a 3-by-3 matrix $U(\varepsilon)$, where

$$\varphi'(x') = U(\varepsilon)\varphi(x) \,, \tag{5a}$$

or in component notation

$$\varphi_i'(x') = U_{ij}(\varepsilon)\varphi_j(x) \,. \tag{5b}$$

The repeated index is summed from 1 to 3, and generalizations to different numbers or kinds of fields are obvious. The parameter $\varepsilon$ is continuous, and as $\varepsilon$ approaches zero, $U(\varepsilon)$ becomes the unit matrix. The dependence of $x'$ on $x$ and $\varepsilon$ is discussed below. The continuous transformation $U(\varepsilon)$ is called linear since $\varphi_j(x)$ occurs linearly on the right-hand side of Eq. 5. (Nonlinear transformations also have an important role in particle physics, but this discussion of the standard model will primarily involve linear transformations except for the vector-boson fields, which have a slightly different transformation law, described in Note 5.) For $N$ independent transformations, there will be a set of parameters $\varepsilon_a$, where the index $a$ takes on values from 1 to $N$.

For these continuous transformations we can expand $\varphi'(x')$ in a Taylor series about $\varepsilon_a = 0$; by keeping only the leading term in the expansion, Eq. 5 can be rewritten in infinitesimal form as

$$\delta\varphi(x) \equiv \varphi'(x) - \varphi(x) = i\varepsilon^a T_a\varphi(x) \,, \tag{6a}$$

where $T_a$ is the first term in the Taylor expansion,

$$i\varepsilon^a T_a = \varepsilon^a \left[ \frac{\partial U(\varepsilon)}{\partial \varepsilon_a} \right]_{\varepsilon=0} - \delta x^\mu \partial_\mu \,, \tag{6b}$$

with $\delta x = x' - x$. The $T_a$ are the "generators" of the symmetry transformations of $\varphi(x)$. (We note that $\delta\varphi(x)$ in Eq. 6a is a small symmetry transformation, not to be confused with the field variations $\delta\varphi$ in Eq. 2.)

The space-time point $x'$ is, in general, a function of $x$. In the case where $x' = x$, Eq. 5 is called an internal transformation. Although our primary focus will be on internal transformations, space-time symmetries have many applications. For example, all theories we describe here have Poincaré symmetry, which means that these theories are invariant under transformations in which $x' = \Lambda x + b$, where $\Lambda$ is a 4-by-4 matrix representing a Lorentz transformation that acts on a four-component column vector $x$ consisting of time and the three space components, and $b$ is the four-component column vector of the parameters of a space-time translation. A spinless field transforms under Poincaré transformations as $\varphi'(x') = \varphi(x)$ or $\delta\varphi = -b^\mu\partial_\mu\varphi(x)$. Upon solving Eq. 6b, we find the infinitesimal translation is represented by $i\partial_\mu$. The components of fields with spin are rearranged by Poincaré transformations according to a matrix that depends on both the $\varepsilon$'s and the spin of the field.

We now restrict attention to internal transformations where the space-time point is unchanged; that is, $\delta x^\mu = 0$. If $\varepsilon_a$ is an infinitesimal, arbitrary function of $x$, $\varepsilon_a(x)$, then Eqs. 5 and 6a are called local transformations. If the $\varepsilon_a$ are restricted to being constants in space-time, then the transformation is called global.

Before beginning a lengthy development of the symmetries of various Lagrangians, we give examples in which each of these kinds of linear transformations are, indeed, symmetries of physical theories. An example of a global, internal symmetry is strong isospin, as discussed briefly in "Particle Physics and the Standard Model." (Actually, strong isospin is not an exact symmetry of Nature, but it is still a good example.) All theories we discuss here have global Lorentz invariance, which is a space-time symmetry. Electrodynamics has a local phase symmetry that is an internal symmetry. For a charged spinless field the infinitesimal form of a local phase transformation is $\delta\varphi(x) = i\varepsilon(x)\varphi(x)$ and $\delta\varphi^\dagger(x) = -i\varepsilon(x)\varphi^\dagger(x)$, where $\varphi(x)$ is a complex field. Larger sets of local internal symmetry transformations are fundamental in the standard model of the weak and strong interactions. Finally, Einstein's gravity makes essential use of local space-time Poincaré transformations. This complicated case is not discussed here. It is quite remarkable how many types of transformations like Eqs. 5 and 6 are basic in the formulation of physical theories.

Let us return to the column vector of three real fields $\varphi(x)$ and suppose we have a Lagrangian that is unchanged by Eqs. 5 and 6, where we now restrict our attention to internal transformations. (One such Lagrangian is Eq. 1a, where $\varphi(x)$ is now a column vector and $\varphi^\dagger(x)$ is its transpose.) Not only the Lagrangian, but the Lagrangian density, too, is unchanged by an internal symmetry transformation.

Let us consider the infinitesimal transformation (Eq. 6a) and calculate $\delta\mathcal{L}$ in two different ways. First of all, $\delta\mathcal{L} = 0$ if $\delta\varphi$ is a symmetry identified from the Lagrangian. Moreover, according to the rules of partial differentiation,

$$\delta\mathcal{L} = \frac{\partial\mathcal{L}}{\partial\varphi_i}\delta\varphi_i + \frac{\partial\mathcal{L}}{\partial(\partial_\mu\varphi_i)}\partial_\mu\delta\varphi_i . \tag{7}$$

Then, using the Euler-Lagrange equations (Eq. 3) for the first term and collecting terms, Eq. 7 can be written in an interesting way:

$$\delta\mathcal{L} = \partial_\mu\left(\frac{\partial\mathcal{L}}{\partial(\partial_\mu\varphi_i)}\right)\delta\varphi_i . \tag{8}$$

The next step is to substitute Eq. 6a into Eq. 8. Thus, let us define the current $J_\mu^a(x)$ as

$$J_\mu^a(x) = i\frac{\partial\mathcal{L}}{\partial(\partial_\mu\varphi_i)}T_{ij}^a\varphi_j . \tag{9}$$

Then Eq. 8 plus the requirement that $\delta\varphi$ is a symmetry imply the continuity equation,

$$\partial^\mu J_\mu^a(x) = 0 . \tag{10}$$

We can gain intuition about Eq. 10 from electrodynamics, since the electromagnetic current satisfies a continuity equation. It says that charge is neither created nor destroyed locally: the change in the charge density, $J_0(x)$, in a small region of space is just equal to the current $\mathbf{J}(x)$ flowing out of the region. Equation 10 generalizes this result of electrodynamics to other kinds of charges, and so $J_\mu^a(x)$ is called a current. In particle physics with its many continuous symmetries, we must be careful to identify which current we are talking about.

Although the analysis just performed is classical, the results are usually correct in the quantum theory derived from a classical Lagrangian. In some cases, however, quantum corrections contribute a nonzero term to the right-hand side of Eq. 10; these terms are called anomalies. For global symmetries these anomalies can improve the predictions from Lagrangians that have too much symmetry when compared with data because the anomaly wrecks the symmetry (it was never there in the quantum theory, even though the classical Lagrangian had the symmetry). However, for local symmetries anomalies are disastrous. A quantum field theory is locally symmetric only if its currents satisfy the continuity equation, Eq. 10. Otherwise local symmetry transformations simply change the theory. (Some care is needed to avoid this kind of anomaly in the standard model.) We now show that Eq. 10 can imply the existence of a conserved quantity called the global charge and defined by

$$Q^a(t) = \int d^3\mathbf{x}\, J_0^a(x) , \tag{11}$$

provided the integral over all space in Eq. 11 is well defined; that is,

$J_0^a(x)$ must fall off rapidly enough as $|\mathbf{x}|$ approaches infinity that the integral is finite.

If $Q^a(t)$ is indeed a conserved quantity, then its value does not change in time, which means that its first time derivative is zero. We can compute the time derivative of $Q^a(t)$ with the aid of Eq. 10:

$$\frac{d}{dt}Q^a(t) = \int d^3\mathbf{x}\,\frac{\partial J_0^a(x)}{\partial t} = \int d^3\mathbf{x}\,\nabla\cdot\mathbf{J}^a(x) = \int \mathbf{J}^a\cdot d\mathbf{S} = 0 . \tag{12}$$

The next to the last step is Gauss's theorem, which changes the volume integral of the divergence of a vector field into a surface integral. If $\mathbf{J}^a(x)$ falls off more rapidly than $1/|\mathbf{x}|^2$ as $|\mathbf{x}|$ becomes very large, then the surface integral must be zero. It is not a always true that $\mathbf{J}^a(x)$ falls off so rapidly, but when it does, $Q^a(t) = Q^a$ is a constant in time. One of the most important experimental tests of a Lagrangian is whether the conserved quantities it predicts are, indeed, conserved in elementary-particle interactions.

The Lagrangian for the complex scalar field defined by Eq. 1 has an internal global symmetry, so let us practice the above steps and identify the conserved current and charge. It is easily verified that the global phase transformation

$$\varphi'(x) = e^{i\varepsilon}\varphi(x) \tag{13}$$

leaves the Lagrangian density invariant. For example, the first term of Eq. 1 by itself is unchanged: $\partial_\mu\varphi^\dagger\partial^\mu\varphi$ becomes $\partial_\mu(e^{-i\varepsilon}\varphi^\dagger)\partial^\mu(e^{i\varepsilon}\varphi) = \partial_\mu\varphi^\dagger\partial^\mu\varphi$, where the last equality follows only if $\varepsilon$ is constant in space-time. (The case of local phase transformations is treated in Note 5.) The next step is to write the infinitesimal form of Eq. 13 and substitute it into Eq. 9. The conserved current is

$$J_\mu(x) = i[(\partial_\mu\varphi^\dagger)\varphi - (\partial_\mu\varphi)\varphi^\dagger] , \tag{14}$$

where the sum in Eq. 9 over the fields $\varphi(x)$ and $\varphi^\dagger(x)$ is written out explicitly.

If $m^2 > 0$ in Eq. 1, then all the charge can be localized in space and time and made to vanish as the distance from the charge goes to infinity. The steps in Eq. 12 are then rigorous, and a conserved charge exists. The calculation was done here for classical fields, but the same results hold for quantum fields: the conservation law implied by Eq. 12 yields a conserved global charge equal to the number of $\varphi$ particles minus the number of $\varphi$ antiparticles. This number must remain constant in any interaction. (We will see in Note 3 that if $m^2 < 0$, the charge distribution is spread out over all space-time, so the global charge is no longer conserved even though the continuity equation remains valid.)

Identifying the transformations of the fields that leave the Lagrangian invariant not only satisfies our sense of symmetry but also leads to important predictions of the theory without solving the equations of motion. In Note 4 we will return to the example of three real scalar fields to introduce larger global symmetries, such as SU(2), that interrelate different fields.

# ③ Spontaneous Breaking of a Global Symmetry

It is possible for the vacuum or ground state of a physical system to have less symmetry than the Lagrangian. This possibility is called spontaneous symmetry breaking, and it plays an important role in the standard model. The simplest example is the complex scalar field theory of Eq. 1a with $m^2 < 0$.

In order to identify the classical fields with particles in the quantum theory, the classical field must approach zero as the number of particles in the corresponding quantum-mechanical state approaches zero. Thus the quantum-mechanical vacuum (the state with no particles) corresponds to the classical solution $\varphi(x) = 0$. This might seem automatic, but it is not. Symmetry arguments do not necessarily imply that $\varphi(x) = 0$ is the lowest energy state of the system. However, if we rewrite $\varphi(x)$ as a function of new fields that do vanish for the lowest energy state, then the new fields may be directly identified with particles. Although this prescription is simple, its justification and analysis of its limitations require extensive use of the details of quantum field theory.

The energy of the complex scalar theory is the sum of kinetic and potential energies of the $\varphi(x)$ and $\varphi^\dagger(x)$ fields, so the energy density is

$$\mathscr{H} = \partial^\mu\varphi^\dagger\partial_\mu\varphi + m^2\varphi^\dagger\varphi + \lambda(\varphi^\dagger\varphi)^2 , \tag{15}$$

with $\lambda > 0$. Note that $\partial^\mu\varphi^\dagger\partial_\mu\varphi$ is nonnegative and is zero if $\varphi$ is a constant. For $\varphi = 0$, $\mathscr{H} = 0$. However, if $m^2 < 0$, then there are nonzero values of $\varphi(x)$ for which $\mathscr{H} < 0$. Thus, there is a nonzero field configuration with lowest energy. A graph of $\mathscr{H}$ as a function of $|\varphi|$ is shown in Fig. 1. In this example $\mathscr{H}$ is at its lowest value when both the kinetic and potential energies ($V = m^2\varphi^\dagger\varphi + \lambda(\varphi^\dagger\varphi)^2$) are at their lowest values. Thus, the vacuum solution for $\varphi(x)$ is found by solving the equation $\partial V/\partial\varphi = 0$, or

$$\varphi^\dagger(x)\varphi(x) = -\frac{m^2}{2\lambda} = \frac{1}{2}|\varphi_0|^2 > 0 . \tag{16}$$

Next we find new fields that vanish when Eq. 16 is satisfied. For example, we can set

$$\varphi(x) = \frac{1}{\sqrt{2}}\Big[\rho(x) + \varphi_0\Big]\exp\Big[i\pi(x)/\varphi_0\Big] , \tag{17}$$

where the real fields $\rho(x)$ and $\pi(x)$ are zero when the system is in the lowest energy state. Thus $\rho(x)$ and $\pi(x)$ may be associated with particles. Note, however, that $\varphi_0$ is not completely specified; it may lie at any point on the circle in field space defined by Eq. 16, as shown in Fig. 2.

Suppose $\varphi_0$ is real and given by

$$\varphi_0 = (-m^2/\lambda)^{1/2} . \tag{18}$$

Then the Lagrangian is still invariant under the phase transformations in Eq. 13, but the choice of the vacuum field solution is changed

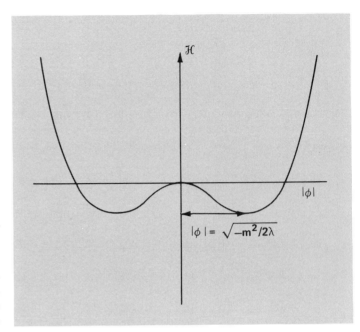

***Fig. 1. The Hamiltonian $\mathscr{H}$ defined by Eq. 15 has minima at nonzero values of the field $\varphi$.***

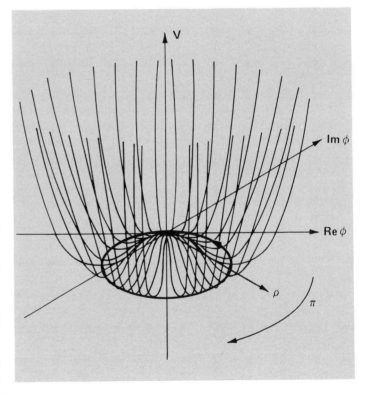

***Fig. 2. The closed curve is the location of the minimum of V in the field space $\varphi$.***

by the phase transformation. Thus, the vacuum solution is not invariant under the phase transformations, so the phase symmetry is spontaneously broken. The symmetry of the Lagrangian is *not* a symmetry of the vacuum. (For $m^2 > 0$ in Eq. 1, the vacuum and the Lagrangian both have the phase symmetry.)

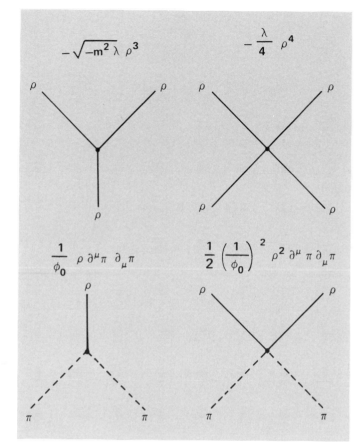

$$-\sqrt{-m^2\,\lambda}\ \rho^3 \qquad\qquad -\frac{\lambda}{4}\ \rho^4$$

$$\frac{1}{\phi_0}\ \rho\,\partial^\mu\pi\ \partial_\mu\pi \qquad\qquad \frac{1}{2}\left(\frac{1}{\phi_0}\right)^2\rho^2\,\partial^\mu\pi\,\partial_\mu\pi$$

*Fig. 3. A graphic representation of the last four terms of Eq. 20, the interaction terms. Solid lines denote the ρ field and dotted lines the π field. The interaction of three ρ(x) fields at a single point is shown as three solid lines emanating from a single point. In perturbation theory this so-called vertex represents the lowest order quantum-mechanical amplitude for one particle to turn into two. All possible configurations of these vertices represent the quantum-mechanical amplitudes defined by the theory.*

We now rewrite the Lagrangian in terms of the particle fields $\rho(x)$ and $\pi(x)$ by substituting Eq. 17 into Eq. 1. The Lagrangian becomes

$$\mathscr{L} = \frac{1}{2}\,\partial^\mu\rho\partial_\mu\rho + \frac{1}{2}(1+\rho/\varphi_0)^2\partial^\mu\pi\partial_\mu\pi$$
$$-\frac{m^2}{2}(\rho+\varphi_0)^2 - \frac{\lambda}{4}(\rho+\varphi_0)^4 . \tag{19}$$

To estimate the masses associated with the particle fields $\rho(x)$ and $\pi(x)$, we substitute Eq. 18 for the constant $\varphi_0$ and expand $\mathscr{L}$ in powers of the fields $\pi(x)$ and $\rho(x)$, obtaining

$$\mathscr{L} = \frac{1}{2}\,\partial^\mu\rho\partial_\mu\rho + \frac{1}{2}\,\partial^\mu\pi\partial_\mu\pi + \frac{m^4}{4\,\lambda} + m^2\rho^2 - (-\lambda m^2)^{1/2}\rho^3 - \frac{\lambda}{4}\rho^4$$
$$+\frac{1}{\varphi_0}\rho\partial^\mu\pi\partial_\mu\pi + \frac{1}{2\varphi_0^2}\,\rho^2\partial^\mu\pi\partial_\mu\pi . \tag{20}$$

This Lagrangian has the following features.

○ The fields $\rho(x)$ and $\pi(x)$ have standard kinetic energy terms.

○ Since $m^2 < 0$, the term $m^2\rho^2$ can be interpreted as the mass term for the $\rho(x)$ field. The $\rho(x)$ field thus describes a particle with mass-squared equal to $|m^2|$, *not* $-|m^2|$.

○ The $\pi(x)$ field has no mass term. (This is obvious from Fig. 2, which shows that $\mathscr{L}(\rho,\pi)$ has no curvature (that is, $\partial^2\mathscr{L}/\partial\pi^2 = 0$) in the $\pi(x)$ direction.) Thus, $\pi(x)$ corresponds to a massless particle. This result is unchanged when all the quantum effects are included.

○ The phase symmetry is hidden in $\mathscr{L}$ when it is written in terms of $\rho(x)$ and $\pi(x)$. Nevertheless, $\mathscr{L}$ has phase symmetry, as is proved by working backward from Eq. 20 to Eq. 16 to recover Eq. 1a.

○ In theories without gravity, the constant term $V \propto m^4/\lambda$ can be ignored, since a constant overall energy level is not measurable. The situation is much more complicated for gravitational theories, where terms of this type contribute to the vacuum energy-momentum tensor and, by Einstein's equations, modify the geometry of space-time.

○ The $\rho$ field interacts with the $\pi$ field only through derivatives of $\pi$. The interaction terms in Eq. 20 may be pictured as in Fig. 3.

Although this model might appear to be an idle curiosity, it is an example of a very general result known as Goldstone's theorem. This theorem states that in any field theory there is a zero-mass spinless particle for each independent global continuous symmetry of the Lagrangian that is spontaneously broken. The zero-mass particle is called a Goldstone boson. (This general result does not apply to local symmetries, as we shall see.)

There has been one very important physical application of spontaneously broken global symmetries in particle physics, namely, theories of pion dynamics. The pion has a surprisingly small mass compared to a nucleon, so it might be understood as a zero-mass particle resulting from spontaneous symmetry breaking of a global symmetry. Since the pion mass is not exactly zero, there must also be some small but explicit terms in the Lagrangian that violate the global symmetry. The feature of pion dynamics that justifies this procedure is that the interactions of pions with nucleons and other pions are similar to the interactions (see Fig. 3) of the $\pi(x)$ field with the $\rho(x)$ field and with itself in the Lagrangian of Eq. 20. Since the pion has three (electric) charge states, it must be associated with a larger global symmetry than the phase symmetry, one where three independent symmetries are spontaneously broken. The usual choice of symmetry is global SU(2) × SU(2) spontaneously broken to the SU(2) of the strong-interaction isospin symmetry (see Note 4 for a discussion of SU(2)). This description accounts reasonably well for low-energy pion physics.

Perhaps we should note that only spinless fields can acquire a vacuum value. Fields carrying spin are not invariant under Lorentz transformations, so if they acquire a vacuum value, Lorentz invariance will be spontaneously broken, in disagreement with experiment. Spinless particles trigger the spontaneous symmetry breaking in the standard model.

# Lagrangians with Larger Global Symmetries

(4)

In a theory with a single complex scalar field the phase transformation in Eq. 13 defines the "largest" possible internal symmetry since the only possible symmetries must relate $\varphi(x)$ to itself. Here we will discuss global symmetries that interrelate different fields and group them together into "symmetry multiplets." Strong isospin, an approximate symmetry of the observed strongly interacting particles, is an example. It groups the neutron and the proton into an isospin doublet, reflecting the fact that the neutron and proton have nearly the same mass and share many similarities in the way that they interact with other particles. Similar comments hold for the three pion states ($\pi^+$, $\pi^0$, and $\pi^-$), which form an isospin triplet.

We will derive the structure of strong isospin symmetry by examining the invariance of a specific Lagrangian for the three real scalar fields $\varphi_i(x)$ already described in Note 2. (Although these fields could describe the pions, the Lagrangian will be chosen for simplicity, not for its capability to describe pion interactions.)

We are about to discover a symmetry by deriving it from a Lagrangian; however, in particle physics the symmetries are often discovered from phenomenology. Moreover, since there can be many Lagrangians with the same symmetry, the predictions following from the symmetry are viewed as more general than the predictions of a specific Lagrangian with the symmetry. Consequently, it becomes important to abstract from specific Lagrangians the general features of a symmetry; see the comments later in this note.

A general linear transformation law for the three real fields can be written

$$\varphi_i'(x) = [\exp(i\varepsilon^a T_a)]_{ij}\varphi_j(x) , \tag{21}$$

where the sum on $j$ runs from 1 to 3. One reason for choosing this form of $U(\varepsilon)$ is that it explicitly approaches the identity as $\varepsilon$ ap-

proaches zero.

To identify the generators $T_a$ with matrix elements $(T_a)_{ij}$, we use a specific Lagrangian,

$$\mathscr{L} = \frac{1}{2}\partial^\mu\varphi_i\partial_\mu\varphi_i - \frac{1}{2}\,m^2\varphi_i\varphi_i - \lambda\,(\varphi_i\varphi_i)^2 . \tag{22}$$

Let us place primes on the fields in Eq. 22 and substitute Eq. 21 into it. Then $\mathscr{L}$ written in terms of the new $\varphi(x)$ is exactly the same as Eq. 22 if

$$[\exp(i\varepsilon^a T_a)]_{ij}\,[\exp(i\varepsilon^b T_b)]_{ik} = \delta_{jk} , \tag{23}$$

where $\delta_{jk}$ are the matrix elements of the 3-by-3 identity matrix. (In the notation of Eq. 5a, Eq. 23 is $U(\varepsilon)U^T(\varepsilon) = I$.) Equation 23 can be expanded in $\varepsilon_a$, and the linear term then requires that $T_a$ be an antisymmetric matrix. Moreover, $\exp(i\varepsilon^a T_a)$ must be a real matrix so that $\varphi(x)$ remains real after the transformation. This implies that all elements of the $T_a$ are imaginary. These constraints are solved by the three imaginary antisymmetric 3-by-3 matrices with elements

$$(T_a)_{ij} = -i\varepsilon_{aij} , \tag{24}$$

where $\varepsilon_{123} = +1$ and $\varepsilon_{abc}$ is antisymmetric under the interchange of any two indices (for example, $\varepsilon_{321} = -1$). (It is a coincidence in this example that the number of fields is equal to the number of independent symmetry generators. Also, the parameter $\varepsilon_a$ with one index should not be confused with the tensor $\varepsilon_{abc}$ with three indices.)

The conditions on $U(\varepsilon)$ imply that it is an orthogonal matrix; 3-by-3 orthogonal matrices can also describe rotations in three spatial dimensions. Thus, the three components of $\varphi_i$ transform in the same way under isospin rotation as a spatial vector $\mathbf{x}$ transforms under a rotation. Since the rotational symmetry is SU(2), so is the isospin symmetry. (Thus "isospin" is like spin.) The $T_a$ matrices satisfy the SU(2) commutation relations

$$[T_a, T_b] \equiv T_a T_b - T_b T_a = i\varepsilon_{abc} T_c. \tag{25}$$

Although the explicit matrices of Eq. 24 satisfy this relation, the $T_a$ can be generalized to be quantum-mechanical operators. In the example of Eqs. 21 and 22, the isospin multiplet has three fields. Drawing on angular momentum theory, we can learn other possibilities for isospin multiplets. Spin-$J$ multiplets (or representations) have $2J + 1$ components, where $J$ can be any nonnegative integer or half integer. Thus, multiplets with isospin of ½ have two fields (for example, neutron and proton) and isospin-3/2 multiplets have four fields (for example, the $\Delta^{++}$, $\Delta^+$, $\Delta^0$, and $\Delta^-$ baryons of mass $\sim 1232 \text{ GeV}/c^2$).

The basic structure of all continuous symmetries of the standard model is completely analogous to the example just developed. In fact, part of the weak symmetry is called weak isospin, since it also has the same mathematical structure as strong isospin and angular momentum. Since there are many different applications to particle theory of given symmetries, it is often useful to know about symmetries and their multiplets. This mathematical endeavor is called group theory, and the results of group theory are often helpful in recognizing patterns in experimental data.

Continuous symmetries are defined by the algebraic properties of their generators. Group transformations can always be written in the form of Eq. 21. Thus, if $Q_a$ ($a = 1, \ldots, N$) are the generators of a symmetry, then they satisfy commutation relations analogous to Eq. 25:

$$[Q_a, Q_b] = i f_{abc} Q_c, \tag{26}$$

where the constants $f_{abc}$ are called the structure constants of the Lie algebra. The structure constants are determined by the multiplication rules for the symmetry operations, $U(\varepsilon_1)U(\varepsilon_2) = U(\varepsilon_3)$, where $\varepsilon_3$ depends on $\varepsilon_1$ and $\varepsilon_2$. Equation 26 is a basic relation in defining a Lie algebra, and Eq. 21 is an example of a Lie group operation. The $Q_a$, which generate the symmetry, are determined by the "group" structure. The focus on the generators often simplifies the study of Lie groups. The generators $Q_a$ are quantum-mechanical operators. The $(T_a)_{ij}$ of Eqs. 24 and 25 are matrix elements of $Q_a$ for some symmetry multiplet of the symmetry.

The general problem of finding all the ways of constructing equations like Eq. 25 and Eq. 26 is the central problem of Lie-group theory. First, one must find all sets of $f_{abc}$. This is the problem of finding all the Lie algebras and was solved many years ago. The second problem is, given the Lie algebra, to find all the matrices that represent the generators. This is the problem of finding all the representations (or multiplets) of a Lie algebra and is also solved in general, at least when the range of values of each $\varepsilon_a$ is finite. Lie group theory thus offers an orderly approach to the classification of a huge number of theories.

Once a symmetry of the Lagrangian is identified, then sets of $n$ fields are assigned to $n$-dimensional representations of the symmetry group, and the currents and charges are analyzed just as in Note 2. For instance, in our example with three real scalar fields and the Lagrangian of Eq. 22, the currents are

$$J_\mu^a(x) = \varepsilon^{aij}(\partial_\mu \varphi_i)\varphi_j \tag{27}$$

and, if $m^2 > 0$, the global symmetry charge is

$$Q^a = \int d^3\mathbf{x} \, \varepsilon^{aij} \frac{\partial \varphi_i}{\partial t} \varphi_j, \tag{28}$$

where the quantum-mechanical charges $Q_a$ satisfy the commutation relations

$$[Q_a, Q_b] = i\varepsilon_{abc} Q_c. \tag{29}$$

(The derivation of Eq. 29 from Eq. 28 requires the canonical commutation relations of the quantum $\varphi_i(x)$ fields.)

The three-parameter group SU(2) has just been presented in some detail. Another group of great importance to the standard model is SU(3), which is the group of 3-by-3 unitary matrices with unit determinant. The inverse of a unitary matrix $U$ is $U^\dagger$, so $U^\dagger U = I$. There are eight parameters and eight generators that satisfy Eq. 26 with the structure constants of SU(3). The low-dimensional representations of SU(3) have 1, 3, 6, 8, 10, ... fields, and the different representations are referred to as **1**, **3**, **3̄**, **6**, **6̄**, **8**, **10**, **10̄**, and so on.

# Local Phase Invariance and Electrodynamics

The theories that make up the standard model are all based on the principle of local symmetry. The simplest example of a local symmetry is the extension of the global phase invariance discussed at the end of Note 2 to local phase invariance. As we will derive below, the requirement that a theory be invariant under local phase transformations implies the existence of a gauge field in the theory that mediates or carries the "force" between the matter fields. For electrodynamics the gauge field is the electromagnetic vector potential $A_\mu(x)$ and its quantum particle is the massless photon. In addition, in the standard model the gauge fields mediating the strong interactions between the quarks are the massless gluon fields and the gauge fields mediating the weak interactions are the fields for the massive $Z^0$ and $W^\pm$ weak bosons.

To illustrate these principles we extend the global phase invariance of the Lagrangian of Eq. 1 to a theory that has local phase invariance. Thus, we require $\mathscr{L}$ to have the same form for $\varphi'(x)$ and $\varphi(x)$, where the local phase transformation is defined by

$$\varphi'(x) = e^{i\varepsilon(x)}\varphi(x) . \tag{30}$$

The potential energy,

$$V(\varphi,\varphi^\dagger) = m^2\varphi^\dagger\varphi + \lambda(\varphi^\dagger\varphi)^2 , \tag{31}$$

already has this symmetry, but the kinetic energy, $\partial^\mu\varphi^\dagger\partial_\mu\varphi$, clearly does not, since

$$\partial_\mu\varphi'(x) = e^{i\varepsilon(x)}\left[\partial_\mu\varphi + i(\partial_\mu\varepsilon)\varphi\right] . \tag{32}$$

$\mathscr{L}$ does not have local phase invariance if the Lagrangian of the transformed fields depends on $\varepsilon(x)$ or its derivatives. The way to eliminate the $\partial_\mu\varepsilon$ dependence is to add a new field $A_\mu(x)$ called the gauge field and then require the local symmetry transformation law for this new field to cancel the $\partial_\mu\varepsilon$ term in Eq. 32. The gauge field can be added by generalizing the derivative $\partial_\mu$ to $D_\mu$, where

$$D_\mu = \partial_\mu - ieA_\mu(x) . \tag{33}$$

This is just the minimal-coupling procedure of electrodynamics. We can then make a kinetic energy term of the form $(D^\mu\varphi)^\dagger(D_\mu\varphi)$ if we require that

$$D'_\mu\varphi'(x) = e^{i\varepsilon(x)}D_\mu\varphi(x) . \tag{34}$$

When written out with Eq. 33, Eq. 34 becomes an equation for $A'_\mu(x)$ in terms of $A_\mu(x)$, which is easily solved to give

$$A'_\mu(x) = A_\mu(x) + \frac{1}{e}\,\partial_\mu\varepsilon(x) . \tag{35}$$

Equation 35 prescribes how the gauge field transforms under the local phase symmetry.

Thus the first step to modifying Eq. 1 to be a theory with local phase invariance is simply to replace $\partial_\mu$ by $D_\mu$ in $\mathscr{L}$. (A slightly generalized form of this trick is used in the construction of all the theories in the standard model.) With this procedure the dominant interaction of the gauge field $A^\mu(x)$ with the matter field $\varphi$ is in the form of a current times the gauge field, $eJ^\mu A_\mu$, where $J_\mu$ is the current defined in Eq. 14.

# Spontaneous Breaking of Local Phase Invariance

We now show that spontaneous breaking of local symmetry implies that the associated vector boson has a mass, in spite of the fact that $A^\mu A_\mu$ by itself is not locally phase invariant. Much of the calculation in Note 3 can be translated to the Lagrangian of Eq. 38. In fact, the calculation is identical from Eq. 16 to Eq. 18, so the first new step is to substitute Eq. 17 into Eq. 38. The only significantly new part of the calculation is replacing $\partial^\mu\varphi^\dagger\partial_\mu\varphi$ by $(D^\mu\varphi)^\dagger(D_\mu\varphi)$. However, instead of simply substituting Eq. 17 for $\varphi$ and computing $(D^\mu\varphi)^\dagger(D_\mu\varphi)$ directly, it is convenient to make a local phase transformation first:

$$\varphi'(x) = \frac{1}{\sqrt{2}}[\rho(x) + \varphi_0] \exp[i\pi(x)/\varphi_0] , \tag{41}$$

where $\varphi(x) = [\rho(x) + \varphi_0]/\sqrt{2}$. (The local phase invariance permits us to remove the phase of $\varphi(x)$ at every space-time point.) We emphasize the difference between Eqs. 17 and 41: Eq. 17 defines the $\rho(x)$ and $\pi(x)$ fields; Eq. 41 is a local phase transformation of $\varphi(x)$ by angle $\pi(x)$. Don't be fooled by the formal similarity of the two equations. Thus, we may write Eq. 38 in terms of $\varphi(x) = [\rho(x) + \varphi_0]/\sqrt{2}$ and obtain

This leaves a problem. If we simply replace $\partial_\mu\varphi$ by $D_\mu\varphi$ in the Lagrangian and then derive the equations of motion for $A_\mu$, we find that $A_\mu$ is proportional to the current $J_\mu$. The $A_\mu$ field equation has no space-time derivatives and therefore $A_\mu(x)$ does not propagate. If we want $A_\mu$ to correspond to the electromagnetic field potential, we must add a kinetic energy term for it to $\mathscr{L}$.

The problem then is to find a locally phase invariant kinetic energy term for $A_\mu(x)$. Note that the combination of covariant derivatives $D_\mu D_\nu - D_\nu D_\mu$, when acting on any function, contains no derivatives of the function. We define the electromagnetic field tensor of electrodynamics as

$$F_{\mu\nu} \equiv \frac{i}{e}\left[D_\mu, D_\nu\right] = \partial_\mu A_\nu - \partial_\nu A_\mu . \tag{36}$$

It contains derivatives of $A_\mu$. Its transformation law under the local symmetry is

$$F'_{\mu\nu} = F_{\mu\nu} . \tag{37}$$

Thus, it is completely trivial to write down a term that is quadratic in the derivatives of $A_\mu$, which would be an appropriate kinetic energy term. A fully phase invariant generalization of Eq. 1a is

$$\mathscr{L} = -\frac{1}{4} F^{\mu\nu}F_{\mu\nu} + (D^\mu\varphi)^\dagger (D_\mu\varphi) - m^2\varphi^\dagger\varphi - \lambda(\varphi^\dagger\varphi)^2 . \tag{38}$$

We should emphasize that $\mathscr{L}$ has no mass term for $A_\mu(x)$. Thus, when the fields correspond directly to the particles in Eq. 38, the vector particles described by $A_\mu(x)$ are massless. In fact, $A^\mu A_\mu$ is not invariant under the gauge transformation in Eq. 35, so it is not obvious how the $A_\mu$ field can acquire a mass if the theory does have local phase invariance. In Note 6 we will show how the gauge field becomes massive through spontaneous symmetry breaking. This is

the key to understanding the electroweak theory.

We now rediscover the Lagrangian of electrodynamics for the interaction of electrons and photons following the same procedure that we used for the complex scalar field. We begin with the kinetic energy term for a Dirac field of the electron $\psi$, replace $\partial_\mu$ by $D_\mu$ defined in Eq. 33, and then add $-\frac{1}{4}F^{\mu\nu}F_{\mu\nu}$, where $F^{\mu\nu}$ is defined in Eq. 36. The Lagrangian for a free Dirac field is

$$\mathscr{L}_{\text{Dirac}} = \bar{\psi}(i\gamma^\mu\partial_\mu - m)\psi , \tag{39}$$

where $\gamma^\mu$ are the four Dirac $\gamma$ matrices and $\bar{\psi} = \psi^\dagger\gamma_0$. Straightening out the definition of the $\gamma^\mu$ matrices and the components of $\psi$ is the problem of describing a spin-½ particle in a theory with Lorentz invariance. We leave the details of the Dirac theory to textbooks, but note that we will use some of these details when we finally write down the interactions of the quarks and leptons. The interaction of the electron field $\psi$ with the electromagnetic field follows by replacing $\partial_\mu$ by $D_\mu$. The electrodynamic Lagrangian is

$$\mathscr{L} = -\frac{1}{4} F^{\mu\nu}F_{\mu\nu} + \bar{\psi}(i\gamma^\mu D_\mu - m)\psi , \tag{40a}$$

where the interaction term in $i\bar{\psi}\gamma^\mu D_\mu\psi$ has the form

$$\mathscr{L}_{\text{interaction}} = e\bar{\psi}\gamma_\mu\psi A^\mu = eJ^{\text{em}}_\mu A^\mu , \tag{40b}$$

where $J^{\text{em}}_\mu = \bar{\psi}\gamma_\mu\psi$ is the electromagnetic current of the electron. What is amazing about the standard model is that all the electroweak and strong interactions between fermions and vector bosons are similar in form to Eq. 40b, and much phenomenology can be understood in terms of such interaction terms as long as we can approximate the quantum fields with the classical solutions.

$$\mathscr{L} = -\frac{1}{4} F^{\mu\nu}F_{\mu\nu} + \frac{1}{2} \partial^\mu\rho\partial_\mu\rho + \frac{e^2}{2}(\rho + \varphi_0)^2 A^\mu A_\mu$$
$$- \frac{m^2}{2}(\rho + \varphi_0)^2 - \frac{\lambda}{4}(\rho + \varphi_0)^4 . \tag{42}$$

(At the expense of a little algebra, the calculation can be done the other way. First substitute Eq. 17 for $\varphi$ in Eq. 38. One then finds an $A^\mu\partial_\mu\pi$ term in $\mathscr{L}$ that can be removed using the local phase transformation $A'_\mu = A_\mu - [1/(e\varphi_0)]\partial_\mu\pi$, $\rho' = \rho$, and $\pi' = 0$. Equation 42 then follows, although this method requires some effort. Thus, a reason for doing the calculation in the order of Eq. 41 is that the algebra gets messy rather quickly if the local symmetry is not used early in the calculation of the electroweak case. However, in principle it makes little difference.)

The Lagrangian in Eq. 42 is an amazing result: the $\pi$ field has

vanished from $\mathscr{L}$ altogether (according to Eq. 41, it was simply a gauge artifact), and there is a term $\frac{1}{2}e^2\varphi_0^2 A^\mu A_\mu$ in $\mathscr{L}$, which is a mass term for the vector particle. Thus, the massless particle of the global case has become the longitudinal mode of a massive vector particle, and there is only one scalar particle $\rho$ left in the theory. In somewhat more picturesque language the vector boson has eaten the Goldstone boson and become heavy from the feast. However, the existence of the vector boson mass terms should not be understood in isolation: the phase invariance of Eq. 42 determines the form of the interaction of the massive $A_\mu$ field with the $\rho$ field.

This calculation makes it clear that it can be tricky to derive the spectrum of a theory with local symmetry and spontaneous symmetry breaking. Theoretical physicists have taken great care to confirm that this interpretation is correct and that it generalizes to the full quantum field theory.

# 7 Yang-Mills Theories

The standard model possesses symmetries of the type described in Note 4, except that they are local. Thus, we need to carry out the calculations of Note 5 for Lie-group symmetries. As the reader might expect, this requires replacing $\varepsilon(x)$ of Eq. 13 by a matrix or, equivalently, the matrix of Eq. 21 by a matrix function of $x$, $\varepsilon^a(x)T_a$. The Yang-Mills Lagrangian can be derived by mimicking with matrix functions Eqs. 34 to 38.

The internal, local transformation of the $\varphi$ field ($\varphi$ is a column vector with components $\varphi_i$, where $i$ runs from 1 to $n$) is

$$\varphi'(x) = e^{i\varepsilon(x)}\varphi(x) , \tag{43}$$

which is formally identical to Eq. 30, except that $\varepsilon(x)$ is now an $n$-by-$n$ matrix. Thus,

$$\varepsilon(x) = \varepsilon^a(x)T_a , \tag{44}$$

where the sum on $a$ is over the $N$ independent symmetries. Equation 43 is a symmetry of the potential energy

$$V = \mu^2\varphi^\dagger\varphi + \lambda(\varphi^\dagger\varphi)^2 , \tag{45}$$

if $\varepsilon(x)$ in Eq. 44 is a Hermitian matrix (that is, if $T_a = T_a^\dagger$ and the $\varepsilon^a(x)$ are real functions). The kinetic energy $(\partial^\mu\varphi)^\dagger(\partial_\mu\varphi)$ can be made phase invariant by extending $\partial_\mu$ to $D_\mu$, analogous to Eq. 33 for electrodynamics:

$$D_\mu = \partial_\mu - ieA_\mu , \tag{46a}$$

where

$$A_\mu = A_\mu^a T_a , \tag{46b}$$

so that $A_\mu$ is an $n$-by-$n$ matrix that acts on the $\varphi$ vector. Just as for Eq. 35, the transformation properties of $A_\mu$ are derived from the equation

$$D'_\mu\varphi'(x) = e^{i\varepsilon(x)} D_\mu\varphi(x) . \tag{47}$$

After some matrix manipulation one finds the solution of Eq. 47 for $A'_\mu(x)$ in terms of $A_\mu(x)$ to be

$$A'_\mu(x) = e^{i\varepsilon(x)} A_\mu(x)e^{-i\varepsilon(x)} - \frac{1}{e} \partial_\mu\varepsilon(x) , \tag{48}$$

where $e^{-i\varepsilon(x)}$ is the inverse of the matrix $e^{i\varepsilon(x)}$. With these requirements, it is easily seen that $(D^\mu\varphi)^\dagger(D_\mu\varphi)$ is invariant under the group of local transformations.

The calculation of the field tensor is formally identical to Eq. 36, except we must take into account that $A_\mu(x)$ is a matrix. Thus, we define a matrix $F_{\mu\nu}$ field tensor as

$$F_{\mu\nu} \equiv \frac{i}{e} [D_\mu , D_\nu] = \partial_\mu A_\nu - \partial_\nu A_\mu - ie [A_\mu , A_\nu] . \tag{49}$$

There is a field tensor for each group generator, and some further matrix manipulation plus Eq. 26 gives the components,

$$F_{\mu\nu}^a = \partial_\mu A_\nu^a - \partial_\nu A_\mu^a + ef^{abc} A_{\mu b} A_{\nu c} . \tag{50}$$

The transformation law for the matrix $F_{\mu\nu}$ is

$$F'_{\mu\nu} = e^{i\varepsilon(x)} F_{\mu\nu} e^{-i\varepsilon(x)} . \tag{51}$$

Thus, we can write down a kinetic energy term in analogy to electrodynamics:

$$\mathcal{L}_{\text{kinetic energy}} = -\frac{1}{4} F_{\mu\nu}^a F_a^{\mu\nu} . \tag{52}$$

The locally invariant Yang-Mills Lagrangian for spinless fields coupled to the vector bosons is

$$\mathcal{L} = -\frac{1}{4} F_{\mu\nu}^a F_a^{\mu\nu} + (D^\mu\varphi)^\dagger(D_\mu\varphi) - \mu^2\varphi^\dagger\varphi - \lambda(\varphi^\dagger\varphi)^2 . \tag{53}$$

Just as in electrodynamics, we can add fermions to the theory in the form

$$\mathcal{L}_{\text{fermion}} = \bar{\psi}(i\gamma^\mu D_\mu - m)\psi , \tag{54}$$

where $D_\mu$ is defined in Eq. 46 and $\psi$ is a column vector with $n_{\text{f}}$ entries ($n_{\text{f}}$ = number of fermions). The matrices $T_a$ in $D_\mu$ for the fermion covariant derivative are usually different from the matrices for the spinless fields, since there is no requirement that $\varphi$ and $\psi$ need to belong to the same representation of the group. It is, of course, necessary for the sets of $T_a$ matrices to satisfy the commutation relations of Eq. 26 with the same set of structure constants.

We will not look at the general case of spontaneous symmetry breaking in a Yang-Mills theory, which is a messy problem mathematically. There is spontaneous symmetry breaking in the electroweak sector of the standard model, and we will work out the steps analogous to Eqs. 41 and 42 for this particular case in the next Note.

# (8) The SU(2) × U(1) Electroweak Model

The main emphasis in these Notes has been on developing just those aspects of Lagrangian field theory that are needed for the standard model. We have now come to the crucial step: finding a Lagrangian that describes the electroweak interactions. It is rather difficult to be systematic. The historical approach would be complicated by the rather late discovery of the weak neutral currents, and a purely phenomenological development is not yet totally logical because there are important aspects of the standard model that have not yet been tested experimentally. (The most important of these are the details of the spontaneous symmetry breaking.) Although we will write down the answer without excessive explanation, the reader should not forget the critical role that experimental data played in the development of the theory.

The first problem is to identify the local symmetry group. Before the standard model was proposed over twenty years ago, the electromagnetic and charge-changing weak interactions were known. The smallest continuous group that can describe these is SU(2), which has a doublet representation. If the weak interactions can change electrons to electron neutrinos, which are electrically neutral, it is not possible to incorporate electrodynamics in SU(2) alone unless a heavy positively charged electron is added to the electron and its neutrino to make a triplet, because the sum of charges in an SU(2) multiplet is zero. Various schemes of this sort have been tried but do not agree with experiment. The only way to leave the electron and electron neutrino in a doublet and include electrodynamics is to add an extra U(1) interaction to the theory. The hypothesis of the extra U(1) factor was challenged many times until the discovery of the weak neutral current. That discovery established that the local symmetry of the electroweak theory had to be at least as large as SU(2) × U(1).

Let us now interpret the physical meaning of the four generators of SU(2) × U(1). The three generators of the SU(2) group are $I^+$, $I_3$, and $I^-$, and the generator of the U(1) group is called $Y$, the weak hypercharge. (The weak SU(2) and U(1) groups are distinguished from other SU(2) and U(1) groups by the label "W.") $I^+$ and $I^-$ are associated with the weak charge-changing currents (the general definition of a current is described in Note 2), and the charge-changing currents couple to the $W^+$ and $W^-$ charged weak vector bosons in analogy to Eq. 40b. Both $I_3$ and $Y$ are related to the electromagnetic current and the weak neutral current. In order to assign the electron and its neutrino to an SU(2) doublet, the electric charge $Q^{em}$ is defined by

$$Q^{em} = I_3 + Y/2 , \tag{55}$$

so the sum of electric charges in an $n$-dimensional multiplet is $nY/2$. The charge of the weak neutral current is a different combination of $I_3$ and $Y$, as will be described below.

The Lagrangian includes many pieces. The kinetic energies of the vector bosons are described by $\mathscr{L}_{Y-M}$, in analogy to the first term in Eq. 38. The three weak bosons have masses acquired through spontaneous symmetry breaking, so we need to add a scalar piece $\mathscr{L}_{scalar}$ to the Lagrangian in order to describe the observed symmetry breaking (also see Eq. 38). The fermion kinetic energy $\mathscr{L}_{fermion}$ includes the fermion-boson interactions, analogous to the electromagnetic interactions derived in Eqs. 39 and 40. Finally, we can add terms that couple the scalars with the fermions in a term $\mathscr{L}_{Yukawa}$. One physical significance of the Yukawa terms is that they provide for masses of the quarks and charged leptons.

The standard model is then a theory with a very long Lagrangian with many fields. The electroweak Lagrangian has the terms

$$\mathscr{L}_{electroweak} = \mathscr{L}_{Y-M} + \mathscr{L}_{scalar} + \mathscr{L}_{fermion} + \mathscr{L}_{Yukawa} . \tag{56}$$

(The reader may find this construction to be ad hoc and ugly. If so, the motivation will be clear for searching for a more unified theory from which this Lagrangian can be derived. However, it is important to remember that, at present, the standard model is the pinnacle of success in theoretical physics and describes a broader range of natural phenomena than any theory ever has.)

The Yang-Mills kinetic energy term has the form given by Eq. 52 for the SU(2) bosons, plus a term for the U(1) field tensor similar to electrodynamics (Eqs. 36 and 38).

$$\mathscr{L}_{Y-M} = -\frac{1}{4} F_a^{\mu\nu} F_{\mu\nu}^a - \frac{1}{4} F^{\mu\nu} F_{\mu\nu} , \tag{57}$$

where the U(1) field tensor is

$$F_{\mu\nu} = \partial_\mu B_\nu - \partial_\nu B_\mu \tag{58}$$

and the SU(2) Yang-Mills field tensor is

$$F_{\mu\nu}^a = \partial_\mu W_\nu^a - \partial_\nu W_\mu^a + g\varepsilon_{abc} W_\mu^b W_\nu^c , \tag{59}$$

where the $\varepsilon_{abc}$ are the structure constants for SU(2) defined in Eq. 24 and the $W_\mu^a$ are the Yang-Mills fields.

# ⑧ continued

SU(2) × U(1) has two factors, and there is an independent coupling constant for each factor. The coupling for the SU(2) factor is called $g$, and it has become conventional to call the U(1) coupling $g'/2$. The two couplings can be written in several ways. The U(1) of electrodynamics is generated by a linear combination of $I_3$ and $Y$, and the coupling is, as usual, denoted by $e$. The other coupling can then be parameterized by an angle $\theta_W$. The relations among $g$, $g'$, $e$, and $\theta_W$ are

$$e \equiv gg'/\sqrt{g^2+g'^2} \quad \text{and} \quad \tan\theta_W \equiv g'/g. \tag{60}$$

These definitions will be motivated shortly. In the electroweak theory both couplings must be evaluated experimentally and cannot be calculated in the standard model.

The scalar Lagrangian requires a choice of representation for the scalar fields. The choice requires that the field with a nonzero vacuum value is electrically neutral, so the photon remains massless, but it must carry nonzero values of $I_3$ and $Y$ so that the weak neutral boson (the $Z_\mu^0$) acquires a mass from spontaneous symmetry breaking. The simplest assignment is

$$\varphi = \begin{pmatrix} \varphi^+ \\ \varphi^0 \end{pmatrix} \quad \text{and} \quad \varphi^\dagger = (-\varphi^-, (\varphi^0)^\dagger), \tag{61}$$

where $\varphi^+$ has $I_3 = \frac{1}{2}$ and $Y = 1$, and $\varphi^0$ has $I_3 = -\frac{1}{2}$ and $Y = 1$. Since $\varphi$ does not have $Y = -1$ fields, it is necessary to make $\varphi$ a complex doublet, so $(\varphi^+)^\dagger = -\varphi^-$ has $I_3 = -\frac{1}{2}$ and $Y = -1$, and $(\varphi^0)^\dagger$ has $I_3 = \frac{1}{2}$ and $Y = -1$. Then we can write down the Lagrangian of the scalar fields as

$$\mathcal{L}_{\text{scalar}} = (D^\mu\varphi)^\dagger(D_\mu\varphi) - m^2\varphi^\dagger\varphi - \lambda(\varphi^\dagger\varphi)^2, \tag{62}$$

where

$$D_\mu\varphi = \partial_\mu\varphi - i\frac{g'}{2}B_\mu\varphi - i\frac{g}{2}\tau_a W_\mu^a\varphi \tag{63}$$

is the covariant derivative. The 2-by-2 matrices $\tau_a$ are the Pauli matrices. The factor of $\frac{1}{2}$ is required because the doublet representation of the SU(2) generators is $\tau_a/2$. The factor of $\frac{1}{2}$ in the $B_\mu$ term is due to the convention that the U(1) coupling is $g'/2$ and the

assignment that the $\varphi$ doublet has $Y = 1$. After the spontaneous symmetry breaking, three of the four scalar degrees of freedom are "eaten" by the weak bosons. Thus just one scalar escapes the feast and should be observable as an independent neutral particle, called the Higgs particle. It has not yet been observed experimentally, and it is perhaps the most important particle in the standard model that does not yet have a firm phenomenological basis. (The minimum number of scalar fields in the standard model is four. Experimental data could eventually require more.)

We now carry out the calculation for the spontaneous symmetry breaking of SU(2) × U(1) down to the U(1) of electrodynamics. Just as in the example worked out in Note 6, spontaneous symmetry breaking occurs when $m^2 < 0$ in Eq. 62. In contrast to the simpler case, it is rather important to set up the problem in a clever way to avoid an inordinate amount of computation. As in Eq. 41, we write the four degrees of freedom in the complex scalar doublet so that it looks like a local symmetry transformation times a simple form of the field:

$$\varphi(x) = \exp[i\pi^a(x)\tau_a/2\varphi_0]\begin{pmatrix} 0 \\ [\rho(x) + \varphi_0]/\sqrt{2} \end{pmatrix}. \tag{64}$$

We can then write the scalar fields in a new gauge where the phases of $\varphi(x)$ are removed:

$$\varphi'(x) = \exp[-i\pi^a(x)\tau_a/2\varphi_0]\varphi(x) = \begin{pmatrix} 0 \\ [\rho(x) + \varphi_0]/\sqrt{2} \end{pmatrix}, \tag{65}$$

where we have used the freedom of making local symmetry transformations to write $\varphi'(x)$ in a very simple form. This choice, called the unitary gauge, will make it easy to write out Eq. 63 in explicit matrix form. Let us drop all primes on the fields in the unitary gauge and redefine $W_\mu^a$ by the equation

$$\tau_a W_\mu^a = \begin{pmatrix} W_\mu^3 & W_\mu^1 - iW_\mu^2 \\ W_\mu^1 + iW_\mu^2 & -W_\mu^3 \end{pmatrix} = \begin{pmatrix} W_\mu^3 & \sqrt{2}\,W_\mu^+ \\ \sqrt{2}\,W_\mu^- & -W_\mu^3 \end{pmatrix}, \tag{66}$$

where the definition of the Pauli matrices is used in the first step, and the $W^\pm$ fields are defined in the second step with a numerical factor that guarantees the correct normalization of the kinetic energy of the charged weak vector bosons.

Next, we write out the $D_\mu\varphi$ in explicit matrix form, using Eqs. 63, 65, and 66:

$$D_\mu\varphi = \frac{1}{\sqrt{2}}\begin{pmatrix} -i\sqrt{2}gW_\mu^+(\rho + \varphi_0)/2 \\ \partial_\mu\rho - i(g'B_\mu - gW_\mu^3)(\rho + \varphi_0)/2 \end{pmatrix}. \tag{67}$$

Finally, we substitute Eqs. 65 and 67 into Eq. 62 and obtain

$$\mathscr{L}_{\text{scalar}} = \frac{g^2}{4} W^{\mu}_{-} W^{+}_{\mu}(\rho + \varphi_0)^2 + \frac{1}{2} \partial^{\mu}\rho\partial_{\mu}\rho$$

$$+ \frac{1}{8} (g'B^{\mu} - gW^{\mu}_3)(g'B_{\mu} - gW^3_{\mu})(\rho + \varphi_0)^2$$

$$+ \frac{m^2}{2} (\rho + \varphi_0)^2 + \frac{\lambda}{4} (\rho + \varphi_0)^4 , \tag{68}$$

where $\rho$ is the, as yet, unobserved Higgs field.

It is clear from Eq. 68 that the $W$ fields will acquire a mass equal to $g\varphi_0/2$ from the term quadratic in the $W$ fields, $(g^2/4)\varphi_0^2 W^{\mu}_{-} W^{+}_{\mu}$. The combination $g'B_{\mu} - gW^3_{\mu}$ will also have a mass. Thus, we "rotate" the $B_{\mu}$ and $W^3_{\mu}$ fields to the fields $Z^0_{\mu}$ for the weak neutral boson and $A_{\mu}$ for the photon so that the photon is massless.

$$\begin{pmatrix} Z^0_{\mu} \\ A_{\mu} \end{pmatrix} = \begin{pmatrix} \sin\theta_W & -\cos\theta_W \\ \cos\theta_W & \sin\theta_W \end{pmatrix} \begin{pmatrix} B_{\mu} \\ W^3_{\mu} \end{pmatrix} , \tag{69}$$

where

$$\cos\theta_W = g/\sqrt{g^2 + g'^2} \quad \text{and} \quad \sin\theta_W = g'/\sqrt{g^2+g'^2} . \tag{70}$$

Upon substituting Eqs. 69 and 70 into Eq. 68, we find that the $Z^0_{\mu}$ mass is $\frac{1}{2}\varphi_0\sqrt{g^2 + g'^2}$, so the ratio of the $W$ and $Z$ masses is

$$M_W/M_Z = \cos\theta_W . \tag{71}$$

Values for $M_W$ and $M_Z$ have recently been measured at the CERN proton-antiproton collider: $M_W = (80.8 \pm 2.7)$ GeV/$c^2$ and $M_Z = (92.9 \pm 1.6)$ GeV/$c^2$. The ratio $M_W/M_Z$ calculated with these values agrees well with that given by Eq. 71. (The angle $\theta_W$ is usually expressed as $\sin^2\theta_W$ and is measured in neutrino-scattering experiments to be $\sin^2\theta_W = 0.224 \pm 0.015$.) The photon field $A_{\mu}$ does not appear in $\mathscr{L}_{\text{scalar}}$, so it does not become massive from spontaneous symmetry breaking. Note, also, that the $\pi^a(x)$ fields appear nowhere in the Lagrangian; they have been eaten by three weak vector bosons, which have become massive from the feast.

The next term in Eq. 56 is $\mathscr{L}_{\text{fermion}}$. Its form is analogous to Eqs. 39 and 40 for electrodynamics:

$$\mathscr{L}_{\text{fermion}} = i\bar{\psi}\gamma^{\mu}D_{\mu}\psi . \tag{72}$$

The physical problem is to assign the left- and right-handed fermions to multiplets of SU(2); the assignments rely heavily on experimental data and are listed in "Particle Physics and the Standard Model."

Our purpose here will be to write out Eq. 72 explicitly for the assignments.

Consider the electron and its neutrino. (The quark and remaining lepton contributions can be worked out in a similar fashion.) The left-handed components are assigned to a doublet and the right-handed components are singlets. (Since a neutral singlet has no weak charge, the right-handed component of the neutrino is invisible to weak, electromagnetic, or strong interactions. Thus, we can neglect it here, whether or not it actually exists.) We adopt the notation

$$\psi_L = \begin{pmatrix} \nu_L \\ e_L \end{pmatrix} \quad \text{and} \quad \psi_R = (e^-_R) , \tag{73}$$

where L and R denote left- and right-handed. Then the explicit statement of Eq. 72 requires constructing $D_{\mu}$ for the left- and right-handed leptons.

$$\mathscr{L}_{\text{lepton}} = i\bar{\psi}_R\gamma^{\mu}(\partial_{\mu} + ig'B_{\mu})\psi_R$$

$$+ i\bar{\psi}_L\gamma^{\mu}[\partial_{\mu} + \frac{i}{2} (g'B_{\mu} - g\tau_a W^a_{\mu})]\psi_L . \tag{74}$$

The weak hypercharge of the right-handed electron is $-2$ so the coefficient of $B_{\mu}$ in the first term of Eq. 74 is $(-g'/2) \times (-2) = g'$. We leave it to the reader to check the rest of Eq. 74. The absence of a mass term is not an error. Mass terms are of the form $\bar{\psi}\psi = \bar{\psi}_L\psi_R + \bar{\psi}_R\psi_L$. Since $\psi_L$ is a doublet and $\bar{\psi}_R$ is a singlet, an electron mass term must violate the SU(2) $\times$ U(1) symmetry. We will see later that the electron mass will reappear as a result of modification of $\mathscr{L}_{\text{Yukawa}}$ due to spontaneous symmetry breaking.

The next task is exciting, because it will reveal how the vector bosons interact with the leptons. The calculation begins with Eq. 74 and requires the substitution of explicit matrices for $\tau_a W^a_{\mu}$, $\psi_R$, and $\psi_L$. We use the definitions in Eqs. 66, 69, and 73. The expressions become quite long, but the calculation is very straightforward. After simplifying some expressions, we find that $\mathscr{L}_{\text{lepton}}$ for the electron lepton and its neutrino is

$$\mathscr{L}_{\text{lepton}} = i\bar{e}\gamma^{\mu}\partial_{\mu}e + i\bar{\nu}_L\gamma^{\mu}\partial_{\mu}\nu_L - e \bar{e}\gamma^{\mu}eA_{\mu}$$

$$+ \frac{g}{\sqrt{2}} (\bar{\nu}_L\gamma^{\mu}e_L W^+_{\mu} + \bar{e}_L\gamma^{\mu}\nu_L W^-_{\mu})$$

$$- \frac{g^2}{2\sqrt{g^2 + g'^2}} [\tan^2\theta_W(2\bar{e}_R\gamma^{\mu}e_R + \bar{e}_L\gamma^{\mu}e_L) - \bar{e}_L\gamma^{\mu}e_L]Z_{\mu}$$

$$- \frac{1}{2} \sqrt{g^2 + g'^2} \, \bar{\nu}_L\gamma^{\mu}\nu_L Z_{\mu} . \tag{75}$$

# ⑧ continued

The first two terms are the kinetic energies of the electron and the neutrino. (Note that $e = e_L + e_R$.) The third term is the electromagnetic interaction (cf. Eq. 40) with electrons of charge $-e$, where $e$ is defined in Eq. 60. The coupling of $A_\mu$ to the electron current does not distinguish left from right, so electrodynamics does not violate parity. The fourth term is the interaction of the $W^\pm$ bosons with the weak charged current of the neutrinos and electrons. Note that these bosons are blind to right-handed electrons. This is the reason for maximal parity violation in beta decay. The final terms predict how the weak neutral current of the electron and that of the neutrino couple to the neutral weak vector boson $Z^0$.

If the left- and right-handed electron spinors are written out explicitly, with $e_L = \frac{1}{2}(1 - \gamma_5)e$, the interaction of the weak neutral current of the electron with the $Z^0$ is proportional to $\bar{e}\gamma^\mu[(1 - 4\sin^2\theta_W) - \gamma_5]eZ_\mu$. This prediction provided a crucial test of the standard model. Recall from Eq. 71 that $\sin^2\theta_W$ is very nearly $\frac{1}{4}$, so that the weak neutral current of the electron is very nearly a purely axial current, that is, a current of the form $\bar{e}\gamma^\mu\gamma_5 e$. This crucial prediction was tested in deep inelastic scattering of polarized electrons and in atomic parity-violation experiments. The results of these experiments went a long way toward establishing the standard model. The tests also ruled out models quite similar to the standard model. We could discuss many more tests and predictions of the model based on the form of the weak currents, but this would greatly lengthen our discussion. The electroweak currents of the quarks will be described in the next section.

We now discuss the last term in Eq. 56, $\mathscr{L}_{\text{Yukawa}}$. In a locally symmetric theory with scalars, spinors, and vectors, the interactions between vectors and scalars, vector and spinors, and vectors and vectors are determined from the local invariance by replacing $\partial_\mu$ by $D_\mu$. In contrast, $\mathscr{L}_{\text{Yukawa}}$, which is the interaction between the scalars and spinors, has the same form for both local and global symmetries:

$$\mathscr{L}_{\text{Yukawa}} = G_Y \bar{\psi}\varphi\psi$$
$$= G_Y(\bar{\psi}_L\varphi\psi_R + \bar{\psi}_R\varphi^\dagger\psi_L) . \qquad (76)$$

This form for $\mathscr{L}_{\text{Yukawa}}$ is rather schematic; to make it explicit we must specify the multiplets and then arrange the component fields so that the form of $\mathscr{L}_{\text{Yukawa}}$ does not change under a local symmetry transformation.

Let us write Eq. 76 explicitly for the part of the standard model we have examined so far: $\varphi$ is a complex doublet of scalar fields that has the form in the unitary gauge given by Eq. 65. The fermions include the electron and its neutrino. If the neutrino has no right-handed component, then it is not possible to insert it into Eq. 76. Since the neutrino has no mass term in $\mathscr{L}_{\text{lepton}}$, the neutrino remains massless in this theory. (If $\nu_R$ is included, then the neutrino mass is a free parameter.) The Yukawa terms for the electron are

$$\mathscr{L}_{\text{Yukawa}} = G_Y\left[ (\bar{\nu}_L, \bar{e}_L)\begin{pmatrix} 0 \\ (\rho + \varphi_0)/\sqrt{2} \end{pmatrix}(e_R) \right.$$

$$\left. + (\bar{e}_R)(0, (\rho + \varphi_0)/\sqrt{2})\begin{pmatrix} \nu_L \\ e_L \end{pmatrix}\right]$$

$$= \frac{1}{\sqrt{2}} G_Y \bar{e}e(\rho + \varphi_0) , \qquad (77)$$

where we have used the fact that $\bar{e}_Le_L = \bar{e}_Re_R = 0$, and $e = e_L + e_R$ is the electron Dirac spinor. Note that Eq. 77 includes an electron mass term,

$$m_e = \frac{1}{\sqrt{2}} G_Y\varphi_0 , \qquad (78)$$

so the electron mass is proportional to the vacuum value of the scalar field. The Yukawa coupling is a free parameter, but we can use the measured electron mass to evaluate it. Recall that

$$M_W = \frac{g\varphi_0}{2} = \frac{e\varphi_0}{2 \sin \theta_W} \simeq 81 \text{ GeV} ,$$

where $e^2/4\pi \simeq 1/137$. This implies that $\varphi_0 = 251$ GeV. Since $m_e = 0.000511$ GeV, $G_Y = 2.8 \times 10^{-6}$ for the electron. There are more than five Yukawa couplings, including those for the $\mu$ and $\tau$ leptons and the three quark doublets as well as terms that mix different quarks of the same electric charge. The standard model in no way determines the values of these Yukawa coupling constants. Thus, the study of fermion masses may turn out to have important hints on how to extend the standard model.

# Quarks

Discovery of the fundamental fields of the strong interactions was not straightforward. It took some years to realize that the hadrons, such as the nucleons and mesons, are made up of subnuclear constituents, primarily quarks. Quarks originated from an effort to provide a simple physical picture of the "Eightfold Way," which is the SU(3) symmetry proposed by M. Gell-Mann and Y. Ne'eman to generalize strong isotopic spin. The hadrons could not be classified by the fundamental three-dimensional representations of this SU(3) but instead are assigned to eight- and ten-dimensional representations. These larger representations can be interpreted as products of the three-dimensional representations, which suggested to Gell-Mann and G. Zweig that hadrons are composed of constituents that are assigned to the three-dimensional representations: the *u* (up), *d* (down), and *s* (strange) quarks. At the time of their conception, it was not clear whether quarks were a physical reality or a mathematical trick for simplifying the analysis of the Eightfold-Way SU(3). The major breakthrough in the development of the present theory of strong interactions came with the realization that, in addition to electroweak and Eightfold-Way quantum numbers, quarks carry a new quantum number, referred to as color. This quantum number has yet to be observed experimentally.

We begin this lecture with a description of the Lagrangian of a strong-interaction theory of quarks formulated in terms of their color quantum numbers. Called quantum chromodynamics, or QCD, it is a Yang-Mills theory with local color-SU(3) symmetry in which each quark belongs to a three-dimensional color multiplet. The eight color-SU(3) generators commute with the electroweak SU(2) × U(1) generators, and they also commute with the generators of the Eightfold Way, which is a different SU(3). (Like SU(2), SU(3) is a recurring symmetry in physics, so its various roles need to be distinguished. Hence we need the label "color.") We conclude with a discussion of the weak interactions of the quarks.

**The QCD Lagrangian.** The interactions among the quarks are mediated by eight massless vector bosons (called gluons) that are required to make the SU(3) symmetry local. As we have already seen,

the assumption of local symmetry leads to a Lagrangian whose form is highly restricted. As far as we know, only the quark and gluon fields are necessary to describe the strong interactions, and so the most general Lagrangian is

$$\mathcal{L}_{\text{QCD}} = -\frac{1}{4} F^a_{\mu\nu} F^{\mu\nu}_a + i\sum_i \bar{\psi}_i \gamma^\mu D_\mu \psi_i + \sum_{i,j} \bar{\psi}_i M_{ij} \psi_j , \qquad (79)$$

where

$$F^a_{\mu\nu} = \partial_\mu A^a_\nu - \partial_\nu A^a_\mu + g_s f_{abc} A^b_\mu A^c_\nu . \qquad (80)$$

The sum on *a* in the first term is over the eight gluon fields $A^a_\mu$. The second term represents the coupling of each gluon field to an SU(3) current of the quark fields, called a color current. This term is summed over the index *i*, which labels each quark type and is independent of color. Since each quark field $\psi_i$ is a three-dimensional column vector in color space, $D_\mu$ is defined by

$$D_\mu \psi_i = \partial_\mu \psi_i - \frac{1}{2} i g_s A^a_\mu \lambda_a \psi_i , \qquad (81)$$

where $\lambda_a$ is a generalization of the three 2-by-2 Pauli matrices of SU(2) to the eight 3-by-3 Gell-Mann matrices of SU(3), and $g_s$ is the QCD coupling. Thus, the color current of each quark has the form $\bar{\psi}\lambda_a \gamma^\mu \psi$. The left-handed quark fields couple to the gluons with exactly the same strength as the right-handed quark fields, so parity is conserved in the strong interactions.

The gluons are massless because the QCD Lagrangian has no spinless fields and therefore no obvious possibility of spontaneous symmetry breaking. Of course, if motivated for experimental reasons, one can add scalars to the QCD Lagrangian and spontaneously break SU(3) to a smaller group. This modification has been used, for example, to explain the reported observation of fractionally charged particles. The experimental situation, however, still remains murky, so it is not (yet) necessary to spontaneously break SU(3) to a smaller group. For the remainder of the discussion, we assume that QCD is not spontaneously broken.

The third term in Eq. 79 is a mass term. In contrast to the electroweak theory, this mass term is now allowed, even in the absence of spontaneous symmetry breaking, because the left- and right-handed quarks are assigned to the same multiplet of SU(3). The numerical coefficients $M_{ij}$ are the elements of the quark mass matrix; they can connect quarks of equal electric charge. The $\mathcal{L}_{\text{QCD}}$ of Eq. 79 permits us to redefine the QCD quark fields so that $M_{ij} = m_i \delta_{ij}$. The

# ⑨ continued

mass matrix is then diagonal and each quark has a definite mass, which is an eigenvalue of the mass matrix. We will reappraise this situation below when we describe the weak currents of the quarks.

After successfully extracting detailed predictions of the electroweak theory from its complicated-looking Lagrangian, we might be expected to perform a similar feat for the $\mathscr{L}_{\text{QCD}}$ of Eq. 79 without too much difficulty. This is not possible. Analysis of the electroweak theory was so simple because the couplings $g$ and $g'$ are always small, regardless of the energy scale at which they are measured, so that a classical analysis is a good first approximation to the theory. The quantum corrections to the results in Note 8 are, for most processes, only a few percent.

In QCD processes that probe the *short-distance* structure of hadrons, the quarks inside the hadrons interact weakly, and here the classical analysis is again a good first approximation because the coupling $g_s$ is small. However, for Yang-Mills theories in general, the renormalization group equations of quantum field theory require that $g_s$ increases as the momentum transfer decreases until the momentum transfer equals the masses of the vector bosons. Lacking spontaneous symmetry breaking to give the gluons mass, QCD contains no mechanism to stop the growth of $g_s$, and the quantum effects become more and more dominant at larger and larger distances. Thus, analysis of the *long-distance* behavior of QCD, which includes deriving the hadron spectrum, requires solving the full quantum theory implied by Eq. 79. This analysis is proving to be very difficult.

Even without the solution of $\mathscr{L}_{\text{QCD}}$, we can, however, draw some conclusions. The quark fields $\psi_i$ in Eq. 79 must be determined by experiment. The Eightfold Way has already provided three of the quarks, and phenomenological analyses determine their masses (as they appear in the QCD Lagrangian). The mass of the $u$ quark is nearly zero (a few MeV/$c^2$), the $d$ quark is a few MeV/$c^2$ heavier than the $u$, and the mass of the $s$ quark is around 300 MeV/$c^2$. If these results are substituted into Eq. 79, we can derive a beautiful result from the QCD Lagrangian. In the limit that the quark mass differences can be ignored, Eq. 79 has a global SU(3) symmetry that is identical to the Eightfold-Way SU(3) symmetry. Moreover, in the limit that the $u$, $d$, and $s$ masses can be ignored, the left-handed $u$, $d$,

and $s$ quarks can be transformed by one SU(3) and the right-handed $u$, $d$, and $s$ quarks by an independent SU(3). Then QCD has the "chiral" SU(3) $\times$ SU(3) symmetry that is the basis of current algebra. The sums of the corresponding SU(3) generators of chiral SU(3) $\times$ SU(3) generate the Eightfold-Way SU(3). Thus, the QCD Lagrangian incorporates in a very simple manner the symmetry results of hadronic physics of the 1960s. The more recently discovered $c$ (charmed) and $b$ (bottom) quarks and the conjectured $t$ (top) quark are easily added to the QCD Lagrangian. Their masses are so large and so different from one another that the SU(3) and SU(3) $\times$ SU(3) symmetries of the Eightfold-Way and current algebra cannot be extended to larger symmetries. (The predictions of, say, SU(4) and chiral SU(4) $\times$ SU(4) do not agree well with experiment.)

It is important to note that the quark masses are undetermined parameters in the QCD Lagrangian and therefore must be derived from some more complete theory or indicated phenomenologically. The Yukawa couplings in the electroweak Lagrangian are also free parameters. Thus, we are forced to conclude that the standard model alone provides no constraints on the quark masses, so they must be obtained from experimental data.

The mass term in the QCD Lagrangian (Eq. 79) has led to new insights about the neutron-proton mass difference. Recall that the quark content of a neutron is $udd$ and that of a proton is $uud$. If the $u$ and $d$ quarks had the same mass, then we would expect the proton to be more massive than the neutron because of the electromagnetic energy stored in the $uu$ system. (Many researchers have confirmed this result.) Since the masses of the $u$ and $d$ quarks are arbitrary in both the QCD and the electroweak Lagrangians, they can be adjusted phenomenologically to account for the fact that the neutron mass is 1.293 MeV/$c^2$ greater than the proton mass. This experimental constraint is satisfied if the mass of the $d$ quark is about 3 MeV/$c^2$ greater than that of the $u$ quark. In a way, this is unfortunate, because we must conclude that the famous puzzle of the $n$-$p$ mass difference will not be solved until the standard model is extended enough to provide a theory of the quark masses.

**Weak Currents.** We turn now to a discussion of the weak currents of the quarks, which are determined in the same way as the weak currents of the leptons in Note 8. Let us begin with just the $u$ and $d$ quarks. Their electroweak assignments are as follows: the left-handed components $u_L$ and $d_L$ form an SU(2) doublet with $Y = \frac{1}{3}$, and the right-handed components $u_R$ and $d_R$ are SU(2) singlets with $Y = 4/3$

and $-\frac{2}{3}$, respectively (recall Eq. 55).

The steps followed in going from Eq. 73 to Eq. 75 will yield the electroweak Lagrangian of quarks. The contribution to the Lagrangian due to interaction of the weak neutral current $J_\mu^{(nc)}$ of the $u$ and $d$ quarks with $Z^0$ is

$$\mathscr{L}^{(nc)} = \frac{e}{\sin\theta_W \cos\theta_W} \, J_\mu^{(nc)} \, Z_0^\mu \,, \qquad (82)$$

where

$$J_\mu^{(nc)} = \left(\frac{1}{2} - \frac{2}{3}\sin^2\theta_W\right) \bar{u}_L \gamma_\mu u_L - \frac{2}{3}\sin^2\theta_W \bar{u}_R \gamma_\mu u_R$$

$$+ \left(-\frac{1}{2} + \frac{1}{3}\sin^2\theta_W\right) \bar{d}_L \gamma_\mu d_L + \frac{1}{3}\sin^2\theta_W \bar{d}_R \gamma_\mu d_R \,. \qquad (83)$$

The reader will enjoy deriving this result and also deriving the contribution of the weak charged current of the quarks to the electroweak Lagrangian. Equation 83 will be modified slightly when we include the other quarks.

So far we have emphasized in Notes 8 and 9 the construction of the QCD and electroweak Lagrangians for just one lepton-quark "family" consisting of the electron and its neutrino together with the $u$ and $d$ quarks. Two other lepton-quark families are established experimentally: the muon and its neutrino along with the $c$ and $s$ quarks and the $\tau$ lepton and its neutrino along with the $t$ and $b$ quarks. Just like $(\nu_e)_L$ and $e_L$, $(\nu_\mu)_L$ and $\mu_L$ and $(\nu_\tau)_L$ and $\tau_L$ form weak-SU(2) doublets; $e_R$, $\mu_R$ and $\tau_R$ are each SU(2) singlets with a weak hypercharge of $-2$. Similarly, the weak quantum numbers of $c$ and $s$ and of $t$ and $b$ echo those of $u$ and $d$: $c_L$ and $s_L$ form a weak-SU(2) doublet as do $t_R$ and $b_L$. Like $u_R$ and $d_R$, the right-handed quarks $c_R$, $s_R$, $t_R$, and $b_R$ are all weak-SU(2) singlets.

This triplication of families cannot be explained by the standard model, although it may eventually turn out to be a critical fact in the development of theories of the standard model. The quantum numbers of the quarks and leptons are summarized in Tables 2 and 3 in "Particle Physics and the Standard Model."

All these quark and lepton fields must be included in a Lagrangian that incorporates both the electroweak and QCD Lagrangians. It is quite obvious how to do this: the standard model Lagrangian is simply the sum of the QCD and electroweak Lagrangians, except that the terms occurring in both Lagrangians (the quark kinetic energy terms $i\bar{\psi}_i\gamma^\mu\partial_\mu\psi_i$ and the quark mass terms $\bar{\psi}_i M_{ij}\psi_j$) are included just once. Only the mass term requires comment.

The quark mass terms appear in the electroweak Lagrangian in the form $\mathscr{L}_{Yukawa}$ (Eq. 77). In the electroweak theory quarks acquire masses only because SU(2) $\times$ U(1) is spontaneously broken. However, when there are three quarks of the same electric charge (such as $d$, $s$, and $b$), the general form of the mass terms is the same as in Eq. 79, $\bar{\psi}_i M_{ij}\psi_j$, because there can be Yukawa couplings between $d$ and $s$, $d$ and $b$, and $s$ and $b$. The problem should already be clear: when we speak of quarks, we think of fields that have a definite mass, that is, fields for which $M_{ij}$ is diagonal. Nevertheless, there is no reason for the fields obtained directly from the electroweak symmetry breaking to be mass eigenstates.

The final part of the analysis takes some care: the problem is to find the most general relation between the mass eigenstates and the fields occurring in the weak currents. We give the answer for the case of two families of quarks. Let us denote the quark fields in the weak currents with primes and the mass eigenstates without primes. There is freedom in the Lagrangian to set $u = u'$ and $c = c'$. If we do so, then the most general relationship among $d$, $s$, $d'$, and $s'$ is

$$\begin{pmatrix} d' \\ s' \end{pmatrix} = \begin{pmatrix} \cos\theta_C & -\sin\theta_C \\ \sin\theta_C & \cos\theta_C \end{pmatrix} \begin{pmatrix} d \\ s \end{pmatrix}. \qquad (84)$$

The parameter $\theta_C$, the Cabibbo angle, is not determined by the electroweak theory (it is related to ratios of various Yukawa couplings) and is found experimentally to be about $13°$. (When the $b$ and $t$ ($= t'$) quarks are included, the matrix in Eq. 84 becomes a 3-by-3 matrix involving four parameters that are evaluated experimentally.) The correct weak currents are then given by Eq. 83 if all quark families are included and primes are placed on all the quark fields. The weak currents can be written in terms of the quark mass eigenstates by substituting Eq. 84 (or its three-family generalization) into the primed version of Eq. 83. The ratio of amplitudes for $s \rightarrow u$ and $d \rightarrow u$ is $\tan\theta_C$; the small ratio of the strangeness-changing to non-strangeness-changing charged-current amplitudes is due to the smallness of the Cabibbo angle. It is worth emphasizing again that the standard model alone provides no understanding of the value of this angle. ◯

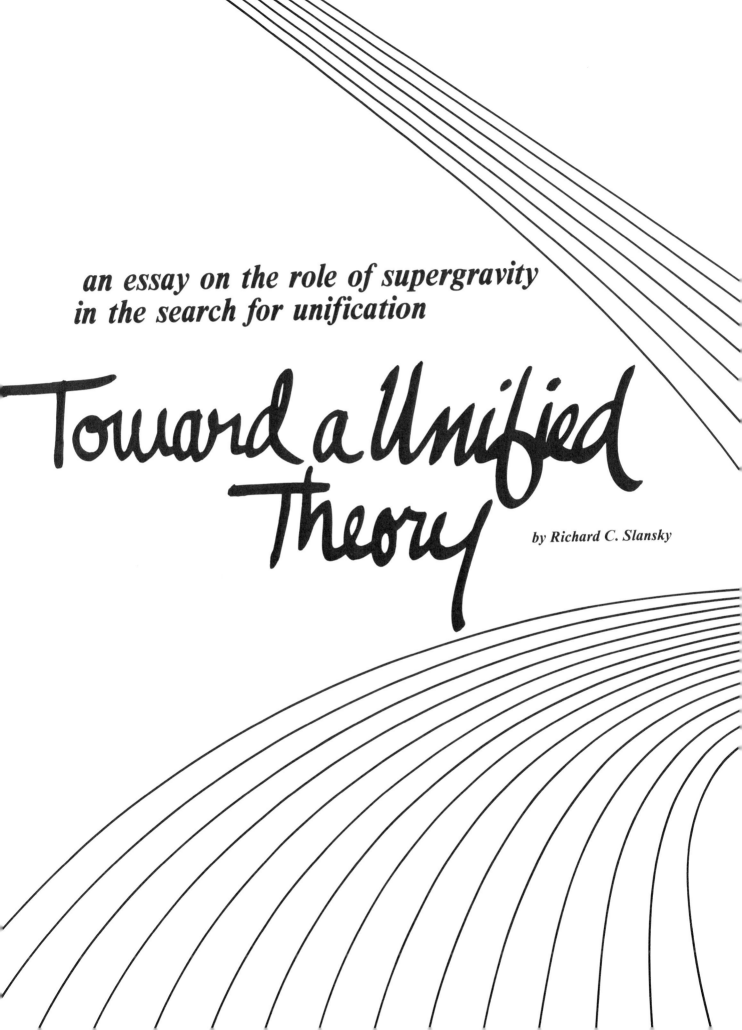

an essay on the role of supergravity
in the search for unification

# Toward a Unified Theory

by Richard C. Slansky

All throughout his history man has wanted to know the dimensions of his world and his place in it. Before the advent of scientific instruments the universe did not seem very large or complicated. Anything too small to detect with the naked eye was not known, and the few visible stars might almost be touched if only there were a higher hill nearby.

Today, with high-energy particle accelerators the frontier has been pushed down to distance intervals as small as $10^{-16}$ centimeter and with super telescopes to cosmological distances. These explorations have revealed a multifaceted universe; at first glance its diversity appears too complicated to be described in any unified manner. Nevertheless, it has been possible to incorporate the immense variety of experimental data into a small number of quantum field theories that describe four basic interactions—weak, strong, electromagnetic, and gravitational. Their mathematical formulations are similar in that each one can be derived from a local symmetry. This similarity has inspired hope for even greater progress: perhaps an extension of the present theoretical framework will provide a single unified description of all natural phenomena.

This dream of unification has recurred again and again, and there have been many successes: Maxwell's unification of electricity and magnetism; Einstein's unification of gravitational phenomena with the geometry of space-time; the quantum-mechanical unification of Newtonian mechanics with the wave-like behavior of matter; the quantum-mechanical generalization of electrodynamics; and finally the recent unification of electromagnetism with the weak force. Each of these advances is a crucial component of the present efforts to seek a more complete physical theory.

Before the successes of the past inspire too much optimism, it is important to note that a unified theory will require an unprecedented extrapolation. The present optimism is generated by the discovery of theories successful at describing phenomena that take place over distance intervals of order $10^{-16}$ centimeter or larger. These theories may be valid to much shorter distances, but that remains to be tested experimentally. A fully unified theory will have to include gravity and therefore will probably have to describe spatial structures as small as $10^{-33}$ centimeter, the fundamental length (determined by Newton's gravitational constant) in the theory of gravity. History suggests cause for further caution: the record shows many failures resulting from attempts to unify the wrong, too few, or too many physical phenomena. The end of the 19th century saw a huge but unsuccessful effort to unify the description of all Nature with thermodynamics. Since the second law of thermodynamics cannot be derived from Newtonian mechanics, some physicists felt it must have the most fundamental significance and sought to derive the rest of physics from it. Then came a period of belief in the combined use of Maxwell's electrodynamics and Newton's mechanics to explain all natural phenomena. This effort was also doomed to failure: not only did these theories lack consistency (Newton's equations are consistent with particles traveling faster than the speed of light, whereas the Lorentz invariant equations of Maxwell are not), but also new experimental results were emerging that implied the quantum structure of matter. Further into this century came the celebrated effort by Einstein to formulate a unified field theory of gravity and electromagnetism. His failure notwithstanding, the mathematical form of his classical theory has many remarkable similarities to the modern efforts to unify all known fundamental interactions. We must be wary that our reliance on quantum field theory and local symmetry may be similarly misdirected, although we suppose here that it is not.

Two questions will be the central themes of this essay. First, should we believe that the theories known today are the correct components of a truly unified theory? The component theories are now so broadly accepted that they have become known as the "standard model." They include the electroweak theory, which gives a unified description of quantum electrodynamics (QED) and the weak interactions, and quantum chromodynamics (QCD), which is an attractive candidate theory for the strong interactions. We will argue that, although Einstein's theory of gravity (also called general relativity) has a somewhat different status among physical theories, it should also be included in the standard model. If it is, then the standard model incorporates all observed physical phenomena—from the shortest distance intervals probed at the highest energy accelerators to the longest distances seen by modern telescopes. However, despite its experimental successes, the standard model remains unsatisfying; among its shortcomings is the presence of a large number of arbitrary constants that require explanations. It remains to be seen whether the next level of unification will provide just a few insights into the standard model or will unify all natural phenomena.

The second question examined in this essay is twofold: What are the possible strategies for generalizing and extending the standard model, and how nearly do models based on these strategies describe Nature? A central problem of theoretical physics is to identify the features of a theory that should be abstracted, extended, modified, or generalized. From among the bewildering array of theories, speculations, and ideas that have grown from the standard model, we will describe several that are currently attracting much attention.

We focus on two extensions of established concepts. The first is called supersymmetry; it enlarges the usual space-time symmetries of field theory, namely, Poincaré invariance, to include a symmetry among the bosons (particles of integer spin) and fermions (particles of half-odd integer spin). One of the intriguing features of supersymmetry is that it can be extended to include internal symmetries (see Note 2 in "Lecture Notes—From Simple Field Theories to the Standard Model). In the standard model internal local symmetries play a crucial role, both for classifying elementary particles and for de-

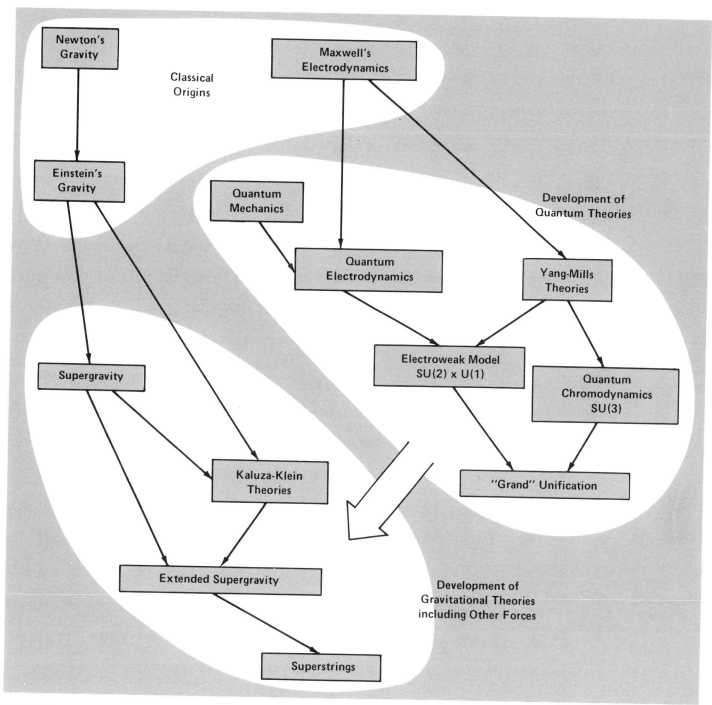

Fig. 1. Evolution of fundamental theories of Nature from the classical field theories of Newton and Maxwell to the grandest theoretical conjectures of today. The relationships among these theories are discussed in the text. Solid lines indicate a direct and well-established extension, or theoretical generalization. The wide arrow symbolizes the goal of present research, the unification of quantum field theories with gravity.

termining the form of the interactions among them. The electroweak theory is based on the internal local symmetry group $SU(2) \times U(1)$ (see Note 8) and quantum chromodynamics on the internal local symmetry group $SU(3)$. Gravity is based on space-time symmetries: general coordinate invariance and local Poincaré symmetry. It is tempting to try to unify all these symmetries with supersymmetry.

Other important implications of supersymmetry are that it enlarges the scope of the classification schemes of the basic particles to include fields of different spins in the same multiplet, and it helps to solve some technical problems concerning large mass ratios that plague certain efforts to derive the standard model. Most significantly, if supersymmetry is made to be a local symmetry, then it automatically implies a theory of gravity, called supergravity, that is a generalization of Einstein's theory. Supergravity theories require the unification of gravity with other kinds of interactions, which may be, in some future version, the electroweak and strong interactions. The near successes of this approach are very encouraging.

The other major idea described here is the extension of the space-time manifold to more than four dimensions, the extra dimensions having, so far, escaped observation. This revolutionary idea implies that particles are grouped into larger symmetry multiplets and the basic interactions have a geometrical origin. Although the idea of extending space-time beyond four dimensions is not new, it becomes natural in the context of supergravity theories because these complicated theories in four dimensions may be derived from relatively simple-looking theories in higher dimensions.

We will follow these developments one step further to a generalization of the field concept: instead of depending on space-time, the fields may depend on paths in space-time. When this generalization is combined with supersymmetry, the resulting theory is called a superstring theory. (The whimsicality of the name is more than matched by the theory's complexity.) Superstring the-

ories are encouraging because some of them reduce, in a certain limit, to the only supergravity theories that are likely to generalize the standard model. Moreover, whereas supergravity fails to give the standard model exactly, a superstring theory might succeed. It seems that superstring theories can be formulated only in ten dimensions.

Figure 1 provides a road map for this essay, which journeys from the origins of the standard model in classical theory to the extensions of the standard model in supergravity and superstrings. These extensions may provide extremely elegant ways to unify the standard model and are therefore attracting enormous theoretical interest. It must be cautioned, however, that at present no experimental evidence exists for supersymmetry or extra dimensions.

## Review of the Standard Model

We now review the standard model with particular emphasis on its potential for being unified by a larger theory. Over the last several decades relativistic quantum field theories with local symmetry have succeeded in describing all the known interactions down to the smallest distances that have been explored experimentally, and they may be correct to much shorter distances.

**Electrodynamics and Local Symmetry.** Electrodynamics was the first theory with local symmetry. Maxwell's great unification of electricity and magnetism can be viewed as the discovery that electrodynamics is described by the simplest possible local symmetry, local phase invariance. Maxwell's addition of the displacement current to the field equations, which was made in order to insure conservation of the electromagnetic current, turns out to be equivalent to imposing local phase invariance on the Lagrangian of electrodynamics, although this idea did not emerge until the late 1920s.

A crucial feature of locally symmetric quantum field theories is this: typically, for each independent internal local symmetry

there exists a gauge field and its corresponding particle, which is a vector boson (spin-1 particle) that mediates the interaction between particles. Quantum electrodynamics has just one independent local symmetry transformation, and the photon is the vector boson (or gauge particle) mediating the interaction between electrons or other charged particles. Furthermore, the local symmetry dictates the exact form of the interaction. The interaction Lagrangian must be of the form $eJ^\mu(x)A_\mu(x)$, where $J^\mu(x)$ is the current density of the charged particles and $A_\mu(x)$ is the field of the vector bosons. The coupling constant $e$ is defined as the strength with which the vector boson interacts with the current. The hypothesis that all interactions are mediated by vector bosons or, equivalently, that they originate from local symmetries has been extended to the weak and then to the strong interactions.

**Weak Interactions.** Before the present understanding of weak interactions in terms of local symmetry, Fermi's 1934 phenomenological theory of the weak interactions had been used to interpret many data on nuclear beta decay. After it was modified to include parity violation, it contained all the crucial elements necessary to describe the low-energy weak interactions. His theory assumed that beta decay (e.g., $n \rightarrow p + e^- + \bar{\nu}_e$) takes place at a single space-time point. The form of the interaction amplitude is a product of two currents $J^\mu J_\mu$, where each current is a product of fermion fields, and $J^\mu J_\mu$ describes four fermion fields acting at the point of the beta-decay interaction. This amplitude, although yielding accurate predictions at low energies, is expected to fail at center-of-mass energies above 300 GeV, where it predicts cross sections that are larger than allowed by the general principles of quantum field theory.

The problem of making a consistent (renormalizable) quantum field theory to describe the weak interactions was not solved until the 1960s, when the electromagnetic and weak interactions were combined into a locally symmetric theory. As outlined in Fig.

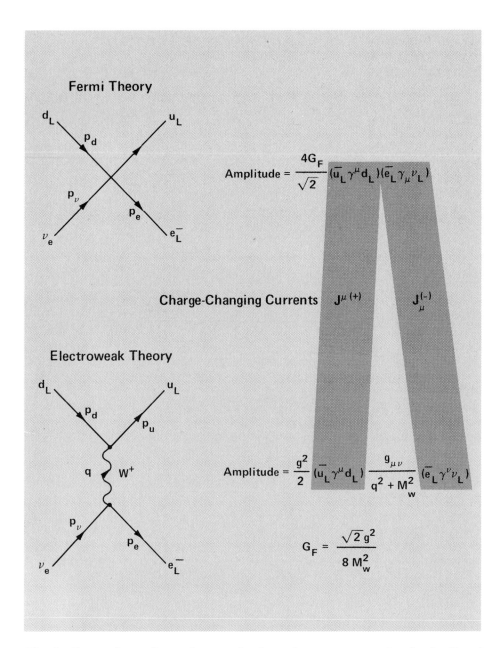

**Fermi Theory**

$$\text{Amplitude} = \frac{4G_F}{\sqrt{2}} (\overline{u}_L \gamma^\mu d_L)(\overline{e}_L \gamma_\mu \nu_L)$$

**Charge-Changing Currents** $\quad J^{\mu\,(+)} \qquad J_\mu^{(-)}$

**Electroweak Theory**

$$\text{Amplitude} = \frac{g^2}{2} (\overline{u}_L \gamma^\mu d_L) \frac{g_{\mu\nu}}{q^2 + M_w^2} (\overline{e}_L \gamma^\nu \nu_L)$$

$$G_F = \frac{\sqrt{2}\, g^2}{8 M_w^2}$$

*Fig. 2. Comparison of neutrino-quark charged-current scattering in the Fermi theory and the modern SU(2) × U(1) electroweak theory. (The bar indicates the Dirac conjugate.) The point interaction of the Fermi theory leads to an inconsistent quantum theory. The $W^+$ boson exchange in the electroweak theory spreads out the weak interactions, which then leads to a consistent (renormalizable) quantum field theory. $J_\mu^{(+)}$ and $J_\mu^{(-)}$ are the charge-raising and charge-lowering currents, respectively. The amplitudes given by the two theories are nearly equal as long as the square of the momentum transfer, $q^2 = (p_u - p_d)^2$, is much less than the square of the mass of the weak boson, $M_W^2$.)*

2, the vector bosons associated with the electroweak local symmetry serve to spread out the interaction of the Fermi theory in space-time in a way that makes the theory consistent. Technically, the major problem with the Fermi theory is that the Fermi coupling constant, $G_F$, is not dimensionless ($G_F = (293\,\text{GeV})^{-2}$), and therefore the Fermi theory is not a renormalizable quantum field theory. This means that removing the infinities from the theory strips it of all its predictive power.

In the gauge theory generalization of Fermi's theory, beta decay and other weak interactions are mediated by heavy weak vector bosons, so the basic interaction has the form $gW^\mu J_\mu$ and the current-current interaction looks pointlike only for energies much less than the rest energy of the weak bosons. (The coupling $g$ is dimensionless, whereas $G_F$ is a composite number that includes the masses of the weak vector bosons.) The theory has four independent local symmetries, including the phase symmetry that yields electrodynamics. The local symmetry group of the electroweak theory is SU(2) × U(1), where U(1) is the group of phase transformations, and SU(2) has the same structure as rotations in three dimensions. The one phase angle and the three independent angles of rotation in this theory imply the existence of four vector bosons, the photon plus three weak vector bosons, $W^+$, $Z^0$, and $W^-$. These four particles couple to the four SU(2) × U(1) currents and are responsible for the "electroweak" interactions.

The idea that all interactions must be derived from local symmetry may seem simple, but it was not at all obvious how to apply this idea to the weak (or the strong) interactions. Nor was it obvious that electrodynamics and the weak interactions should be part of the same local symmetry since, experimentally, the weak bosons and the photon do not share much in common: the photon has been known as a physical entity for nearly eighty years, but the weak vector bosons were not observed until late 1982 and early 1983 at the CERN proton-antiproton collider in the highest energy accelerator experiments ever performed; the mass of the photon is consistent with zero, whereas the weak vector bosons have huge masses (a little less than 100 GeV/$c^2$); electromagnetic interactions can take place over very large distances, whereas the weak interactions take place on a distance scale of about $10^{-16}$ centimeter; and finally, the photon has no electric charge, whereas the weak vector bosons carry the electric and weak charges of the electroweak interactions. Moreover, in the early days of gauge theories, it was generally believed, although incorrectly, that local symmetry of a Lagrangian implies masslessness for the vector bosons.

How can particles as different as the photon and the weak bosons possibly be unified by local symmetry? The answer is explained in detail in the Lecture Notes; we mention here merely that if the vacuum of a locally symmetric theory has a nonzero symmetry charge density due to the presence of a spinless field, then the vector boson associated with that symmetry acquires a mass. Solutions to the equations of motion in which the vacuum is not invariant under symmetry transformations are called spontaneously broken solutions, and the vector boson mass can be arbitrarily large without upsetting the symmetry of the Lagrangian.

In the electroweak theory spontaneous symmetry breaking separates the weak and electromagnetic interactions and is the most important mechanism for generating masses of the elementary particles. In the theories dicussed below, spontaneous symmetry breaking is often used to distinguish interactions that have been unified by extending symmetries (see Note 8).

The range of validity of the electroweak theory is an important issue, especially when considering extensions and generalizations to a theory of broader applicability. "Range of validity" refers to the energy (or distance) scale over which the predictions of a theory are valid. The old Fermi theory gives a good account of the weak interactions for energies less than 50 GeV, but at higher energies, where the effect of the weak bosons is to

## Table 1

**Review of fundamental interactions.**

| Interaction | | |
|---|---|---|
| Example | Name | Local Symmetry |
| Any Charged Particle — Photon — Any Charged Particle | Electromagnetic (QED) | U(1) |
| Quark — Gluon — Quark | Strong (QCD) | SU(3) |
| $\nu_e$, $e^-$ — $W^+$ | Electroweak | SU(2) × U(1) |
| Any Massive Particle — Graviton — Any Massive Particle | Gravity | Poincaré |
| $\bar{d}$, $u$ — X — $e^+$, $u$ (Proton Decay) | Conjectured Strong-Electroweak Unification | SU(5) |

**Local Symmetry:** The generator of the electromagnetic U(1) is a linear combination of the generators of the electroweak U(1) and the diagonal generator of the electroweak SU(2). The general coordinate invariance of gravity permits several formulations of gravity in which different local symmetries can be emphasized.

**Range of Force:** The electromagnetic and gravitational forces fall off as $1/r^2$. Of course, the electromagnetic part of the electroweak force is long range.

**Relative Strength at Low Energy:** The strength of the strong interactions is extremely energy-dependent. At low energy hadronic amplitudes are typically 100 times stronger than electromagnetic amplitudes.

**Number of Vector Bosons:** The graviton can be viewed as the gauge particle for translations, and as a consequence it has a spin of 2. After all the symmetries of gravity are taken into account, the graviton is massless and has only two degrees of freedom with helicities (spin components) ±2.

| Number of Vector Bosons | Range of Force | Relative Strength at Low Energy | Mass Scale |
|---|---|---|---|
| 1 (photon) | Infinite | 1/137 | |
| 8 (gluons) | $10^{-13}$ cm | 1 | $\mu = 200$ MeV/$c^2$ |
| 4 (3 weak bosons, 1 photon) | $10^{-15}$ cm (weak) | $10^{-5}$ | $G_F^{-1/2} = 290$ GeV/$c^2$ |
| (Graviton) | Infinite | $10^{-38}$ | $G_N^{-1/2} = 1.2 \times 10^{19}$ GeV/$c^2$ |
| 24 | $10^{-29}$ cm | $10^{-32}$ | $10^{15}$ GeV/$c^2$ |

**Mass Scale**: There is no universal definition of mass scale in particle physics. It is, however, possible to select a mass scale of physical significance for each of these theories. For example, in the electroweak and SU(5) theories the mass scale is associated with the spontaneous symmetry breaking. In both cases the vacuum value of a scalar field (which has dimensions of mass) has a nonzero value. In the weak interactions $G_F$ is related directly to this vacuum value (see Fig. 2) and, at the same time, to the masses of the weak bosons. Similarly, the scale of the SU(5) model is related to the proton-decay rate and to the vacuum value of a different scalar field. In the Fermi theory $G_F$ is the strength of the weak interaction in the same way that $G_N$ is the strength of the gravitational interaction. However, in gravity theory, with its massless graviton, the origin of the large value of $G_N$ is not well understood. (It might be related to a vacuum value but not in precisely the way that $G_F$ is.) The QCD mass scale is defined in a completely different way. Aside from the quark masses, the classical QCD Lagrangian has no mass scales and no scalar fields. However, in quantum field theory the coupling of a gluon to a quark current depends on the momentum carried by the gluon, and this coupling is found to be large for momentum transfers below 200 MeV/$c$. It is thus customary to select $\mu = 200$ MeV/$c^2$ (where $\mu$ is the parameter governing the scale of asymptotic freedom) as the mass scale for QCD.

spread out the weak interactions in space-time, the Fermi theory fails. The electroweak theory remains a consistent quantum field theory at energies far above a few hundred GeV and reduces to the Fermi theory (with the modification for parity violation) at lower energies. Moreover, it correctly predicts the masses of the weak vector bosons. In fact, until experiment proves otherwise, there are no logical impediments to extending the electroweak theory to an energy scale as large as desired. Recall the example of electrodynamics and its quantum-mechanical generalization. As a theory of light in the mid-19th century, it could be tested to about $10^{-5}$ centimeter. How could it have been known that QED would still be valid for distance scales ten orders of magnitude smaller? Even today it is not known where quantum electrodynamics breaks down.

**Strong Interactions.** Quantum chromodynamics is the candidate theory of the strong interactions. It, too, is a quantum field theory based on a local symmetry; the symmetry, called color SU(3), has eight independent kinds of transformations, and so the strong interactions among the quark fields are mediated by eight vector bosons, called gluons. Apparently, the local symmetry of the strong interaction theory is not spontaneously broken. Although conceptually simpler, the absence of symmetry breaking makes it harder to extract experimental predictions. The exact SU(3) color symmetry may imply that the quarks and gluons, which carry the SU(3) color charge, can never be observed in isolation. There seem to be no simple relationships between the quark and gluon fields of the theory and the observed structure of hadrons (strongly interacting particles). The quark model of hadrons has not been rigorously derived from QCD.

One of the main clues that quantum chromodynamics is correct comes from the results of "deep" inelastic scattering experiments in which leptons are used to probe the structure of protons and neutrons at very short distance intervals. The theory predicts that at very high momentum transfers or, equivalently, at very short distances ($<10^{-13}$ centimeter) the quark and gluon fields that make up the nucleons have a direct and fundamental interpretation: they are almost noninteracting, point-like particles. Deep inelastic electron, muon, and neutrino experiments have tested the short-distance structure of protons and neutrons and have confirmed qualitatively this short-distance prediction of quantum chromodynamics. At relatively long distance intervals of $10^{-13}$ centimeter or greater, the theory must account for the existence of the observed hadrons, which are complicated composites of the quark and gluon fields. Until progress is made in deriving the list of hadrons from quantum chromodynamics, we will not know whether it is the correct theory of the strong interactions. This is a rather peculiar situation: the validity of QCD at energies above a few GeV is established (and there is no experimental or theoretical reason to limit the range of validity of the theory at even higher energies), but the long-distance (low-energy) structure of the theory, including the hadron spectrum, has not yet been calculated. Perhaps the huge computational effort now being devoted to testing the theory will resolve this question soon.

**Gravity.** Gravity theory (and by this is meant Einstein's theory of general relativity) should be added to the standard model, although it has a different status from the electroweak and strong theories. The energy scale at which gravity becomes strong, according to Einstein's (or Newton's) theory, is far above the electroweak scale: it is given by the Planck mass, which is defined as $(\hbar c/G_N)^{1/2}$, where $G_N$ is Newton's gravitational constant, and is equal to $1.2 \times 10^{19}$ GeV/$c^2$. (In quantum theories distance is inversely proportional to energy; the Planck mass corresponds to a length (the Planck length) of $1.6 \times 10^{-33}$ centimeter.) Large mass scales are typically associated with small interaction rates, so gravity has a negligible effect on high-energy particle physics at present accelerator energies. The reason we feel the

effect of this very weak interaction so readily in everyday life is that the graviton, which mediates the interaction, is massless and has long-range interactions like the photon. Moreover, the gravitational force has always been found to be attractive; matter in bulk cannot be "gravitationally neutral" in the way that it is typically electrically neutral.

At present there are no experimental reasons that compel us to include gravity in the standard model; present particle phenomenology is explained without it. Moreover, its theoretical standing is shaky, since all attempts to formulate Einstein's gravity as a consistent quantum field theory have failed. The problem is similar to that of the Fermi theory: Newton's constant has dimensions of (energy)$^{-2}$ so the theory is not renormalizable. However, like the Fermi theory, it is valid up to an energy that is a substantial fraction of its energy scale of $10^{19}$ GeV. This is the only known serious inconsistency in the standard model when gravity is included. Thus, including gravity in the standard model seems to pose many problems. Yet, there is a good reason to attempt this unification: there exist theoretical models (as we discuss later) that suggest that the electroweak and strong theories may cure the ills of gravitational theory, and unification with gravity may require a theory that predicts the phenomenological inputs of the electroweak and strong theories.

The mathematical structure of gravity theory provides another reason for its inclusion in the standard model. Like the other interactions, gravity is based on a local symmetry, the Poincaré symmetry, which includes Lorentz transformations and space-time translations. In this case, however, not all the generators of the symmetry group give rise to particles that mediate the gravitational interaction. In particular, Einstein's theory has no kinetic energy terms in the Lagrangian for the gauge fields corresponding to the six independent symmetries of the Lorentz group. The space-time translations have associated with them the gauge field called the graviton that mediates the gravitational interaction. The graviton field has a spin of 2 and is

denoted by $e_\mu^a(x)$, where the vector index $\mu$ on the usual boson field is combined with the space-time translation index $a$ to form a spin of 2. The metric tensor is, essentially, the square of $e_\mu^a(x)$. The massless graviton has two helicities (spin projections along the direction of motion) of values $\pm 2$. In some ways these are merely technical differences, and gravity is like the other interactions. Nevertheless, these differences are crucial in the search for theories that unify gravity with the other interactions.

**Summary.** Let us summarize why the standard model including gravity may be the correct set of component theories of a truly unified theory.

○ The standard model (with its phenomenologically motivated symmetries, choice of fields, and Lagrangian) correctly accounts for all elementary-particle data.

○ The standard model contains no known mathematical inconsistencies up to an energy scale near $10^{19}$ GeV, and then only gravity gives difficulty.

○ All components of the standard model have similar mathematical structures. Essentially, they are local gauge theories, which can be derived from a principle of local symmetry.

○ There are no logical or phenomenological requirements that force the addition of further components to describe phenomena at scales greater than $10^{-16}$ centimeter. Thus, we are free to seek theories with a range of validity that may transcend the present experimental frontier.

We still have to cope with the huge extrapolation, by seventeen orders of magnitude, in energy scale necessary to include gravity in the theory. At best it appears reckless to begin the search for such a unification, in spite of the good luck historically with quantum electrodynamics. However, even if we ignore gravity, the energy scales encountered in attempts to unify just the electroweak and strong interactions are surprisingly close to the Planck mass. These more

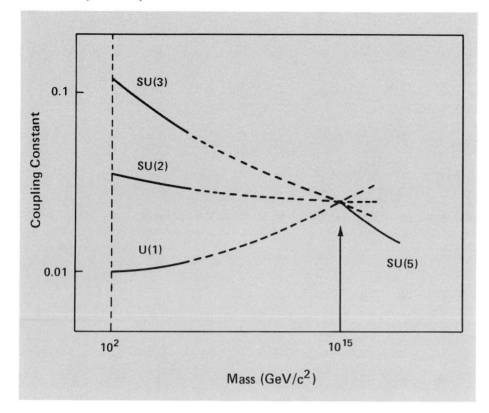

**Fig. 3. Unification in the SU(5) model. The values of the SU(2), U(1), and SU(3) couplings in the SU(5) model are shown as functions of mass scale. These values are calculated using the renormalization group equations of quantum field theory. At the unification energy scale the proton-decay bosons begin to contribute to the renormalization group equations; at higher energies, the ratios track together along the solid curve. If the high-mass bosons were not included in the calculation, the couplings would follow the dashed curves.**

modest efforts to unify the fundamental interactions may be an important step toward including gravity. Moreover, these efforts require the belief that local gauge theories are correct to distance intervals around $10^{-29}$ centimeter, and so they have made theorists more "comfortable" when considering the extrapolation to gravity, which is only four orders of magnitude further. Whether this outlook has been misleading remains to be seen. The components of the standard model are summarized in Table 1.

## Electroweak-Strong Unification without Gravity

The $SU(2) \times U(1) \times SU(3)$ local theory is a detailed phenomenological framework in which to analyze and correlate data on electroweak and strong interactions, but the choice of symmetry group, the charge assignments of the scalars and fermions, and the values of many masses and couplings must be deduced from experimental data. The problem is to find the simplest extension of this part of the standard model that also unifies (at least partially) the interactions,

assignments, and parameters that must be put into it "by hand." Total success at unification is not required at this stage because the range of validity will be restricted by gravitational effects.

One extension is to a local symmetry group that includes $SU(2) \times U(1) \times SU(3)$ and interrelates the transformations of the standard model by further internal symmetry transformations. The simplest example is the group $SU(5)$, although most of the comments below also apply to other proposals for electroweak-strong unification. The $SU(5)$ local symmetry implies new constraints on the fields and parameters in the theory. However, the theory also includes new interactions that mix the electroweak and strong quantum numbers; in $SU(5)$ there are vector bosons that transform quarks to leptons and quarks to antiquarks. These vector bosons provide a mechanism for proton decay.

If the $SU(5)$ local symmetry were exact, all the couplings of the vector bosons to the symmetry currents would be equal (or related by known factors), and consequently the proton decay rate would be near the weak

decay rates. Spontaneous symmetry breaking of $SU(5)$ is introduced into the theory to separate the electroweak and strong interactions from the other $SU(5)$ interactions as well as to provide a huge mass for the vector bosons mediating proton decay and thereby reduce the predicted decay rate. To satisfy the experimental constraint that the proton lifetime be at least $10^{31}$ years, the masses of the heavy vector bosons isn the $SU(5)$ model must be at least $10^{14}$ GeV/$c^2$. Thus, experimental facts already determine that the electroweak-strong unification must introduce masses into the theory that are within a factor of $10^5$ of the Planck mass.

It is possible to calculate the proton lifetime in the $SU(5)$ model and similar unified models from the values of the couplings and masses of the particles in the theory. The couplings of the standard model (the two electroweak couplings and the strong coupling) have been measured in low-energy processes. Although the ratios of the couplings are predicted by $SU(5)$, the symmetry values are accurate only at energies where $SU(5)$ looks exact, which is at energies above the masses of the vector bosons mediating proton decay. In general, the strengths of the couplings depend on the mass scale at which they are measured. Consequently, the $SU(5)$ ratios cannot be directly compared with the values measured at low energy. However, the renormalization group equations of field theory prescribe how they change with the mass scale. Specifically, the change of the coupling at a given mass scale depends only on all the elementary particles with masses less than that mass scale. Thus, as the mass scale is lowered below the mass of the proton-decay bosons, the latter must be omitted from the equations, so the ratios of the couplings change from the $SU(5)$ values. If we assume that the only elementary fields contributing to the equations are the low-mass fields known experimentally and if the proton-decay bosons have a mass of $10^{14}$ GeV/$c^2$ (see Fig. 3), then the low-energy experimental ratios of the standard model couplings are predicted correctly by the renormalization group equations but the proton lifetime prediction is a little less than the experimental lower bound. However, adding a few more "low-mass" (say, less than $10^{12}$ GeV/$c^2$) particles to the equations lengthens the lifetime predictions, which can thereby be pushed well beyond the limit attainable in present-day experiments.

Thus, using the proton-lifetime bound directly and the standard model couplings at low mass scale, we have seen that electroweak-strong unification implies mass scales close to the scale where gravity must be included. Even if it turns out that the electroweak-strong unification is not exactly correct, it has encouraged the extrapolation of present theoretical ideas well beyond the energies available in present accelerators.

Electroweak-strong unified models such as SU(5) achieve only a partial unification. The vector bosons are fully unified in the sense that they and their interactions are determined by the choice of SU(5) as the local symmetry. However, this is only a partial unification. The choice of fermion and scalar multiplets and the choice of symmetry-breaking patterns are left to the discretion of the physicist, who makes his selections based on low-energy phenomenology. Thus, the "unification" in SU(5) (and related local symmetries) is far from complete, except for the vector bosons. (This suggests that theories in which all particles are more closely related to the vector bosons might remove some of the arbitrariness; this will prove to be the case for supergravity.)

In summary, strong experimental evidence for electroweak-strong unification, such as proton decay, would support the study of quantum field theories at energies just below the Planck mass. From the vantage of these theories, the electroweak and strong interactions should be the low-energy limit of the unifying theory, where "low energy" corresponds to the highest energies available at accelerators today! Only future experiments will help decide whether the standard model is a complete low-energy theory, or whether we are repeating the age-old error of omitting some low-energy interactions that are not yet discovered. Never-

theless, the quest for total unification of the laws of Nature is exciting enough that these words of caution are not sufficient to delay the search for theories incorporating gravity.

## Toward Unification with Gravity

Let us suppose that the standard model including gravity is the correct set of theories to be unified. On the basis of the previous discussion, we also accept the hypothesis that quantum field theory with local symmetry is the correct theoretical framework for extrapolating physical theory to distances perhaps as small as the Planck length. Quantum field theory assumes a mathematical model of space-time called a manifold. On large scales a manifold can have many different topologies, but at short enough distance scale, a manifold always looks like a flat (Minkowski) space, with space and time infinitely divisible. This might not be the structure of space-time at very small distances, and the manifold model of space-time might fail. Nevertheless, all progress at unifying gravity and the other interactions described here is based on theories in which space-time is assumed to be a manifold.

Einstein's theory of gravity has fascinated physicists by its beauty, elegance, and correct predictions. Before examining efforts to extend the theory to include other interactions, let us review its structure. Gravity is a "geometrical" theory in the following sense. The shape or geometry of the manifold is determined by two types of tensors, called curvature and torsion, which can be constructed from the gravitational field. The Lagrangian of the gravitational field depends on the curvature tensor. In particular, Einstein's brilliant discovery was that the curvature scalar, which is obtained from the curvature tensor, is essentially a unique choice for the kinetic energy of the gravitational field. The gravitational field calculated from the equations of motion then determines the geometry of the space-time manifold. Particles travel along "straight lines" (or geodesics) in this space-time. For

example, the orbits of the planets are geodesics of the space-time whose geometry is determined by the sun's gravitational field.

In Einstein's gravity all the remaining fields are called matter fields. The Lagrangian is a sum of two terms:

$$\mathscr{L} = \mathscr{L}_{\text{gravity}} + \mathscr{L}_{\text{matter}}, \qquad (1)$$

where the curvature scalar $\mathscr{L}_{\text{gravity}}$ is the kinetic energy of the graviton, and $\mathscr{L}_{\text{matter}}$ contains all the other fields and their interactions with the gravitational field. The interaction term in the Lagrangian, which couples the gravitational field (the metric tensor) to the energy-momentum tensor, has a form almost identical to the term that couples the electromagnetic field to the electromagnetic current. Newton's constant, which has dimensions of (mass)$^{-2}$, appears in the ratio of the two terms in Eq. 1 as a coupling analogous to the Fermi coupling in the weak theory. This complicates the quantum generalization, just as it did in Fermi's weak interaction theory, and it is not possible to formulate a consistent quantum theory with Eq. 1. Actually, the situation is even worse, because $\mathscr{L}_{\text{gravity}}$ alone does not lead to a consistent quantum theory either, although the inconsistencies are not as bad as when $\mathscr{L}_{\text{matter}}$ is included.

This suggests that our efforts to unify gravity with the other interactions might solve the problems of gravity: perhaps we can join the matter fields together with the gravitational field in something like a curvature scalar and thereby eliminate $\mathscr{L}_{\text{matter}}$. In addition, generalizing the graviton field in this way might lead to a consistent (renormalizable) quantum theory of gravity. There are reasons to hope that the problem of finding a renormalizable theory of gravity is solved by superstrings, although the proof is far from complete. For now, we discuss the unification of the graviton with other fields without concern for renormalizability.

We will discuss several ways to find manifolds for which the curvature scalar depends on many fields, not just the gravitational

apply to symmetries such as supersymmetry, with its anticommuting generators.

These two loopholes in the assumptions of the theorem have suggested two directions of research in the attempt to unify gravity with the other interactions. First, we might suppose that the dimensionality of space-time is greater than four, and that spontaneous symmetry breaking of the Poincaré invariance of this larger space separates 4-dimensional space-time from the other dimensions. The symmetries of the extra dimensions can then correspond to internal symmetries, and the symmetries of the states in four dimensions need not imply an unsatisfactory infinity of states. A second approach is to extend the Poincaré symmetry to supersymmetry, which then requires additional fermionic fields to accompany the graviton. A combination of these approaches leads to the most interesting theories.

## Higher Dimensional Space-Time

If the dimensionality of space-time is greater than four, then the geometry of space-time must satisfy some strong observational constraints. In a 5-dimensional world the fourth spatial direction must be invisible to present experiments. This is possible if at each 4-dimensional space-time point the additional direction is a little circle, so that a tiny person traveling in the new direction would soon return to the starting point. Theories with this kind of vacuum geometry are generically called Kaluza-Klein theories.[1]

It is easy to visualize this geometry with a two-dimensional analogue, namely, a long pipe. The direction around the pipe is analogous to the extra dimension, and the location along the pipe is analogous to a location in 4-dimensional space-time. If the means for examining the structure of the pipe are too coarse to see distance intervals as small as its diameter, then the pipe appears 1-dimensional (Fig. 4). If the probe of the structure is sensitive to shorter distances, the pipe is a 2-dimensional structure with one dimension wound up into a circle.

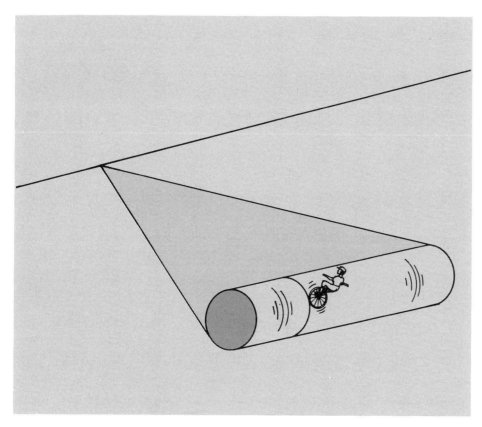

*Fig. 4. Two-dimensional analogue of the vacuum geometry of a Kaluza-Klein theory. From great distances the geometry looks one-dimensional, but up close the second dimension, which is wound up in a circle, becomes visible. If space-time has more than four dimensions, then the extra dimensions could have escaped detection if each is wound into a circle whose radius is less than $10^{-16}$ centimeter.*

field. This generally requires extending the 4-dimensional space-time manifold. The fields and manifold must satisfy many constraints before this can be done. All the efforts to unify gravity with the other interactions have been formulated in this way, but progress was not made until the role of spontaneous symmetry breaking was appreciated. As we now describe, it is crucial for the solutions of the theory to have less symmetry than the Lagrangian has.

In the standard model the generators of the space-time Poincaré symmetry commute with (are independent of) the generators of the internal symmetries of the electroweak and strong interactions. We might look for a

local symmetry that interrelates the space-time and internal symmetries, just as SU(5) interrelates the electroweak and strong internal symmetries. Unfortunately, if this enlarged symmetry changes simultaneously the internal and space-time quantum numbers of several states of the same mass, then a theorem of quantum field theory requires the existence of an infinite number of particles of that mass. However, this seemingly catastrophic result does not prevent the unification of space-time and internal symmetries for two reasons: first, all symmetries of the Lagrangian need not be symmetries of the states because of spontaneous symmetry breaking; and second, the theorem does not

The physically interesting solutions of Einstein's 4-dimensional gravity are those in which, if all the matter is removed, space-time is flat. The 4-dimensional space-time we see around us is flat to a good approximation; it takes an incredibly massive hunk of high-density (much greater than any density observed on the earth) matter to curve space. However, it might also be possible to construct a higher dimensional theory in which our 4-dimensional space-time remains flat in the absense of identifiable matter, and the extra dimensions are wound up into a "little ball." We must study the generalizations of Einstein's equations to see whether this can happen, and if it does, to find the geometry of the extra dimensions.

**The Cosmological Constant Problem.** Before we examine the generalizations of gravity in more detail, we must raise a problem that pervades all gravitational theories. Einstein's equations state that the Einstein tensor (which is derived from the curvature scalar in finding the equations of motion from the Lagrangian) is proportional to the energy-momentum tensor. If, in the absence of all matter and radiation, the energy-momentum tensor is zero, then Einstein's equations are solved by flat space-time and zero gravitational field. In 4-dimensional classical general relativity, the curvature of space-time and the gravitational field result from a nonzero energy-momentum tensor due to the presence of physical particles.

However, there are many small effects, such as other interactions and quantum effects, not included in classical general relativity, that can radically alter this simple picture. For example, recall that the electroweak theory is spontaneously broken, which means that the scalar field has a nonzero vacuum value and may contribute to the vacuum value of the energy-momentum tensor. If it does, the solution to the Einstein equations in vacuum is no longer flat space but a curved space in which the curvature increases with increasing vacuum energy. Thus, the constant value of the potential energy, which had no effect on the

weak interactions, has a profound effect on gravity.

At first glance, we can solve this difficulty in a trivial manner: simply add a constant to the Lagrangian that cancels the vacuum energy, and the universe is saved. However, we may then wish to compute the quantum-mechanical corrections to the electroweak theory or add some additional fields to the theory; both may readjust the vacuum energy. For example, electroweak-strong unification and its quantum corrections will contribute to the vacuum energy. Almost all the details of the theory must be included in calculating the vacuum energy. So, we could repeatedly readjust the vacuum energy as we learn more about the theory, but it seems artificial to keep doing so unless we have a good theoretical reason. Moreover, the scale of the vacuum energy is set by the mass scale of the interactions. This is a dilemma. For example, the quantum corrections to the electroweak interactions contribute enough vacuum energy to wind up our 4-dimensional space-time into a tiny ball about $10^{-13}$ centimeter across, whereas the scale of the universe is more like $10^{28}$ centimeters. Thus, the observed value of the cosmological constant is smaller by a factor of $10^{82}$ than the value suggested by the standard model. Other contributions can make the theoretical value even larger. This problem has the innocuous-sounding name of "the cosmological constant problem." At present there are no principles from which we can impose a zero or nearly zero vacuum energy on the 4-dimensional part of the theory, although this problem has inspired much research effort. Without such a principle, we can safely say that the vacuum-energy prediction of the standard model is wrong. At best, the theory is not adequate to confront this problem.

If we switch now to the context of gravity theories in higher dimensions, the difficult question is not why the extra dimensions are wound up into a little ball, but why our 4-dimensional space-time is so nearly flat, since it would appear that a large cosmological constant is more natural than a

small one. Also, it is remarkable that the vacuum energy winding the extra dimensions into a little ball is conceptually similar to the vacuum charge of a local symmetry providing a mass for the vector bosons. However, in the case of the vacuum geometry, we have no experimental data that bear on these speculations other than the remarkable flatness of our 4-dimensional space-time. The remaining discussion of unification with gravity must be conducted in ignorance of the solution to the cosmological constant problem.

## Internal Symmetries from Extra Dimensions

The basic scheme for deriving local symmetries from higher dimensional gravity was pioneered by Kaluza and Klein[1] in the 1920s, before the weak and strong interactions were recognized as fundamental. Their attempts to unify gravity and electrodynamics in four dimensions start with pure gravity in five dimensions. They assumed that the vacuum geometry is flat 4-dimensional space-time with the fifth dimension a little loop of definite radius at each space-time point, just as in the pipe analogy of Fig. 4. The Lagrangian consists of the curvature scalar, constructed from the gravitational field in five dimensions with its five independent components. The relationship of a higher dimensional field to its 4-dimensional fields is summarized in Fig. 5 and the sidebar, "Fields and Spin in Higher Dimensions." The infinite spectrum in four dimensions includes the massless graviton (two helicity components of values $\pm 2$), a massless vector boson (two helicity components of $\pm 1$), a massless scalar field (one helicity component of 0), and an infinite series of massive spin-2 pyrgons of increasing masses. (The term "pyrgon" derives from πύργος, the Greek word for tower.) The Fourier expansion for each component of the gravitational field is identical to Eq. 1 of the sidebar. Since the extra dimension is a circle, its symmetry is a phase symmetry just as in electrodynamics.

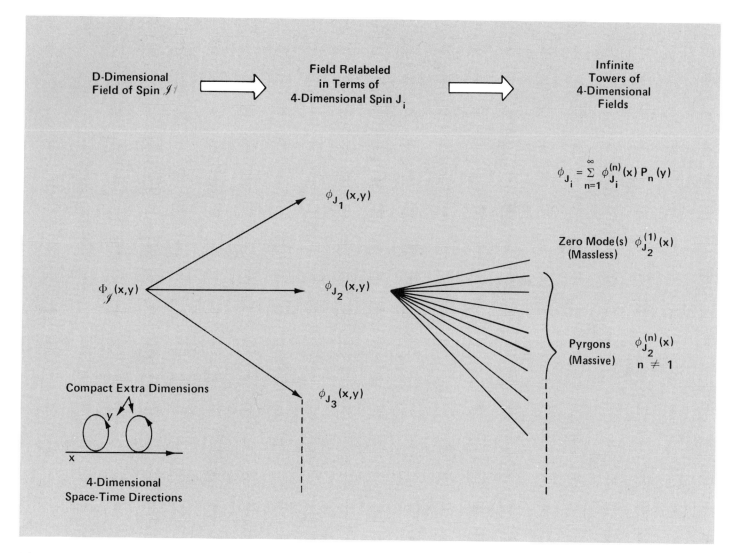

*Fig. 5. A field in D dimensions unifies fields of different spins and masses in four dimensions. In step 1 the spin components of a single higher dimensional spin are resolved into several spins in four dimensions. (The total number of components remains constant.) Mathematically this is achieved by finding the spins $J_1$, $J_2$, ... in four dimensions that are contained in "spin-$\mathcal{J}$" of D dimensions. Step 2 is the harmonic expansion of the 4-dimensional spin components on the extra dimensions, which then resolves a single massless D-dimensional field into an infinite number of 4-dimensional fields of varying masses. When the 4-dimensional mass is zero, the corresponding 4-dimensional field is called a zero mode. The 4-dimensional fields with 4-dimensional mass form an infinite sequence of pyrgons.*

The symmetry of this vacuum state is not the 5-dimensional Poincaré symmetry but the direct product of the 4-dimensional Poincaré group and a phase symmetry.

This skeletal theory should not be taken seriously, except as a basis for generalizing to more realistic theories. The zero modes (massless particles in four dimensions) are electrically neutral. Only the pyrgons carry electric charge. The interaction associated with the vector boson in four dimensions cannot be electrodynamics because there are no low-mass charged particles. (Adding fermions to the 5-dimensional theory does not help, because the resulting 4-dimensional fermions are all pyrgons, which cannot be low mass either.) Nevertheless, the hypothesis that all interactions are conse-

# Fields and Spins in

**Fields in Higher Dimensions.** We describe here how to construct a field in higher dimensions and how such a field is related to fields in the 4-dimensional world in which we live. Higher dimensional fields unify an infinite number of 4-dimensional fields. A typical and simple example of this can be seen from a scalar field (a spin-0 field) in five dimensions. A scalar field has only one component, so it can be written as $\varphi(x,y)$, where $x$ is the 4-dimensional space-time coordinate and $y$ is the coordinate for the fifth dimension. We will assume that the fifth dimension is a little circle with radius $R$, where $R$ is independent of $x$. (After this example, we examine the generalizations to more than five dimensions and to fields carrying nonzero spin in the higher dimensions.)

Functions on a circle can be expanded in a Fourier series; thus, the 5-dimensional scalar field can be written in the form

$$\varphi(x,y) = \sum_{n=-\infty}^{\infty} \varphi_n(x)\exp(iny/R) , \tag{1}$$

where $n$ is an integer, and $\varphi_n(x)$ are 4-dimensional fields. The Fourier series satisfies the requirement that the field is single-valued in the extra dimension, since Eq. 1 has the same value at the identical points $y$ and $y + 2\pi R$. Usually the wave equation of $\varphi(x,y)$ is a straightforward generalization of the 4-dimensional scalar wave equation (that is, the Klein-Gordon equation) in the limit that interactions can be ignored. The 5-dimensional Klein-Gordon equation for a massless 5-dimensional particle is

$$\left(\frac{\partial^2}{\partial t^2} - \nabla^2 - \frac{\partial^2}{\partial y^2}\right)\varphi(x,y) = 0 . \tag{2}$$

The presence of additional terms depends on the details of the Lagrangian, and we ignore them for the present description. It is a simple matter to substitute the Fourier expansion of Eq. 1 into Eq. 2 and use the orthogonality of the expansion functions $\exp(iny/R)$ to rewrite Eq. 2 as an infinite number of equations in four dimensions, one for each $\varphi_n(x)$:

$$\left[\frac{\partial^2}{\partial t^2} - \nabla^2 + \left(\frac{n}{R}\right)^2\right]\varphi_n(x) = 0 . \tag{3}$$

Note the following very important point: for $n = 0$, Eq. 3 is the equation for a massless 4-dimensional scalar field, whereas for $n \neq 0$, Eq. 3 is the wave equation for a particle with mass $|n|/R$. The massless particle, or "zero mode," should correspond to a field observable in our world. The fields with nonzero mass are called "pyrgons," since they are on a "tower" of particles, one for each $n$. If $R$ is near the Planck length ($10^{-33}$ centimeter), then the pyrgons have masses on the order of the Planck mass. However, it is also possible that $R$ can be much larger, say as large as $10^{-16}$ centimeter, without conflicting with experience.

The 4-dimensional form of the Lagrangian depends on an infinite number of fields and is very complicated to analyze. For many purposes it is helpful to truncate the theory, keeping a specially chosen set of fields. For example, 5-dimensional Einstein gravity is simplified by omitting all the pyrgons. This can be achieved by requiring that the fields do not depend on $y$, a procedure called "dimensional reduction." The dimensionally reduced theory should

quences of the symmetries of space-time is so attractive that efforts to generalize the Kaluza-Klein idea have been vigorously pursued. These theories require a more complete discussion of the possible candidate manifolds of the extra dimensions.

The geometry of the extra dimensions in the absence of matter is typically a space with a high degree of symmetry. Symmetry requires the existence of transformations in which the starting point looks like the point reached after the transformation. (For example, the environments surrounding each point on a sphere are identical.) Two of the most important examples are "group manifolds" and "coset spaces," which we briefly describe.

The transformations of a continuous group

are identified by $N$ parameters, where $N$ is the number of independent transformations in the group. For example, $N = 3$ for SU(2) and 8 for SU(3). These parameters are the coordinates of an $N$-dimensional manifold. If the vacuum values of fields are constant on the group manifold, then the vacuum solution is said to be symmetric.

Coset spaces have the symmetry of a group too, but the coordinates are labeled by a subset of the parameters of a group. For example, consider the space SO(3)/SO(2). In this example, SO(3) has three parameters, and SO(2) is the phase symmetry with one parameter, so the coset space SO(3)/SO(2) has three minus one, or two, dimensions. This space is called the 2-sphere, and it has the geometry of the surface of an ordinary

sphere. Spheres can be generalized to any number of dimensions: the $N$-dimensional sphere is the coset space [SO($N$ + 1)]/SO($N$). Many other cosets, or "ratios" of groups, make spaces with large symmetries. It is possible to find spaces with the symmetries of the electroweak and strong interactions. One such space is the group manifold SU(2) $\times$ U(1) $\times$ SU(3), which has twelve dimensions. More interesting is the lowest dimensional space with those symmetries, namely, the coset space [SU(3) $\times$ SU(2) $\times$ U(1)]/[SU(2) $\times$ U(1) $\times$ U(1)], which has dimension $8 + 3 + 1 - 3 - 1 - 1 = 7$. (The SU(2) and the U(1)'s in the denominator differ from those in the numerator, so they cannot be "canceled.") Thus, one might hope that $(4 + 7 = 11)$-dimensional gravity would

# Higher Dimensions

describe the low-energy limit of the theory.

The gravitational field can be generalized to higher ($>5$) dimensional manifolds, where the extra dimensions at each 4-dimensional space-time point form a little ball of finite volume. The mathematics requires a generalization of Fourier series to "harmonic" expansions on these spaces. Each field (or field component if it has spin) unifies an infinite set of pyrgons, and the series may also contain some zero modes. The terms in the series correspond to fields of increasing 4-dimensional mass, just as in the 5-dimensional example. The kinetic energy in the extra dimensions of each term in the series then corresponds to a mass in our space-time. The higher dimensional field quite generally describes mathematically an infinite number of 4-dimensional fields.

**Spin in Higher Dimensions**. The definition of spin in $D$ dimensions depends on the $D$-dimensional Lorentz symmetry; 4-dimensional Lorentz symmetry is naturally embedded in the $D$-dimensional symmetry. Consequently a $D$-dimensional field of a specific spin unifies 4-dimensional fields with different spins.

Conceptually the description of $D$-dimensional spin is similar to that of spin in four dimensions. A massless particle of spin $J$ in four dimensions has helicities $+J$ and $-J$ corresponding to the projections of spin along the direction of motion. These two helicities are singlet multiplets of the 1-dimensional rotations that leave unchanged the direction of a particle traveling at the speed of light. The group of 1-dimensional rotations is the phase symmetry SO(2), and this method for identifying the physical degrees of freedom is called the "light-cone classification." However, the situation is a little more complicated in five dimensions, where there are three directions orthogonal to the direction of the particle. Then the helicity symmetry becomes SO(3) (instead of SO(2)), and the spin multiplets in five dimensions group together sets of 4-dimensional helicity. For example, the graviton in five dimensions has five components. The SO(2) of four dimensions is contained in this SO(3) symmetry, and the 4-dimensional helicities of the 5-dimensional graviton are 2, 1, 0, $-1$, and $-2$.

Quite generally, the light-cone symmetry that leaves the direction of motion of a massless particle unchanged in $D$ dimensions is SO($D-2$), and the $D$-dimensional helicity corresponds to the multiplets (or representations) of SO($D-2$). For example, the graviton has $D(D-3)/2$ independent degrees of freedom in $D$ dimensions; thus the graviton in eleven dimensions belongs to a 44-component representation of SO(9). The SO(2) of the 4-dimensional helicity is inside the SO(9), so the forty-four components of the graviton in eleven dimensions carry labels of 4-dimensional helicity as follows: one component of helicity 2, seven of helicity 1, twenty-eight of helicity 0, seven of helicity $-1$ and one of helicity $-2$. (The components of the graviton in eleven dimensions then correspond to the graviton, seven massless vector bosons, and twenty-eight scalars in four dimensions.)

The analysis for massive particles in $D$ dimensions proceeds in exactly the same way, except the helicity symmetry is the one that leaves a resting particle at rest. Thus, the massive helicity symmetry is SO($D-1$). (For example, SO(3) describes the spin of a massive particle in ordinary 4-dimensional space-time.) These results are summarized in Fig. 5 of the main text.

---

unify all known interactions.

It turns out that the 4-dimensional fields implied by the 11-dimensional gravitational field resemble the solution to the 5-dimensional Kaluza-Klein case, except that the gravitational field now corresponds to many more 4-dimensional fields. There are methods of dimensional reduction for group manifolds and coset spaces, and the zero modes include a vector boson for each symmetry of the extra dimensions. Thus, in the (4 + 7)-dimensional example mentioned above, there is a complete set of vector bosons for the standard model. At first sight this model appears to provide an attractive unification of all the interactions of the standard model; it explains the origins of the local symmetries of the standard model as space-time symmetries of gravity in eleven dimensions.

Unfortunately, this 11-dimensional Kaluza-Klein theory has some shortcomings. Even with the complete freedom consistent with quantum field theory to add fermions, it cannot account for the parity violation seen in the weak neutral-current interactions of the electron. Witten[1] has presented very general arguments that no 11-dimensional Kaluza-Klein theory will ever give the correct electroweak theory.

## Supersymmetry and Gravity in Four Dimensions

We return from our excursion into higher dimensions and discuss extending gravity not by enlarging the space but rather by enlarging the symmetry. The local Poincaré symmetry of Einstein's gravity implies the massless spin-2 graviton; our present goal is to extend the Poincaré symmetry (without increasing the number of dimensions) so that additional fields are grouped together with the graviton. However, this cannot be achieved by an ordinary (Lie group) symmetry: the graviton is the only known elementary spin-2 field, and the local symmetries of the standard model are internal symmetries that group together particles of the same spin. Moreover, gravity has an exceptionally weak interaction, so if the graviton carries quantum numbers of symmetries similar to those of the standard model, it will interact too strongly. We can accommodate these facts if the graviton is a singlet under the internal symmetry, but then its multiplet in this new symmetry must include particles of other spins. Supersymmetry[2] is capable of fulfilling this requirement.

**Four-Dimensional Supersymmetry.** Supersymmetry is an extension of the Poincaré symmetry, which includes the six Lorentz generators $M_{\mu\nu}$ and four translations $P_\mu$. The Poincaré generators are boson operators, so they can change the spin components of a massive field but not the total spin. The simplest version of supersymmetry adds fermionic generators $Q_\alpha$ to the Poincaré generators; $Q_\alpha$ transforms like a spin-$\frac{1}{2}$ field under Lorentz transformations. (The index $\alpha$ is a spinor index.) To satisfy the Pauli exclusion principle, fermionic operators in quantum field theory always satisfy anticommutation relations, and the supersymmetry generators are no exception. In the algebra the supersymmetry generators $Q_\alpha$ anticommute to yield a translation

$$\left\{ Q_\alpha, \bar{Q}_\beta \right\} = \gamma^\mu_{\alpha\beta} P_\mu, \qquad (2)$$

where $P_\mu$ is the energy-momentum 4-vector and the $\gamma^\mu_{\alpha\beta}$ are matrix elements of the Dirac $\gamma$ matrices.

The significance of the fermionic generators is that they change the spin of a state or field by $\pm\frac{1}{2}$; that is, supersymmetry unifies bosons and fermions. A multiplet of "simple" supersymmetry (a supersymmetry with one fermionic generator) in four dimensions is a pair of particles with spins $J$ and $J - \frac{1}{2}$; the supersymmetry generators transform bosonic fields into fermionic fields and vice versa. The boson and fermion components are equal in number in all supersymmetry multiplets relevant to particle theories.

We can construct larger supersymmetries by adding more fermionic generators to the Poincaré symmetry. "$N$-extended" supersymmetry has $N$ fermionic generators. By applying each generator to the state of spin $J$,

we can lower the helicity up to $N$ times. Thus, simple supersymmetry, which lowers the helicity just once, is called $N = 1$ supersymmetry. $N = 2$ supersymmetry can lower the helicity twice, and the $N = 2$ multiplets have spins $J$, $J - \frac{1}{2}$, and $J - 1$. There are twice as many $J - \frac{1}{2}$ states as $J$ or $J - 1$, so that there are equal numbers of fermionic and bosonic states. The $N = 2$ multiplet is made up of two $N = 1$ multiplets: one with spins $J$ and $J - \frac{1}{2}$ and the other with spins $J - \frac{1}{2}$ and $J - 1$.

In principle, this construction can be extended to any $N$, but in quantum field theory there appears to be a limit. There are serious difficulties in constructing simple field theories with spin 5/2 or higher. The largest extension with spin 2 or less has $N = 8$. In $N = 8$ extended supersymmetry, there is one state with helicity of 2, eight with 3/2, twenty-eight with 1, fifty-six with 1/2, seventy with 0, fifty-six with $-1/2$, twenty-eight with $-1$, eight with 3/2 and one with $-2$. This multiplet with 256 states will play an important role in the supersymmetric theories of gravity or supergravity discussed below. Table 2 shows the states of $N$-extended supersymmetry.

**Theories with Supersymmetry.** Rather ordinary-looking Lagrangians can have supersymmetry. For example, there is a Lagrangian with simple global supersymmetry in four dimensions with a single Majorana fermion, which has one component with helicity $+1/2$, one with helicity $-1/2$, and two spinless particles. Thus, there are two bosonic and two fermionic degrees of freedom. The supersymmetry not only requires the presence of both fermions and bosons in the Lagrangian but also restricts the types of interactions, requires that the mass parameters in the multiplet be equal, and relates some other parameters in the Lagrangian that would otherwise be unconstrained.

The model just described, the Wess-Zumino model,[3] is so simple that it can be written down easily in conventional field notation. However, more realistic supersym-

metric Lagrangians take pages to write down. We will avoid this enormous complication and limit our discussion to the spectra of particles in the various theories.

Although supersymmetry may be an exact symmetry of the Lagrangian, it does not appear to be a symmetry of the world because the known elementary particles do not have supersymmetric partners. (The photon and a neutrino cannot form a supermultiplet because their low-energy interactions are different.) However, like ordinary symmetries, the supersymmetries of the Lagrangian do not have to be supersymmetries of the vacuum: supersymmetry can be spontaneously broken. The low-energy predictions of spontaneously broken supersymmetric models are discussed in "Supersymmetry at 100 GeV."

**Local Supersymmetry and Supergravity.** There is a curious gap in the spectrum of the spin values of the known elementary particles. Almost all spins less than or equal to 2 have significant roles in particle theory: spin-1 vector bosons are related to the local internal symmetries; the spin-2 graviton mediates the gravitational interaction; low-mass spin-$\frac{1}{2}$ fermions dominate low-energy phenomenology; and spinless fields provide the mechanism for spontaneous symmetry breaking. All these fields are crucial to the standard model, although there seems to be no relation among the fields of different spin. A spin of 3/2 is not required phenomenologically and is missing from the list. If the supersymmetry is made local, the resulting theory is supergravity, and the spin-2 graviton is accompanied by a "gravitino" with spin 3/2.

Local supersymmetry can be imposed on a theory in a fashion formally similar to the local symmetries of the standard model, except for the additional complications due to the fact that supersymmetry is a space-time symmetry. Extra gauge fields are required to compensate for derivatives of the space-time-dependent parameters, so, just as for ordinary symmetries, there is a gauge particle corresponding to each independent super-

**Table 2**

**The fields of *N*-extended supergravity in four dimensions. Shown are the number of states of each helicity for each possible supermultiplet containing a graviton but with spin ≤ 2. Simple supergravity ($N = 1$) has a graviton and gravitino. $N = 4$ supergravity is the simplest theory with spinless particles. The overlap of the multiplets with the largest (+2) and smallest (−2) helicities gives rise to large additional symmetries in supergravity. $N = 7$ and $N = 8$ supergravities have the same list of helicities because particle-antiparticle symmetry implies that the $N = 7$ theory must have two multiplets (as for $N < 7$), whereas $N = 8$ is the first and last case for which particle-antiparticle symmetry can be satisfied by a single multiplet.**

|           |   |   |   | *N* |   |   |        |
|-----------|---|---|---|---|----|-----|--------|
| Helicity  | 1 | 2 | 3 | 4 | 5  | 6   | 7 or 8 |
| 2         | 1 | 1 | 1 | 1 | 1  | 1   | 1      |
| 3/2       | 1 | 2 | 3 | 4 | 5  | 6   | 8      |
| 1         |   | 1 | 3 | 6 | 10 | 16  | 28     |
| 1/2       |   |   | 1 | 4 | 11 | 26  | 56     |
| 0         |   |   |   | 2 | 10 | 30  | 70     |
| −1/2      |   |   | 1 | 4 | 11 | 26  | 56     |
| −1        |   | 1 | 3 | 6 | 10 | 16  | 28     |
| −3/2      | 1 | 2 | 3 | 4 | 5  | 6   | 8      |
| −2        | 1 | 1 | 1 | 1 | 1  | 1   | 1      |
| Total     | 4 | 8 | 16| 32| 64 | 128 | 256    |

symmetry transformation. However, the gauge particles associated with the supersymmetry generators must be fermions. Just as the graviton has spin 2 and is associated with the local translational symmetry, the gravitino has spin 3/2 and gauges the local supersymmetry. The graviton and gravitino form a simple ($N = 1$) supersymmetry multiplet. This theory is called simple supergravity and is interesting because it succeeds in unifying the graviton with another field.

The Lagrangian of simple supergravity[4] is an extension of Einstein's Lagrangian, and one recovers Einstein's theory when the gravitational interactions of the gravitino are ignored. This model must be generalized to a more realistic theory with vector bosons, spin-½ fermions, and spinless fields to be of much use in particle theory.

The generalization is to Lagrangians with extended local supersymmetry, where the largest spin is 2. The extension is extremely complicated. Nevertheless, without much work we can surmise some features of the extended theory. Table 2 shows the spectrum of particles in *N*-extended supergravity.

We start here with the largest extended supersymmetry and investigate whether it includes the electroweak and strong interactions. In $N = 8$ extended supergravity the spectrum is just the $N = 8$ supersymmetric multiplet of 256 helicity states discussed before. The massless particles formed from these states include one graviton, eight gravi-

tinos, twenty-eight vector bosons, fifty-six fermions, and seventy spinless fields.

$N = 8$ supergravity[5] is an intriguing theory. (Actually, several different $N = 8$ supergravity Lagrangians can be constructed.) It has a remarkable set of internal symmetries, and the choice of theory depends on which of these symmetries have gauge particles associated with them. Nevertheless, supergravity theories are highly constrained and we can look for the standard model in each. We single out one of the most promising versions of the theory, describe its spectrum, and then indicate how close it comes to unifying the electroweak, strong, and gravitational interactions.

In the $N = 8$ supergravity of de Wit-Nicolai theory[6] the twenty-eight vector bosons gauge an SO(8) symmetry found by Cremmer and Julia.[5] Since the standard model needs just twelve vector bosons, twenty-eight would appear to be plenty. In the fermion sector, the eight gravitinos must have fairly large masses in order to have escaped detection. Thus, the local supersymmetry must be broken, and the gravitinos acquire masses by absorbing eight spin-½ fermions. This leaves $56 - 8 = 48$ spin-½ fermion fields. For the quarks and leptons in the standard model, we need forty-five fields, so this number also is sufficient.

The next question is whether the quantum numbers of SO(8) correspond to the electroweak and strong quantum numbers and the spin-½ fermions to quarks and leptons. This is where the problems start: if we separate an SU(3) out of the SO(8) for QCD, then the only other independent interactions are two local phase symmetries of U(1) × U(1), which is not large enough to include the SU(2) × U(1) of the electroweak theory. The rest of the SO(8) currents mix the SU(3) and the two U(1)'s. Moreover, many of the fifty-six spin-½ fermion states (or forty-eight if the gravitinos are massive) have the wrong SU(3) quantum numbers to be quarks and leptons.[7] Finally, even if the quantum numbers for QCD were right and the electroweak local symmetry were present, the weak interactions could still not be accounted for. No mechanism in this theory can guarantee the almost purely axial weak neutral current of the electron. Thus this interpretation of $N = 8$ supergravity cannot be the ultimate theory. Nevertheless, this is a model of unification, although it gave the wrong sets of interactions and particles.

Perhaps the 256 fields do not correspond directly to the observable particles, but we need a more sophisticated analysis to find them. For example, there is a "hidden" local SU(8) symmetry, independent of the gauged SO(8) mentioned above, that could easily contain the electroweak and strong interactions. It is hidden in the sense that the Lagrangian does not contain the kinetic energy terms for the sixty-three vector bosons of SU(8). These sixty-three vector bosons are composites of the elementary supergravity fields, and it is possible that the quantum corrections will generate kinetic energy terms. Then the fields in the Lagrangian do not correspond to physical particles; instead the photon, electron, quarks, and so on, which look elementary on a distance scale of present experiments, are composite. Unfortunately, it has not been possible to work out a logical derivation of this kind of result for $N = 8$ supergravity.[8]

In summary, $N = 8$ supergravity may be correct, but we cannot see how the standard model follows from the Lagrangian. The basic fields seem rich enough in structure to account for the known interactions, but in detail they do not look exactly like the real world. Whether $N = 8$ supergravity is the wrong theory, or is the correct theory and we simply do not know how to interpret it, is not yet known.

## Supergravity in Eleven Dimensions

The apparent phenomenological shortcomings of $N = 8$ supergravity have been known for some time, but its basic mathematical structure is so appealing that many theorists continue to work on it in hope that some variant will give the electroweak and strong interactions. One particularly interesting development is the generalization of $N = 8$ supergravity in four dimensions to simple ($N = 1$) supergravity in eleven dimensions.[9] This generalization combines the ideas of Kaluza-Klein theories with supersymmetry.

The formulation and dimensional reduction of simple supergravity in eleven dimensions requires most of the ideas already described. First we find the fields of 11-dimensional supergravity that correspond to the graviton and gravitino fields in four dimensions. Then we describe the components of each of the 11-dimensional fields. Finally, we use the harmonic expansion on the extra seven dimensions to identify the zero modes and pyrgons. For a certain geometry of the extra dimensions, the dimensionally reduced, 11-dimensional supergravity without pyrgons is $N = 8$ supergravity in four dimensions; for other geometries we find new theories. We now look at each of these steps in more detail.

In constructing the 11-dimensional fields, we begin by recalling that the helicity symmetry of a massless particle is SO(9) and the spin components are classified by the multiplets of SO(9). The multiplets of SO(9) are either fermionic or bosonic, which means that all the four-dimensional helicities are either integers (bosonic) or half-odd integers (fermionic) for all the components in a single multiplet. The generators independent of the SO(2) form an SO(7), which is the Lorentz group for the extra seven dimensions. Thus, the SO(9) multiplets can be expressed in terms of a sum of multiplets of SO(7) $\times$ SO(2), which makes it possible to reduce 11-dimensional spin to 4-dimensional spin.

The fields of 11-dimensional, $N = 1$ supergravity must contain the graviton and gravitino in four dimensions. We have already mentioned in the sidebar that the graviton in eleven dimensions has forty-four bosonic components. The smallest SO(9) multiplet of 11-dimensional spin that yields a helicity of 3/2 in four dimensions for the gravitinos has 128 components, eight components with helicity 3/2, fifty-six with 1/2, fifty-six with

$-1/2$, and eight with $-3/2$. Since by supersymmetry the number of fermionic states is equal to the number of bosonic states, eighty-four bosonic components remain. It turns out that there is a single 11-dimensional spin with eighty-four components, and it is just the field needed to complete the $N = 1$ supersymmetry multiplet in eleven dimensions.

Thus, we have recovered the 256 components of $N = 8$ supergravity in terms of just three fields in eleven dimensions (see Table 3). The Lagrangian is much simpler in eleven dimensions than it is in four dimensions. The three fields are related to one another by supersymmetry transformations that are very similar to the simple supersymmetry transformations in four dimensions. Thus, in many ways the 11-dimensional theory is no more complicated than simple supergravity in four dimensions.

The dimensional reduction of the 11-dimensional supergravity, where the extra dimensions are a 7-torus, gives one version of $N = 8$ supergravity in four dimensions.[5] In this case each of the components is expanded in a sevenfold Fourier series, one series for each dimension just as in Eq. 1 in the sidebar, except that $ny$ is replaced by $\Sigma n_i y_i$. The dimensional reduction consists of keeping only those fields that do not depend on any $y_i$, that is, just the 4-dimensional fields corresponding to $n_1 = n_2 = \ldots = n_7 = 0$. Thus, there is one zero mode (massless field in four dimensions) for each component. The pyrgons are the 4-dimensional fields with any $n_i \neq 0$, and these are omitted in the dimensional reduction.

The 11-dimensional theory has a simple Lagrangian, whereas the 4-dimensional, $N = 8$ Lagrangian takes pages to write down. In fact the $N = 8$ Lagrangian was first derived in this way.[5] It is easy to be impressed by a formalism in which everything looks simple. This is the first of several reasons to take seriously the proposal that the extra dimensions might be physical, not just a mathematical trick.

The seven extra dimensions of the 11-dimensional theory must be wound up into a little ball in order to escape detection. The

## Table 3

The relation of simple ($N = 1$) supergravity in eleven dimensions and $N = 8$ supergravity in four dimensions. The 256 components of the massless fields of 11-dimensional, $N = 1$ supergravity fall into three $n$-member multiplets of SO(9). The members of these multiplets have definite helicities in four dimensions. The count of helicity states is given in terms of the size of SO(7) multiplets, where SO(7) is the Lorentz symmetry of the seven extra dimensions in the 11-dimensional theory.

| | **4-Dimensional Helicity** | | | | | | | | |
|---|---|---|---|---|---|---|---|---|---|
| $n$ | 2 | 3/2 | 1 | 1/2 | 0 | −1/2 | −1 | −3/2 | −2 |
| 44 | 1 | | 7 | | 1+27 | | 7 | | 1 |
| 84 | | | 21 | | 7+35 | | 21 | | |
| 128 | | 8 | | 8+48 | | 8+48 | | 8 | |
| **Total** | 1 | 8 | 28 | 56 | 70 | 56 | 28 | 8 | 1 |

case described above assumes that the little ball is a 7-torus, which is the group manifold made of the product of seven phase symmetries. As a Kaluza-Klein theory, the seven vector bosons in the graviton (Table 3) gauge these seven symmetries. Since the twenty-eight vector bosons of $N = 8$ supergravity can be the gauge fields for a local SO(8), it is interesting to see if we can redo the dimensional reduction so that 11-dimensional supergravity is a Kaluza-Klein theory for SO(8), the de Wit-Nicolai theory. Indeed, this is possible. If the extra dimensions are assumed to be the 7-sphere, which is the coset space SO(8)/SO(7), the vector bosons do gauge SO(8).[10] This is, perhaps, the ultimate Kaluza-Klein theory, although it does not contain the standard model. The main difference between the 7-torus and coset spaces is that for coset spaces there is not necessarily a one-to-one correspondence between components and zero modes. Some components may have several zero modes, while others have none (recall Fig. 5).

There are other manifolds that solve the 11-dimensional supergravity equations, although we do not describe them here. The internal local symmetries are just those of the

extra dimensions, and the fermions and bosons are unified by supersymmetry. Thus, 11-dimensional supergravity can be dimensionally reduced to one of several different 4-dimensional supergravity theories, and we can search through these theories for one that contains the standard model. Unfortunately, they all suffer phenomenological shortcomings.

Eleven-dimensional supergravity contains an additional error. In the solution where the seven extra dimensions are wound up in a little ball, our 4-dimensional world gets just as compacted: the cosmological constant is about 120 orders of magnitude larger than is observed experimentally.[11] This is the cosmological constant problem at its worst. Its solution may be a major breakthrough in the search for unification with gravity. Meanwhile, it would appear that supergravity has given us the worst prediction in the history of modern physics!

## Superstrings

In view of its shortcomings, supergravity is apparently not the unified theory of all

elementary particle interactions. In many ways it is close to solving the problem, but a theory that is correct in all respects has not been found. The weak interactions are not exactly right nor is the list of spin-$\frac{1}{2}$ fermions. There seems to be no good reason that the cosmological constant should be nearly or exactly zero as observed experimentally. The issue of the renormalizability of the quantum theory of gravity also remains unsolved. Supergravity improves the quantum structure of the theory in that the unwanted infinities are not as bad as in Einstein's theory with matter, but troubles still appear. Newton's constant is a fundamental parameter in the theory, and 4-fermion terms similar to those in Fermi's weak interaction theory are still present. In $N = 8$ supergravity, which is the best case, the perturbation solution to the quantum field theory is expected to break down eventually.

In spite of these difficulties we have reasons to be optimistic that supergravity is on the right track. It does unify gravity with some interactions and is almost a consistent quantum field theory. The line of generalization followed so far has led to theories that are enormous improvements, in a mathematical sense, over Einstein's gravity. It would seem reasonable to look for generalizations beyond supergravity.

Superstring theories may answer some of these questions. Just as the progress of supergravity was based on the systematic addition of fields to Einstein's gravity, superstring theory can also be viewed in terms of the systematic addition of fields to supergravity. Although the formulation of superstring theory looks quite different from the formulation of supergravity, this may be partially due to its historical origin.

Superstring theories were born from an early effort to find a theory of the strong interactions. They began as a very efficient means of understanding the long list of hadronic resonances. In particular, hadrons of high spin have been identified experimentally. It is interesting that sets of hadrons of different spins but the same internal quantum numbers can be grouped together into

"Regge trajectories." Figure 6 shows examples of Regge trajectories (plots of spin versus mass-squared) for the first few states of the Δ and $N$ resonances; these resonances for hadrons of different spins fall along nearly straight lines. Such sequences appear to be general phenomena, and so, in the '60s and early '70s, a great effort was made to incorporate these results directly into a theory. The basic idea was to build a set of hadron amplitudes with rising Regge trajectories that satisfied several important constraints of quantum field theory, such as Lorentz invariance, crossing symmetry, the correct analytic properties, and factorization of resonance-pole residues.[12] Although the theory was a prescription for calculating the amplitudes, these constraints are true of quantum field theory and are necessary for the theory to make sense.

The constraints of field theory proved to be too much for this theory of hadrons. Something always went wrong. Some theories predicted particles with imaginary mass (tachyons) or particles produced with negative probability (ghosts), which could not be interpreted. Several theories had no logical difficulties, but they did not look like hadron theories. First of all, the consistency requirements forced them to be in ten dimensions rather than four. Moreover, they predicted massless particles with a spin of 2; no hadrons of this sort exist. These original superstring theories did not succeed in describing hadrons in any detail, but the solution of QCD may still be similar to one of them.

In 1974 Scherk and Schwarz[13] noted that the quantum amplitudes for the scattering of the massless spin-2 states in the superstring are the same as graviton-graviton scattering in the simplest approximation of Einstein's theory. They then boldly proposed throwing out the hadronic interpretation of the superstring and reinterpreting it as a fundamental theory of elementary particle interactions. It was easily found that superstrings are closely related to supergravity, since the states fall into supersymmetry multiplets and massless spin-2 particles are required.[14]

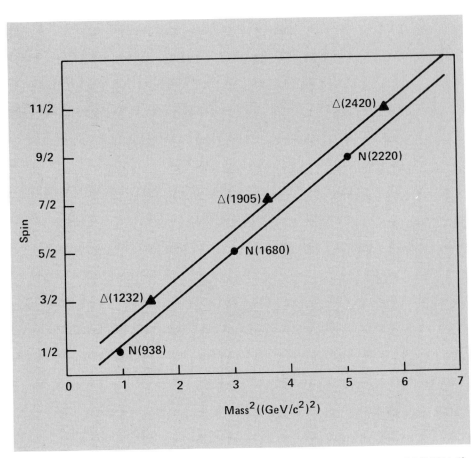

Fig. 6. Regge trajectories in hadron physics. The neutron and proton (N(938)) lie on a linearly rising Regge trajectory with other isospin-½ states: the N(1680) of spin 5/2, the N(2220) of spin 9/2, and so on. This fact can be interpreted as meaning that the N(1680), for example, looks like a nucleon except that the quarks are in an F wave rather than a P wave. Similarly the isospin-3/2 Δ resonance at 1232 MeV lies on a trajectory with other isospin-3/2 states of spins 7/2, 11/2, 15/2, and so on. The slope of the hadronic Regge trajectories is approximately (1 GeV/c²)⁻². The slope of the superstring trajectories must be much smaller

The theoretical development of superstrings is not yet complete, and it is not possible to determine whether they will finally yield the truly unified theory of all interactions. They are the subject of intense research today. Our plan here is to present a qualitative description of superstrings and then to discuss the types and particle spectra of superstring theories.

Recent formulations of superstring theories are generalizations of quantum field theory.[15] The fields of an ordinary field theory, such as supergravity, depend on the space-time point at which the field is evaluated. The fields of superstring theory depend on paths in space-time. At each moment in time, the string traces out a path in space, and as time advances, the string propagates through space forming a surface called the "world sheet." Strings can be closed, like a rubber band, or open, like a broken rubber band. Theories of both types

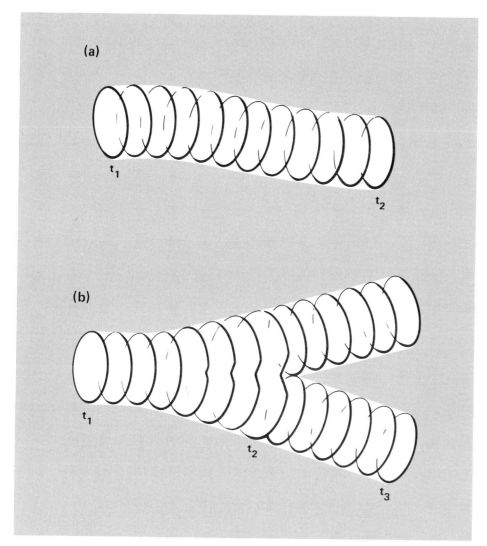

(a)

t₁

t₂

(b)

t₁

t₂

t₃

*Fig. 7. Dynamics of closed strings. The figures show the string configurations at a sequence of times (in two dimensions instead of ten). In Fig. 7(a) a string in motion from times t₁ to t₂ traces out a world sheet. Figure 7(b) shows the three closed string interaction, where one string at t₁ undergoes a change of shape until it pinches off at a point at time t₂ (the interaction time). At time t₃ two strings are propagating away from the interaction region.*

are promising, but the graviton is always associated with closed strings.

Before analyzing the motion of a superstring, we must return to a discussion of space-time. Previously, we described extensions of space-time to more than four dimensions. In all those cases coordinates were numbers that satisfied the rules of ordinary arithmetic. Yet another extension of space-time, which is useful in supergravity and crucial in superstring theory, is the addition to space-time of "supercoordinates" that do not satisfy the rules of ordinary arithmetic. Instead, two supercoordinates $\theta_\alpha$ and

$\theta_\beta$ satisfy anticommutation relations $\theta_\alpha\theta_\beta + \theta_\beta\theta_\alpha = 0$, and consequently $\theta_\alpha\theta_\alpha$ (with no sum on $\alpha$) $= 0$. Spaces with this kind of additional coordinate are called superspaces.[16]

At first encounter superspaces may appear to be somewhat silly constructions. Nevertheless, much of the apparatus of differential geometry of manifolds can be extended to superspaces, so applications in physics may exist. It is possible to define fields that depend on the coordinates of a superspace. Rather naturally, such fields are called superfields.

Let us apply this idea to supergravity, which is a field theory of both fermionic and bosonic fields. The supergravity fields can be further unified if they are written as a smaller number of superfields. Supergravity Lagrangians can then be written in terms of superfields; the earlier formulations are recovered by expanding the superfields in a power series in the supercoordinates. The anticommutation rule $\theta_\alpha\theta_\alpha = 0$ leads to a finite number of ordinary fields in this expansion.

The motion of a superstring is described by the motion of each space-time coordinate and supercoordinate along the string; thus the motion of the string traces out a "world sheet" in superspace. The full theory describes the motions and interactions of superstrings. In particular, Fig. 7 shows the basic form of the three closed superstring interactions. All other interactions of closed strings can be built up out of this one kind of interaction.[15] Needless to say, the existence of only one kind of fundamental interaction would severely restrict theories with only closed strings.

There is a direct connection between the quantum-mechanical states of the string and the elementary particle fields of the theory. The string, whether it is closed or open, is under tension. Whatever its source, this tension, rather than Newton's constant, defines the basic energy scale of the theory. To first approximation each point on the string has a force on it depending on this tension and the relative displacement between it and neighboring points on the string. The prob-

# Table 4

Ground states of Type II superstrings. The 10-dimensional fields are listed according to the multiplets of the SO(8) light-cone symmetry. The 4-dimensional fields are listed in terms of helicity and multiplets of the SO(6) Lorentz group of the extra six dimensions.

| | | | | | Helicity | | | | |
|---|---|---|---|---|---|---|---|---|---|
| | 2 | 3/2 | 1 | 1/2 | 0 | −1/2 | −1 | −3/2 | −2 |
| **Type IIA: Bosons** | | | | | | | | | |
| 1 | | | | | 1 | | | | |
| 28 | | | 6 | | $1 + 15$ | | 6 | | |
| $35_v$ | 1 | | 6 | | $1 + 20'$ | | 6 | | 1 |
| $8_v$ | | | 1 | | 6 | | 1 | | |
| $56_v$ | | | 15 | | $6 + 10 + \overline{10}$ | | 15 | | |
| **Type IIA: Fermions** | | | | | | | | | |
| $8_s$ | | | | 4 | | $\overline{4}$ | | | |
| $8_s$ | | | | $\overline{4}$ | | 4 | | | |
| $56_s$ | | $\overline{4}$ | | $4 + 20$ | | $\overline{4} + \overline{20}$ | | 4 | |
| $56_c$ | | 4 | | $\overline{4} + \overline{20}$ | | $4 + 20$ | | $\overline{4}$ | |
| **Type IIB: Bosons** | | | | | | | | | |
| 1 (twice) | | | | | 1 | | | | |
| 28 (twice) | | | 6 | | $1 + 15$ | | 6 | | |
| $35_v$ | 1 | | 6 | | $1 + 20'$ | | 6 | | 1 |
| $35_c$ | | | $\overline{10}$ | | 15 | | 10 | | |
| **Type IIB: Fermions** | | | | | | | | | |
| $8_s$ (twice) | | | | 4 | | $\overline{4}$ | | | |
| $56_s$ (twice) | | $\overline{4}$ | | $4 + 20$ | | $\overline{4} + \overline{20}$ | | 4 | |

lem of unravelling this infinite number of harmonic oscillators is one of the most famous problems of physics. The amplitudes of the Fourier expansion of the string displacement decouple the infinite set of harmonic oscillators into independent Fourier modes. These Fourier modes then correspond to the elementary-particle fields. The quantum-mechanical ground state of this infinite set of oscillators corresponds to the fields of 10-dimensional supergravity. Ten space-time dimensions are necessary to avoid tachyons and ghosts. The excited modes of the superstring then correspond to the new fields being added to supergravity.

The harmonic oscillator in three dimensions can provide insight into the qualitative features of the superstring. The maximum value of the spin of a state of the harmonic oscillator increases with the level of the excitation. Moreover, the energy necessary to reach a given level increases as the spring constant is increased. The superstring is similar. The higher the excitation of the string, the higher are the possible spin values (now in ten dimensions). The larger

the string tension, the more massive are the states of an excited level.

The consistency requirements restrict superstring theories to two types. Type I theories have 10-dimensional $N = 1$ supersymmetry and include both closed and open strings and five kinds of string interactions. Nothing more will be said here about Type I theories, although they are extremely interesting (see Refs. 14 and 15).

Type II theories have $N = 2$ supersymmetry in ten dimensions and accommodate closed strings only. There are two $N = 2$ supersymmetry multiplets in ten dimensions, and each corresponds to a Type II superstring theory. We will now describe these two superstring theories.

The Type IIA ground-state spectrum is the one that can be derived by dimensional reduction of simple supergravity in eleven dimensions to $N = 2$ supergravity in ten dimensions. Thus, if we continue to reduce from ten to four dimensions with the hypothesis that the extra six dimensions form a 6-torus, we will obtain $N = 8$ supergravity in four dimensions. The superstring

theory adds both pyrgons and Regge recurrences to the 256 $N = 8$ supergravity fields, but it has been possible (and often simpler) to investigate several aspects of supergravity directly from the superstring theory.

The classification of the excited 10-dimensional string states (or elementary fields of the theory) is complicated by the description of spin in ten dimensions. However, the analysis does not differ conceptually from the analysis of spin for 11-dimensional supergravity. The massless states, which form the ground state of the superstring, are classified by multiplets of SO(8), and the excitations of the string are massive fields in ten dimensions that belong to multiplets of SO(9). The ground-state fields of the Type IIA superstring are found in Table 4.

The Type IIB ground-state fields cannot be derived from 11-dimensional supergravity. Instead the theory has a useful phase symmetry in ten dimensions. The fields listed as occurring twice in Table 4 carry nonzero values of the quantum number associated with U(1). So far, the main application of the U(1) symmetry has been the

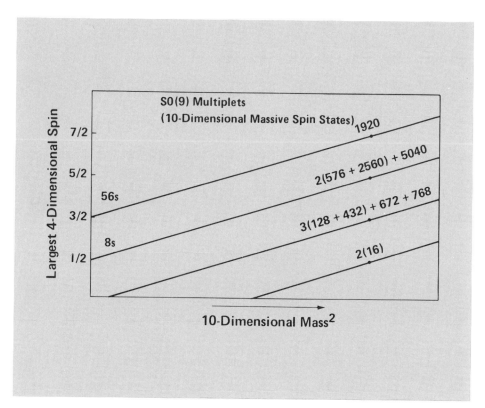

*Fig. 8. The ground state and first Regge recurrence of fermionic states in the 10-dimensional Type IIB superstring theory. There are a total of 256 fermionic and bosonic states in the ground state. (The $56_s$ contains the gravitino.) The first excited states contain 65,536 component fields. Half of these are fermions. (Each representation of the fermions shown above appears twice.)*

derivation of the equations of motion for the ground-state fields.[17] It will certainly have a crucial role in the future understanding of Type IIB superstrings.

The quantum-mechanical excitations of the superstring correspond to the Regge recurrences, which are massive in ten dimensions; they belong to multiplets of SO(9). Thus, it is possible to fill in a diagram similar to Fig. 6, although the huge number of states makes the results look complicated. We give a few results to illustrate the method.

The sets of Regge recurrences in Type IIA and IIB are identical. In Figure 8 we show the first recurrence of the fermion trajectories. (Note that only one-half of the 32,768 fer-

mionic states of this mode are shown. The boson states are even messier.) The first excited level has a total of 65,536 states, and the next two excited levels have 5,308,416 and 235,929,600 states, respectively, counting both fermions and bosons. (Particle physicists seem to show little embarrassment these days over adding a few fields to a theory!)

The component fields in ten dimensions can now be expanded into 4-dimensional fields as was done in supergravity. Besides the zero modes and pyrgons associated with the ground states, there will be infinite ladders of pyrgon fields associated with each of the fields of the excited levels of the superstring.

The zero modes in four dimensions have been investigated only for the 6-torus; in this case all the zero modes come from the ground states. There is one zero mode for each component field, since the dimensional reduction is done as a 6-dimensional Fourier series on the 6-torus. The answers for other geometries are not yet known. It may be that many more fields become zero modes (or have nearly zero mass) in four dimensions when the dimensional reduction is studied for other spaces. An important problem is the analysis of superstrings on curved spaces, which has not yet been definitively studied.

Although not much progress has been made toward understanding the phenomenology of these superstring theories, there has been some formal progress. The theory described here may be a quantum theory of gravity. (It may take all those new fields to obtain a renormalizable theory.) Although local symmetries can be ruined by anomalies, Type II (and several Type I) superstrings satisfy the constraints. Also, the one-loop calculation is finite; there are no candidates for counter terms, so the theory may be finite. Of course, this promising result needs support from higher order calculations.

These results give some encouragement that superstrings may solve some long-standing problems in particle theory; whether they will lead to the ultimate unification of all interactions remains to be seen.

## Postscript

The search for a unified theory may be likened to an old geography problem. Columbus sailed westward to reach India believing the world had no edge. By analogy, we are searching for a unified theory at shorter and shorter distance scales believing the microworld has no edge. Perhaps we are wrong and space-time is not continuous. Or perhaps we are only partly wrong, like Columbus, and will discover something new, but something consistent with what we already know. Then again, we may finally be right on course to a theory that unifies all Nature's interactions. ■

**Richard C. Slansky** has a broad background in physics with more than a taste of metaphysics. He received a B.A. in physics from Harvard in 1962 and then spent the following year as a Rockefeller Fellow at Harvard Divinity School. Dick then attended the University of California, Berkeley, where he received his Ph.D. in physics in 1967. A two-year postdoctoral stint at the California Institute of Technology was followed by five years as Instructor and Assistant Professor at Yale University (1969-1974). Dick joined the Laboratory in 1974 as a Staff Member in the Elementary Particles and Field Theory group of the Theoretical Division, where his interests encompass phenomenology, high-energy physics, and the early universe.

## References

1. For a modern description of Kaluza-Klein theories, see Edward Witten, *Nuclear Physics* B186(1981):412 and A. Salam and J. Strathdee, *Annals of Physics* 141(1982):316.

2. Two-dimensional supersymmetry was discovered in dual-resonance models by P. Ramond, *Physical Review D* 3(1971):2415 and by A. Neveu and J. H. Schwarz, *Nuclear Physics* B31(1971):86. Its four-dimensional form was discovered by Yu. A. Gol'fand and E. P. Likhtman, *Journal of Experimental and Theoretical Physics Letters* 13(1971):323.

3. J. Wess and B. Zumino, *Physics Letters* 49B(1974):52 and *Nuclear Physics* B70(1974):39.

4. Daniel Z. Freedman, P. van Nieuwenhuizen, and S. Farrara, *Physical Review D* 13(1976):3214; S. Deser and B. Zumino, *Physics Letters* 62B(1976):335; Daniel Z. Freedman and P. van Nieuwenhuizen, *Physical Review D* 14(1976):912.

5. E. Cremmer and B. Julia, *Physics Letters* 80B(1982):48 and *Nuclear Physics* B159(1979):141.

6. B. de Wit and H. Nicolai, *Physics Letters* 108B(1982):285 and *Nuclear Physics* B208(1982):323.

7. This shortage of appropriate low-mass particles was noted by M. Gell-Mann in a talk at the 1977 Spring Meeting of the American Physical Society.

8. J. Ellis, M. Gaillard, L. Maiani, and B. Zumino in *Unification of the Fundamental Particle Interactions*, S. Farrara, J. Ellis, and P. van Nieuwenhuizen, editors (New York:Plenum Press, 1980), p. 69.

9. E. Cremmer, B. Julia, and J. Scherk, *Physics Letters* 76B(1978):409. Actually, the $N = 8$ supergravity Lagrangian in four dimensions was first derived by dimensionally reducing the $N = 1$ supergravity Lagrangian in eleven dimensions to $N = 8$ supergravity in four dimensions.

10. M. J. Duff in *Supergravity 81*, S. Farrara and J. G. Taylor, editors (London: Cambridge University Press, 1982), p. 257.

11. Peter G.O. Freund and Mark A. Rubin, *Physics Letters* 97B(1983):233.

12. "Dual Models," *Physics Reports Reprint*, Vol. I, M. Jacob, editor (Amsterdam: North-Holland, 1974).

13. J. Scherk and John H. Schwarz, *Nuclear Physics* B81(1974):118.

14. For a history of this development and a list of references, see John H. Schwarz, *Physics Reports* 89(1982):223 and Michael B. Green, *Surveys in High Energy Physics* 3(1983):127.

15. M. B. Green and J. H. Schwarz, Caltech preprint CALT-68-1090, 1984.

16. For detailed textbook explanations of superspace, superfields, supersymmetry, and supergravity see S. James Gates, Jr., Marcus T. Grisaru, Martin Roček, and Warren Siegel, *Superspace: One Thousand and One Lessons in Supersymmetry* (Reading, Massachusetts: Benjamin/Cummings Publishing Co., Inc., 1983) and Julius Wess and Jonathan Bagger, *Supersymmetry and Supergravity* (Princeton, New Jersey:Princeton University Press, 1983).

17. John H. Schwarz, *Nuclear Physics* B226(1983):269; P. S. Howe and P. C. West, *Nuclear Physics* B238(1984):181.

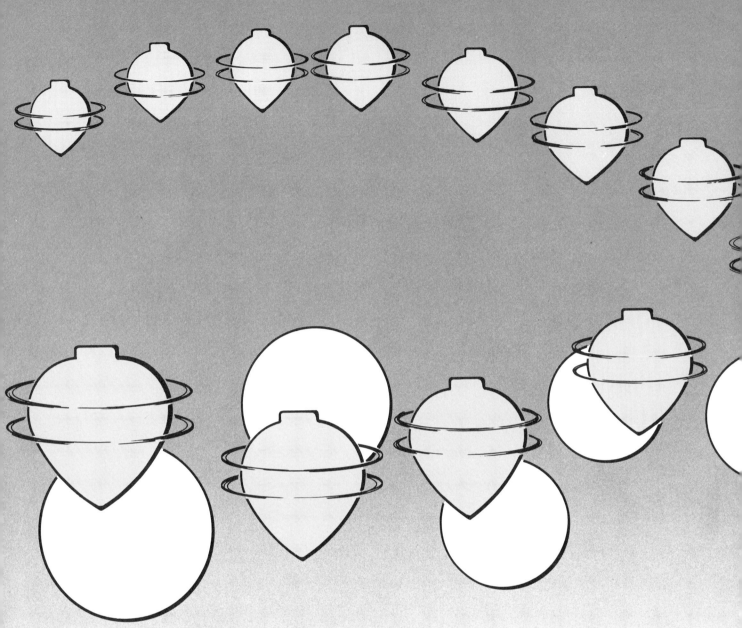

# *Supersymmetry*

$S$upersymmetry is a symmetry that connects particles of integral and half-integral spin. Invented about ten years ago by physicists in Europe and the Soviet Union, supersymmetry was immediately recognized as having amazing dynamical properties. In particular, this symmetry provides a rational framework for unifying *all* the known forces between elementary particles—the strong, weak, electromagnetic, and gravitational. Indeed, it may also unify the separate concepts of matter and force into one comprehensive framework.

*In the supersymmetric world depicted here, each boson pairs with a fermion partner.*

# at 100 GeV

*by Stuart Raby*

There are two types of symmetries in nature: external (or space-time) symmetries and internal symmetries. Examples of internal symmetries are the symmetry of isotopic spin that identifies related energy levels of the nucleons (protons and neutrons) and the more encompassing SU(3) × SU(2) × U(1) symmetry of the standard model (see "Particle Physics and the Standard Model"). Operations with these symmetries do not change the space-time properties of a particle.

External symmetries include translation invariance and invariance under the Lorentz transformations. Lorentz transformations, in turn, include rotations as well as the *special* Lorentz transformations, that is, a "boost" or a change in the velocity of the frame of reference.

Each symmetry defines a particular operation that does not affect the result of any experiment. An example of a spatial translation is to, say, move our laboratory (accelerators and all) from Chicago to New Mexico. We are, of course, not surprised that the result of any experiment is unaffected by the move, and we say that our system is translationally invariant. Rotational invariance is similarly defined with respect to rotating our apparatus about any axis. Invariance under a special Lorentz transformation corresponds to finding our results unchanged when our laboratory, at rest in our reference frame, is replaced by one moving at a constant velocity.

Corresponding to each symmetry operation is a quantity that is conserved. Energy and momentum are conserved because of time and space-translational invariance, respectively. The energy of a particle at rest is its mass ($E = mc^2$). Mass is thus an intrinsic property of a particle that is conserved because of invariance of our system under space-time translations.

**Spin**. Angular momentum conservation is a result of Lorentz invariance (both rotational and special). Orbital angular momentum refers to the angular momentum of a particle in motion, whereas the intrinsic angular momentum of a particle (remaining even at rest) is called spin. (Particle spin is an external symmetry, whereas isotopic spin, which is not based on Lorentz invariance, is not.)

In quantum mechanics spin comes in integral or half-integral multiples of a fundamental unit $\hbar$ ($\hbar = h/2\pi$ where $h$ is Planck's constant). (Orbital angular momentum only comes in integral multiples of $\hbar$.) Particles with integral values of spin ($0, \hbar, 2\hbar, \ldots$) are called bosons, and those with half-integral spins ($\hbar/2, 3\hbar/2, 5\hbar/2, \ldots$) are called fermions. Photons (spin 1), gravitons (spin 2), and pions (spin 0) are examples of bosons. Electrons, neutrinos, quarks, protons, and neutrons—the particles that make up ordinary matter—are all spin-½ fermions.

The conservation laws, such as those of energy, momentum, or angular momentum, are very useful concepts in physics. The following example dealing with spin and the conservation of angular momentum provides one small bit of insight into their utility.

In the process of beta decay, a neutron decays into a proton, an electron, and an antineutrino. The antineutrino is massless (or very close to being massless), has no charge, and interacts only very weakly with other particles. In short, it is practically invisible, and for many years beta decay was thought to be simply

$$n \rightarrow p + e^- .$$

However, angular momentum is not conserved in this process since it is not possible for the initial angular momentum (spin 1/2 for the neutron) to equal the final total angular momentum (spin 1/2 for the proton ± spin 1/2 for the electron ± an integral value for the orbital angular momentum). As a result, W. Pauli predicted that the neutrino must exist because its half-integral spin restores conservation of angular momentum to beta decay.

There is a dramatic difference between the behavior of the two groups of spin-classified particles, the bosons and the fermions. This difference is clarified in the so-called spin-statistics theorem that states that bosons must satisfy commutation relations (the quantum mechanical wave function is symmetric under the interchange of identical bosons) and that fermions must satisfy anticommutation relations (antisymmetric wave functions). The ramification of this simple statement is that an indefinite number of bosons can exist in the same place at the same time, whereas only one fermion can be in any given place at a given time (Fig. 1). Hence "matter" (for example, atoms) is made of fermions. Clearly, if you can't put more than one in any given place at a time, then they must take up space. If they are also observable in some way, then this is exactly our concept of matter. Bosons, on the other hand, are associated with "forces." For example, a large number of photons in the same place form a macroscopically observable electromagnetic field that affects charged particles.

**Supersymmetry**. The fundamental property of supersymmetry is that it is a space-time symmetry. A supersymmetry operation alters particle spin in half-integral jumps, changing bosons into fermions and vice versa. Thus supersymmetry is the first symmetry that can unify matter and force, the basic attributes of nature.

If supersymmetry is an exact symmetry in nature, then for every boson of a given mass there exists a fermion of the same mass and vice versa; for example, for the electron there should be a *scalar* electron (selectron), for the neutrino, a *scalar* neutrino (sneutrino), for quarks, *scalar* quarks (squarks), and so forth. Since no such degeneracies have been observed, supersymmetry cannot be an exact symmetry of nature. However, it might be a symmetry that is inexact or broken. If so, it can be broken in either of two inequivalent ways: *explicit* supersymmetry breaking in which the Lagrangian contains explicit terms that are not supersymmetric, or *spontaneous* supersymmetry breaking in which the Lagrangian is supersymmetric but the vacuum is not (spontaneous symmetry breaking is

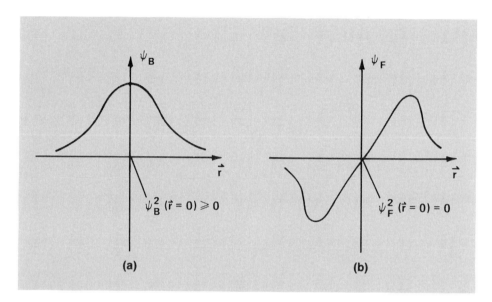

**Fig. 1. (a)** *An example of a symmetric wave function for a pair of bosons and* **(b)** *an antisymmetric wave function for a pair of fermions, where the vector* **r** *represents the distance between each pair of identical particles. Because the boson wave function is symmetric with respect to exchange* ($\psi_B$ (**r**) = $\psi_B$(−**r**)), *there can be a nonzero probablity* ($\psi_B^2$) *for two bosons to occupy the same position in space* (**r** = 0), *whereas for the asymmetric fermion wave function* ($\psi_F$ (**r**) = −$\psi_F$ (−**r**)) *the probability* ($\psi_F^2$) *of two fermions occupying the same position in space* **must be zero.**

explained in Notes 3 and 6 of "Lecture Notes—From Simple Field Theories to the Standard Model"). Either way will lift the boson-fermion degeneracy, but the latter way will introduce (in a somewhat analogous way to the Higgs boson of weak-interaction symmetry breaking) a new particle, the Goldstone fermion. (We develop mathematically some of the ideas of this paragraph in "Supersymmetry and Quantum Mechanics".)

A question of extreme importance is the scale of supersymmetry breaking. This scale can be characterized in terms of the so-called *supergap*, the mass splitting between fermions and their bosonic partners ($\delta^2 = M_B^2 - M_F^2$). Does one expect this scale to be of the order of the weak scale ($\sim 100$ GeV), or is it much larger? We will discuss the first possiblity at length because if supersymmetry is broken on a scale of order 100 GeV

there are many predictions that can be verified in the next generation of high-energy accelerators. The second possibility would not necessarily lead to any new low-energy consequences.

We will also discuss the role gravity has played in the description of low-energy supersymmetry. This connection betweeen physics at the largest mass scale in nature (the Planck scale: $M_{pl} = (\hbar c/G_N)^{1/2} \simeq 1.2 \times 10^{19}$ GeV/$c^2$, where $G_N$ is Newton's gravitational constant) and physics at the low energies of the weak scale ($M_W \simeq 83$ GeV/$c^2$ where $M_W$ is the mass of the $W$ boson responsible for weak interactions) is both novel and exciting.

**Motivations.** Why would one consider supersymmetry to start with?

First, supersymmetry is the largest possible symmetry of nature that can com-

bine internal symmetries and space-time symmetries in a nontrivial way. This combination is *not* a necessary feature of supersymmetry (in fact, it is accomplished by extending the algebra of Eqs. 2 and 3 in "Supersymmetry and Quantum Mechanics" to include more supersymmetry generators and internal symmetry generators). However, an important consequence of such an extension might be that bosons and fermions in different representations of an internal symmetry group are related. For example, quarks (fermions) are in triplets in the strong-interaction group SU(3), whereas the gluons (bosons) are in octets. Perhaps they are all related in an extended supersymmetry, thus providing a unified description of quarks and their forces.

Second, supersymmetry can provide a theory of gravity. If supersymmetry is global, then a given supersymmetry rotation must be the same over all space-time. However, if supersymmetry is local, the system is invariant under a supersymmetry rotation that may be arbitrarily different at every point. Because the various generators (supersymmetry charges, four-momentum translational generators, and Lorentz generators for both rotations and boosts) satisfy a common algebra of commutation and anticommutation relations, consistency *requires* that all the symmetries are local. (In fact, the anticommutator of two supersymmetry generators is a translation generator.) Thus different points in space-time can transform in different ways; put simply, this can amount to acceleration between points, which, in turn, is equivalent to gravity. In fact, the theory of local translations and Lorentz transformations is just general relativity, that is, Einstein's theory of gravity, and a supersymmetric theory of gravity is called supergravity. It is just the theory invariant under local supersymmetry. Thus, supersymmetry allows for a possible unification of all of nature's particles and their interactions.

These two motivations were realized quite soon after the advent of supersymmetry. They are possibilities that unfortunately have not yet led to any reasonable predic-

*continued on page 106*

# Supersymmetry in Quantum Mechanics

I intend to develop here some of the algebra pertinent to the basic concepts of supersymmetry. I will do this by showing an analogy between the quantum-mechanical harmonic oscillator and a bosonic field and a further analogy between the quantum-mechanical spin-½ particle and a fermionic field. One result of combining the two resulting fields will be to show that a "tower" of degeneracies between the states for bosons and fermions is a natural feature of even the simplest of supersymmetry theories.

A supersymmetry operation changes bosons into fermions and vice versa, which can be represented schematically with the operators $Q_\alpha^\dagger$ and $Q_\alpha$ and the equations

$$Q_\alpha^\dagger |\text{boson}\rangle = |\text{fermion}\rangle_\alpha$$

and

$$Q_\alpha |\text{fermion}\rangle = |\text{boson}\rangle_\alpha . \tag{1}$$

In the simplest version of supersymmetry, there are four such operators or generators of supersymmetry ($Q_\alpha$ and the Hermitian conjugate $Q_\alpha^\dagger$ with $\alpha = 1, 2$). Mathematically, the generators are Lorentz spinors satisfying fermionic anticommutation relations

$$\{Q_\alpha^\dagger, Q_\beta\} = p^\mu (\sigma_\mu)_{\alpha\beta} , \tag{2}$$

where $p^\mu$ is the energy-momentum four-vector ($p^0 = H$, $p^i$ = three-momentum) and the $\sigma_\mu$ are two-by-two matrices that include the Pauli spin matrices $\sigma^i$ ($\sigma_\mu = (1, \sigma^i)$ where $i = 1, 2, 3$). Equation 2 represents the unusual feature of this symmetry: the supersymmetry operators combine to generate translation in space and time. For example, the operation of changing a fermion to a boson and back again results in changing the position of the fermion.

If supersymmetry is an invariance of nature, then

$$[H, Q_\alpha] = 0 , \tag{3}$$

that is, $Q_\alpha$ commutes with the Hamiltonian $H$ of the universe. Also, in this case, the vacuum is a supersymmetric singlet ($Q_\alpha |\text{vac}\rangle = 0$).

Equations 1 through 3 are the basic defining equations of supersymmetry. In the form given, however, the supersymmetry is solely an external or space-time symmetry (a supersymmetry operation changes particle spin without altering any of the particle's internal symmetries). An extended supersymmetry that connects external and internal symmetries can be constructed by expanding the number of operators of Eq. 2. However, for our purposes, we need not consider that complication.

**The Harmonic Oscillator.** In order to illustrate the consequences of Eqs. 1 through 3, we first need to review the quantum-mechanical treatment of the harmonic oscillator.

The Hamiltonian for this system is

$$H_{\text{osc}} = \frac{1}{2} (p^2 + \omega^2 q^2) , \tag{4}$$

where $p$ and $q$ are, respectively, the momentum and position coordinates of a nonrelativistic particle with unit mass and a $2\pi/\omega$ period of oscillation. The coordinates satisfy the quantum-mechanical commutation relation

$$[p, q] = (pq - qp) = -i\hbar . \tag{5}$$

The well-known solution to the harmonic oscillator (the set of eigenstates and eigenvalues of $H_{\text{osc}}$) is most conveniently expressed in terms of the so-called raising and lowering operators, $a^\dagger$ and $a$, respectively, which are defined as

$$a^\dagger = \frac{1}{\sqrt{2\omega\hbar}} (p + i\omega q)$$

and $\hspace{10cm}$ (6)

$$a = \frac{1}{\sqrt{2\omega\hbar}} (p - i\omega q) ,$$

and which satisfy the commutation relation

$$[a, a^\dagger] = 1 . \tag{7}$$

In terms of these operators, the Hamiltonian becomes

$$H_{\text{osc}} = \hbar\omega(a^\dagger a + \tfrac{1}{2}) , \tag{8}$$

with eigenstates

$$|n\rangle = N_n(a^\dagger)^n|0\rangle , \tag{9}$$

where $N_n$ is a normalization factor and $|0\rangle$ is the ground state satisfying

$$a|0\rangle = 0$$

and $\hspace{10cm}$ (10)

$$\langle 0|0\rangle = 1 .$$

It is easy to show that

$$a^\dagger|\text{n}\rangle = \sqrt{n + 1} \, |n + 1\rangle$$

and $\hspace{10cm}$ (11)

$$a|n\rangle = \sqrt{n} \, |n - 1\rangle ,$$

hence the names raising operator for $a^\dagger$ and lowering operator for $a$. Also note that $a^\dagger a$ is just a counting operator since $a^\dagger a |n\rangle = n | n\rangle$. Finally, we find that

$$H_{\text{osc}} |n\rangle = \hbar\omega(n + \tfrac{1}{2}) |n\rangle , \tag{12}$$

that is, the states $|n\rangle$ have energy $(n + \tfrac{1}{2}) \hbar\omega$ .

**The Bosonic Field.** There is a simple analogy between the quantum oscillator and the scalar quantum field needed to represent bosons (scalar particles). A free scalar field is quite rigorously described by an infinite set of noninteracting harmonic oscillators $\{a_p^\dagger, a_p\}$, where $p$ is an index labeling the set. The Hamiltonian of the free field can be written as

$$H_{\text{scalar}} = \sum_p \hbar\omega_p \left( a_p^\dagger a_p + \tfrac{1}{2} \right) , \tag{13}$$

with the summation taken over the individual oscillators $p$.

The ground state of the free scalar quantum field is called the vacuum (it contains no scalar particles) and is described mathematically by the conditions

$$a_p |\text{vac}\rangle = 0$$

and $\hspace{10cm}$ (14)

$$\langle \text{vac}|\text{vac}\rangle = 1 .$$

The $a_p^\dagger$ and $a_p$ operators create or annihilate, respectively, a single scalar particle with energy $\hbar\omega_p$ ($\hbar\omega_p = \sqrt{p^2 + m^2}$, where $p$ is the momentum carried by the created particle and $m$ is the mass). A scalar particle is thus an excitation of one particular oscillator mode.

**The Fermionic Field.** The simple quantum-mechanical analogue of a spin-$\tfrac{1}{2}$ field needed to represent fermions is just a quantum particle with spin $\tfrac{1}{2}$. This is necessary because, whereas bosons can be represented by scalar particles satisfying commutation relations, fermions must be represented by spin-$\tfrac{1}{2}$ particles satisfying anticommutation relations.

A spin-$\tfrac{1}{2}$ particle has two spin states: $|0\rangle$ for spin down and $|1\rangle$ for spin up. Once again we define raising and lowering operators, here $b^\dagger$ and $b$, respectively. These operators satisfy the anticommutation relations

$$\{b, b^\dagger\} = (bb^\dagger + b^\dagger b) = 1$$

and (15)

$$\{b^\dagger, b^\dagger\} = \{b, b\} = 0 .$$

If $b|0\rangle = 0$, it is easy to show that

$$b^\dagger |0\rangle = |1\rangle$$

and (16)

$$b |1\rangle = |0\rangle ,$$

where $b^\dagger b$ is again a counting operator satisfying

$$b^\dagger b |1\rangle = |1\rangle$$

and (17)

$$b^\dagger b |0\rangle = 0 .$$

We may define a Hamiltonian

$$H_{\text{spin}} = \hbar\omega(b^\dagger b - \tfrac{1}{2}) , \qquad (18)$$

so that states $|1\rangle$ and $|0\rangle$ will have energy equal to $\tfrac{1}{2}\hbar\omega$ and $-\tfrac{1}{2}\hbar\omega$, respectively.

The analogy between the free quantum-mechanical fermionic field and the simple quantum-mechanical spin-$\tfrac{1}{2}$ particle is identical to the scalar field case. For example, once again we may define an infinite set $\{b_p^\dagger, b_p\}$ of noninteracting spin-$\tfrac{1}{2}$ particles labeled by the index $p$. The vacuum state satisfies

$$b_p |\text{vac}\rangle = 0$$

and (19)

$$\langle\text{vac}|\text{vac}\rangle = 1 .$$

Here $b_p^\dagger$ and $b_p$ are identified as creation and annihilation operators, respectively, of a single fermionic particle. Note that since $\{b_p^\dagger, b_p^\dagger\} = 0$, it is only possible to create *one* fermionic particle in the state $p$. This is the Pauli exclusion principle.

**Supersymmetry.** Let us now construct a simple supersymmetric quantum-mechanical system that includes the bosonic oscillator degrees of freedom ($a^\dagger$ and $a$) and the fermionic spin-$\tfrac{1}{2}$ degrees of freedom ($b^\dagger$ and $b$). We define the anticommuting charges

$$Q = a^\dagger b(\hbar\omega)^{1/2}$$

and (20)

$$Q^\dagger = ab^\dagger(\hbar\omega)^{1/2} .$$

It is then easy to verify that

$$\{Q^\dagger, Q\} = H = H_{\text{osc}} + H_{\text{spin}}$$
$$= \hbar\omega(a^\dagger a + b^\dagger b) , \qquad (21)$$

and

$$[H, Q] = 0 . \qquad (22)$$

Equations 21 and 22 are the direct analogues of Eqs. 2 and 3, respectively. We see that the anticommuting charges $Q$ combine to form the generator of time translation, namely, the Hamiltonian $H$. The ground state of this system is the state $|0\rangle_{\text{osc}}|0\rangle_{\text{spin}} = |0,0\rangle$, where both the oscillator and the spin-$\tfrac{1}{2}$ degrees of freedom are in the lowest energy state. This state is a unique one, satisfying

$$Q|0,0\rangle = Q^\dagger|0,0\rangle = 0 . \qquad (23)$$

The excited states form a tower of degenerate levels (see figure) with energy $(n + \tfrac{1}{2})\hbar\omega \pm \tfrac{1}{2}\hbar\omega$, where the sign of the second term is determined by whether the spin-$\tfrac{1}{2}$ state is $|1\rangle$ (plus) or $|0\rangle$ (minus).

The tower of states illustrates the boson-fermion degeneracy for exact supersymmetry. The bosonic states $|n+1,0\rangle$ (called bosonic in the field theory analogy because they contain no fermions) have the same energy as their fermionic partners $|n,1\rangle$.

Moreover, it is easy to see that the charges $Q$ and $Q^\dagger$ satisfy the relations

$$Q|n,1\rangle = \sqrt{n+1} \,|n+1,0\rangle$$

and

| Energy | States | |
| --- | --- | --- |
| | Boson | Fermion |
| 0 | $|0,0\rangle$ | |
| $\hbar\omega$ | $|1,0\rangle$ | $|0,1\rangle$ |
| $2\hbar\omega$ | $|2,0\rangle$ | $|1,1\rangle$ |
| $3\hbar\omega$ | $|3,0\rangle$ | $|2,1\rangle$ |
| . | . | . |
| . | . | . |
| . | . | . |

*The boson-fermion degeneracy for exact supersymmetry in which the first number in $|n,m\rangle$ corresponds to the state for the oscillator degree of freedom (the scalar, or bosonic, field) and the second number to that for the spin-½ degree of freedom (the fermionic field).*

$$Q^\dagger|n+1,0\rangle = \sqrt{n+1}\ |n,1\rangle\,, \qquad (24)$$

which are analogous to Eq. 1 because they represent the conversion of a fermionic state to a bosonic state and vice versa.

The above example is a simple representation of supersymmetry in quantum mechanics. It is, however, trivial since it describes non-interacting bosons (oscillators) and fermions (spin-½ particles). Non-trivial *interacting* representations of supersymmetry may also be obtained. In some of these representations it it possible to show that the ground state is not supersymmetric even though the Hamiltonian is. This is an example of spontaneous supersymmetry breaking.

**Symmetry Breaking.** If supersymmetry were an exact symmetry of nature, then bosons and fermions would come in degenerate pairs. Since this is not the case, the symmetry must be broken. There are two inequivalent ways in which to do this and thus to have the degeneracy removed.

First we may add a small symmetry breaking term to the Hamiltonian, that is, $H \to H + \varepsilon H'$, where $\varepsilon$ is a small parameter and

$$[H', Q] \neq 0\,. \qquad (25)$$

This mechanism is called *explicit symmetry breaking*. Using it we can give scalars a mass that is larger than that of their fermionic partners, as is observed in nature. Although this breaking mechanism may be perfectly self-consistent (even this is in doubt when one includes gravity), it is totally ad hoc and lacks predictive power.

The second symmetry breaking mechanism is termed *spontaneous symmetry breaking*. This mechanism is characterized by the fact that the Hamiltonian remains supersymmetric,

$$[Q,H] = 0\,, \qquad (26)$$

but the ground state does not,

$$Q|\text{vac}\rangle \neq 0\,. \qquad (27)$$

Supersymmetry can either be a global symmetry, such as the rotational invariance of a ferromagnet, or a local symmetry, such as a phase rotation in electrodynamics. Spontaneous breaking of a *global* symmetry leads to a massless Nambu-Goldstone particle. In supersymmetry we obtain a massless fermion $\tilde{G}$, the goldstino.

Spontaneous breaking of a *local* symmetry, however, results in the gauge particle becoming massive. (In the standard model, the $W$ bosons obtain a mass $M_W = gV$ by "eating" the massless Higgs bosons, where $g$ is the SU(2) coupling constant and $V$ is the vacuum expectation value of the neutral Higgs boson.) The gauge particle of local supersymmetry is called a gravitino. It is the spin-3/2 partner of the graviton; that is, local supersymmetry incorporates Einstein's theory of gravity. When supersymmetry is spontaneously broken, the gravitino obtains a mass

$$m_G = G_N^{1/2}\Lambda_{ss}^2 \qquad (28)$$

by "eating" the goldstino (here $G_N$ is Newton's gravitational constant and $\Lambda_{ss}$ is the vacuum expectation of some field that spontaneously breaks supersymmetry).

Thus, if the ideas of supersymmetry are correct, there is an underlying symmetry connecting bosons and fermions that is "hidden" in nature by spontaneous symmetry breaking. ∎

continued from page 101

tions. Many workers in the field are, however, still pursuing these elegant notions.

Recently a third motivation for supersymmetry has been suggested. I shall describe the motivation and then discuss its expected consequences.

For many years Dirac focused attention on the "problem of large numbers" or, more recently, the "hierarchy problem." There are many extremely large numbers that appear in physics and for which we currently have no good understanding of their origin. One such large number is the ratio of the gravitational and weak-interaction mass scales mentioned earlier ($M_{\text{pl}}/M_W \sim 10^{17}$).

The gravitational force between two particles is proportional to the product of the energy (or mass if the particles are at rest) of the two particles times $G_{\text{N}}$. Thus, since $G_{\text{N}} \propto 1/M_{\text{pl}}^2$, the force between two $W$ bosons at rest is proportional to $M_W^2/M_{\text{pl}}^2 \sim 10^{-34}$. This is to be compared to the electric force between $W$ bosons, which is proportional to $\alpha = e^2/(4\pi\hbar c) \sim 10^{-2}$, where $e$ is the electromagnetic coupling constant. Hence gravitational interactions between all known elementary particles are, at observable energies, at least $10^{32}$ times weaker than their electromagnetic interactions.

The key word is observable, for if we could imagine reaching an energy of order $M_{\text{pl}}c^2$, then the gravitational interactions would become quite strong. In other words, gravitationally bound states can be formed, in principle, with mass of order $M_{\text{pl}} \sim 10^{19}$ GeV. The Planck scale might thus be associated with particles, as yet unobserved, that have strong gravitational interactions.

At a somewhat lower energy, we also have the grand unification scale ($M_{\text{G}} \sim 10^{15}$ GeV or greater), another very large scale with similar theoretical significance. New particles and interactions are expected to become important at $M_{\text{G}}$.

In either case, should these new phenomena exist, we are faced with the question of why there are two such diverse scales, $M_W$ and $M_{\text{pl}}$ (or $M_{\text{G}}$), in nature.

The problem is exacerbated in the context of the standard model. In this mathematical

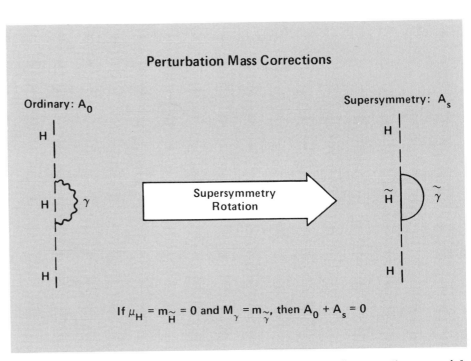

**Fig. 2. If $A_0$ (left) represents a perturbative mass correction for an ordinary particle H due to the creation of a virtual photon $\gamma$, then a supersymmetry rotation of the central region of the diagram will generate a second mass correction $A_s$ (right) involving the supersymmetric partners $\tilde{H}$ and the photino $\tilde{\gamma}$. If supersymmetry is an exact symmetry, then the total mass correction is zero.**

framework, the $W$ boson has a nonzero mass $M_W$ because of spontaneous symmetry breaking and the existence of the scalar particle called the Higgs boson. Moreover, the mass of the $W$ and the mass of the Higgs particle must be approximately equal. Unfortunately scalar masses are typically extremely sensitive to the details of the theory at very high energies. In particular, when one calculates quantum mechanical corrections to the Higgs mass $\mu_H$ in perturbation theory, one finds

$$\mu_H^2 = (\mu_H^0)^2 + \delta\mu^2 , \qquad (1)$$

where

$$\delta\mu^2 \sim \alpha\, M_{\text{large}}^2 . \qquad (2)$$

In these equations $\mu_H^0$ is the zeroth order value of the Higgs boson mass, which can be

zero, and $\delta\mu^2$ is the perturbative correction. The parameter $\alpha$ is a generic coupling constant connecting the low mass states of order $M_W$ and the heavy states of order $M_{\text{large}}$, that is, the largest mass scale in the theory. For example, some of the theorized particles with mass $M_{\text{pl}}$ or $M_{\text{G}}$ will have electric charge and interact with known particles. In this case, $\alpha = e^2/4\pi\hbar c$, a measure of the electromagnetic coupling. Clearly $\mu_H$ is naturally very large here and *not* approximately equal to the mass of the $W$.

Supersymmetry can ameliorate the problem because, in such theories, scalar particles are no longer sensitive to the details at high energies. As a result of miraculous cancellations, one finds

$$\delta\mu^2 \sim \alpha\, (\mu_H^0)^2 \ln (M_{\text{large}}) . \qquad (3)$$

This happens in the following way (Fig. 2).

## Table 1

### The Supersymmetry Doubling of Particles

| Standard Model | | Supersymmetric Partners | |
|---|---|---|---|
| spin-½ quarks | $\begin{pmatrix} u \\ d \end{pmatrix} \quad \bar{u} \quad \bar{d}$ | $\begin{pmatrix} \tilde{u} \\ \tilde{d} \end{pmatrix} \quad \tilde{\bar{u}} \quad \tilde{\bar{d}}$ | spin-0 squarks |
| spin-½ leptons | $\begin{pmatrix} \nu \\ e \end{pmatrix} \quad \bar{e}$ | $\begin{pmatrix} \tilde{\nu} \\ \tilde{e} \end{pmatrix} \quad \tilde{\bar{e}}$ | spin-0 sleptons |

**(There are two other quark-lepton families similar to this one.)**

| | | | |
|---|---|---|---|
| spin-1 gauge bosons | $\gamma, W^{\pm}, Z^0, g$ | $\tilde{\gamma}, \tilde{W}^{\pm}, \tilde{Z}^0, \tilde{g}$ | spin-½ gauginos |
| spin-0 Higgs bosons | $\begin{pmatrix} H^+ \\ H^0 \end{pmatrix} \begin{pmatrix} \bar{H}^- \\ \bar{H}^0 \end{pmatrix}$ | $\begin{pmatrix} \tilde{H}^+ \\ \tilde{H}^0 \end{pmatrix} \begin{pmatrix} \tilde{\bar{H}}^- \\ \tilde{\bar{H}}^0 \end{pmatrix}$ | spin-½ Higgsinos |

#### Global Supersymmetry

| | | | |
|---|---|---|---|
| spin-0 scalar partner | $G$ | $\tilde{G}$ (massless) | spin-½ Goldstino |

#### Local Supersymmetry

| | | | |
|---|---|---|---|
| spin-0 scalar partner | $G$ | | |
| spin-2 graviton | **g** | $\tilde{G}$ (massive) | spin-3/2 gravitino |

**The Particles.** We've discussed a bit of the motivation for supersymmetry. Now let's describe the consequences of the minimal supersymmetric extension of the standard model, that is, the particles, their masses, and their interactions.

The particle spectrum is literally doubled (Table 1). For every spin-½ quark or lepton there is a spin-0 scalar partner (squark or slepton) with the same quantum numbers under the SU(3) × SU(2) × U(1) gauge interactions. (We show only the first family of quarks and leptons in Table 1; the other two families include the *s*, *c*, *b*, and *t* quarks, and, for leptons, the muon and tau and their associated neutrinos.)

The spin-1 gauge bosons (the photon $\gamma$, the weak interaction bosons $W^{\pm}$ and $Z^0$, and the gluons g) have spin-½ fermionic partners, called gauginos.

Likewise, the spin-0 Higgs boson, responsible for the spontaneous symmetry breaking of the weak interaction, should have a spin-½ fermionic partner, called a Higgsino. However, we have included two sets of weak *doublet* Higgs bosons, denoted $H$ and $\bar{H}$, giving a total of four Higgs bosons and four Higgsinos. Although only one weak doublet of Higgs bosons is required for the weak breaking of the standard model, a consistent supersymmetry theory requires the two sets. As a result (unlike the standard model, which predicts one neutral Higgs boson), supersymmetry predicts that we should observe two *charged* and three neutral Higgs bosons.

Finally, other particles, related to symmetry breaking and to gravity, should be introduced. For a global supersymmetry, these particles will be a massless spin-½ Goldstino and its spin-0 partner. However, in the local supersymmetry theory needed for gravity, there will also be a graviton and its supersymmetric partner, the gravitino. We will discuss this point in greater detail later, but local symmetry breaking combines the Goldstino with the gravitino to form a massive, rather than a massless, gravitino.

In many cases the doubling of particles just outlined creates a supersymmetric partner that is absolutely stable. Such a particle

For each ordinary mass correction, there will be a second mass correction related to the first by a supersymmetry rotation (the symmetry operation changes the virtual particles of the ordinary correction into their corresponding supersymmetric partners). Although each correction *separately* is proportional to $\alpha M_{\text{large}}^2$, the sum of the two corrections is given by Eq. 3. In this case, if $\mu_H^0 = 0$, then $\mu_H = 0$ and will remain zero to all orders in perturbation theory as long as supersymmetry remains unbroken. Hence supersymmetry is a symmetry that prevents scalars from getting "large" masses, and one can even imagine a limit in which scalar masses vanish. Under these conditions we say scalars are "naturally" light.

How then do we obtain the spontaneous breaking of the weak interactions and a $W$ boson mass? We remarked that supersymmetry cannot be an exact symmetry of nature; it must be broken. Once supersymmetry is broken, the perturbative correction (Eq. 3) is replaced by

$$\delta\mu^2 \sim \alpha\,(\mu_H^0)^2 \ln(M_{\text{large}}) + \alpha\,\Lambda_{\text{ss}}^2, \qquad (4)$$

where $\Lambda_{\text{ss}}$ is the scale of supersymmetry breaking. If supersymmetry is broken spontaneously, then $\Lambda_{\text{ss}}$ is not sensitive to $M_{\text{large}}$ and could thus have a value that is much less than $M_{\text{large}}$. This correction to the Higgs boson mass can then result in a spontaneous breaking of the weak interactions, with the standard mechanism, at a scale of order $\Lambda_{\text{ss}} \ll M_{\text{large}}$.

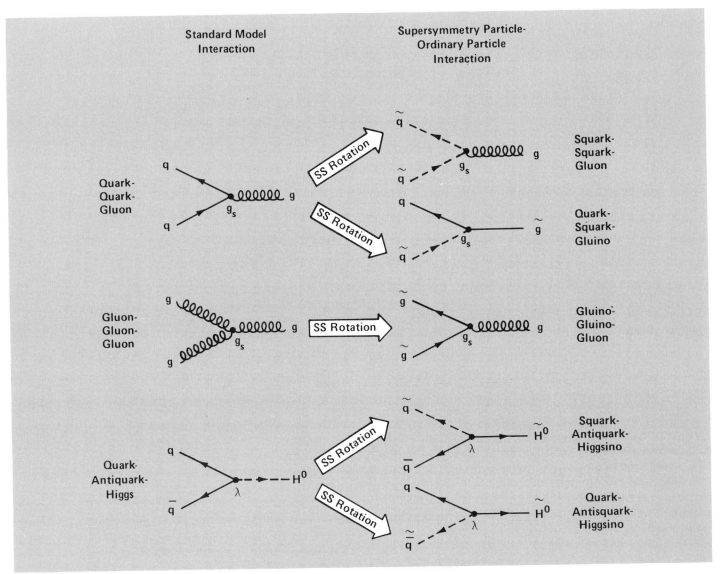

**Fig. 3.** *Examples of interactions between ordinary particles (left) and the corresponding interactions between an ordinary particle and two supersymmetric particles (right)* *obtained by performing a supersymmetry rotation on the first interaction.*

could, in fact, be the dominant form of matter in our universe.

**The Masses.** What is the expected mass for the supersymmetric partners of the ordinary particles? The theory, to date, does not make any firm predictions; we can nevertheless obtain an order-of-magnitude estimate in the

following manner.

Although an unbroken supersymmetry can keep scalars massless, once supersymmetry is broken, all scalars obtain quantum corrections to their masses proportional to the supersymmetry breaking scale $\Lambda_{ss}$, that is

$$\delta\mu^2 \sim \alpha\,\Lambda_{ss}^2\,, \tag{5}$$

which is Eq. 4 with the first negligible term dropped. If we demand the Higgs mass $\mu_H^2 \sim \delta\mu^2$ to be of order $M_W^2$, then $\Lambda_{ss}^2 \sim M_W^2/\alpha$ is at most of order 1000 GeV. Moreover, the mass splitting between all ordinary particles and their supersymmetric partners is again of order $M_W$. We thus conclude that if supersymmetry is responsible for the large ratio

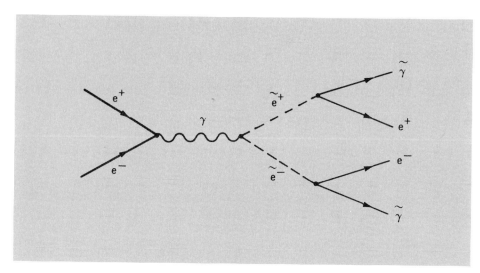

*Fig. 4. A possible interaction involving supersymmetric particles (the selectrons $\tilde{e}^+$ and $\tilde{e}^-$ and the photino $\tilde{\gamma}$) that experimentally would be easily recognizable.*

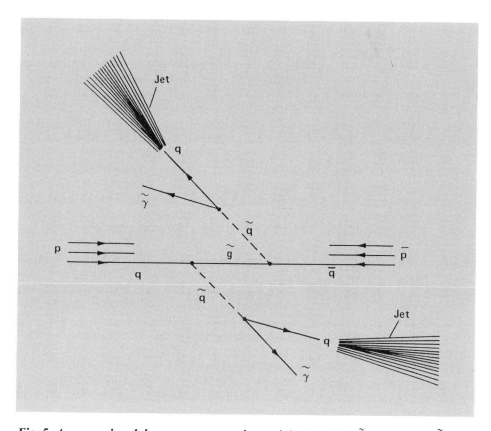

*Fig. 5. A process involving supersymmetric particles (a gluino $\tilde{g}$ and squarks $\tilde{q}$) that generates two hadronic jets.*

$M_{pl}/M_W$, then the new particles associated with supersymmetry will be seen in the next generation of high-energy accelerators.

**The Interactions.** As a result of supersymmetry, the entire low-energy spectrum of particles has been doubled, the masses of the new particles are of order $M_W$, but these masses cannot be predicted with any better accuracy. A reasonable person might therefore ask what properties, if any, *can* we predict. The answer is that we know all the interactions of the new particles with the ordinary ones, of which several examples are shown in Fig. 3. To get an interaction between ordinary and new particles, we can start with an interaction between three ordinary particles and rotate two of these (with a supersymmetry operation) into their supersymmetric partners. The important point is that as a result of supersymmetry the coupling constants remain unchanged.

Since we understand the interactions of the new particles with the ordinary ones, we know how to find these new objects. For example, an electron and a positron can annihilate and produce a pair of selectrons that subsequently decay into an electron-positron pair and two photinos (Fig. 4). This process is easily recognizable and would be a good signal of supersymmetry in high-energy electron-positron colliders.

Supersymmetry is also evident in the process illustrated in Fig. 5. Here one of the three quarks in a proton interacts with one of the quarks in an antiproton; the interaction is mediated by a gluino. The result is the generation of two squarks that decay into quarks and photinos. Because quarks do not exist as free particles, the experimenter should observe two hadronic jets (each jet is a collection of hadrons moving in the same direction as, and as a consequence of, the initial motion of a single quark). The two photinos will generally not interact in the detector, and thus some of the total energy of the process will be "missing".

The theories we have been discussing until now have been a minimal supersymmetric extension of the standard model. There are, however, two further extrapolations that are interesting both theoretically and phenomenologically. The first concerns gravity and the second, grand unified supersymmetry models.

**Gravity**. We have already remarked that supersymmetry may be either a global or a local symmetry. If it is a global symmetry, the Goldstino is massless and the lightest supersymmetric partner. However, if supersymmetry is a local symmetry, it necessarily includes the gravity of general relativity and the Goldstino becomes part of a massive gravitino (the spin-3/2 partner of the graviton) with mass

$$m_G \cong \frac{\Lambda_{ss}^2}{M_{pl}}. \tag{6}$$

With $\Lambda_{ss}$ of order $M_W/\sqrt{\alpha}$ or 1000 GeV, $m_G$ is extremely small ($\sim 10^{-10}$ times the mass of the electron).

Recently it was realized that under certain circumstances $\Lambda_{ss}$ can be much larger than $M_W$, but, at the same time, the perturbative corrections $\delta\mu^2$ can still satisfy the constraint that they be of order $M_W^2$. In these special cases, supersymmetry breaking effects vanish in the limit as some very large mass diverges; that is, we obtain

$$\delta\mu^2 \sim \alpha \left(\frac{\Lambda_{ss}^2}{M_{large}}\right)^2 \tag{7}$$

instead of Eq. 5. An example is already provided by the gravitino mass $m_G$ (where $M_{large} = M_{pl}$). In fact, models have now been constructed in which the gravitino mass is of order $M_W$ and sets the scale of the low-energy supergap $\delta^2$ between bosons and fermions.

In either case (an extremely small or a very large gravitino mass), the observation of a *massive* gravitino is a clear signal of local supersymmetry in nature, that is, the nontrivial extension of Einstein's gravity or supergravity.

**Grand Unification.** Our second extrapolation of supersymmetry has to do with grand

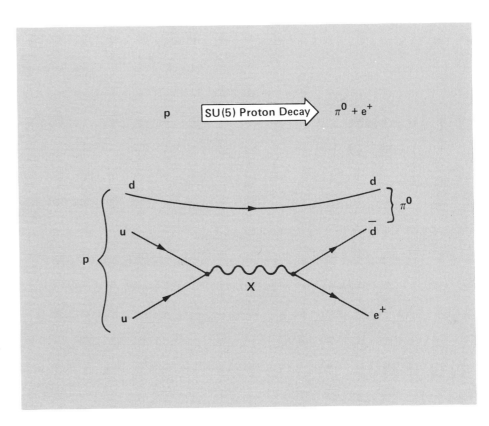

**Fig. 6. The decay mode of the proton predicted by the minimal unification symmetry SU(5). The expected decay products are a neutral pion $\pi^0$ and a positron $e^+$.**

unified theories, which provide a theoretically appealing unification of quarks and leptons and their strong, weak, and electromagnetic interactions. So far there has been one major experimental success for grand unification and two unconfirmed predictions.

The success has to do with the relationship between various coupling constants. In the minimal unification symmetry SU(5), two independent parameters (the coupling constant $g_G$ and the value of the unification mass $M_G$) determine the three independent coupling constants ($g_s$, $g$, and $g'$) of the standard-model SU(3)$\times$SU(2)$\times$U(1) symmetry. As a result, we obtain one prediction, which is typically expressed in terms of the weak-interaction parameter:

$$\sin^2\theta_W = \frac{g'^2}{g'^2 + g^2}. \tag{8}$$

The theory of minimal SU(5) predicts $\sin^2\theta_W = 0.21$, whereas the experimentally observed value is $0.22 \pm 0.01$, in excellent agreement.

The two predictions of SU(5) that have not been verified experimentally are the existence of magnetic monopoles and proton decay. The expected abundance of magnetic monopoles today is crucially dependent on poorly understood processes occurring in the first $10^{-35}$ second of the history of the universe. As a result, if they are not seen, we may ascribe the problem to our poor understanding of the early universe. On the other hand, if proton decay is not observed at the ex-

$$\tau_p \sim \frac{M_G^4}{m_p^5} \frac{\hbar}{c^2} \sim 10^{28 \pm 2} \text{ years} , \qquad (10)$$

where $m_p$ is the proton mass.

Recent experiments, especially sensitive to the decay modes of Eq. 9, have found $\tau_p \geq 10^{32}$ years, in contradiction with the prediction. Hence minimal SU(5) appears to be in trouble. There are, of course, ways to complicate minimal SU(5) so as to be consistent with the experimental values for both $\sin^2\theta_W$ and proton decay. Instead of considering such ad hoc changes, we will discuss the unexpected consequences of making minimal SU(5) globally supersymmetric. The parameter $\sin^2\theta_W$ does not change considerably, whereas $M_G$ increases by an order of magnitude. Hence, the good prediction for $\sin^2\theta_W$ remains intact while the proton lifetime, via the gauge boson exchange process of Fig. 6, naturally increases and becomes unobservable.

It was quickly realized, however, that other processes in supersymmetric SU(5) give the dominant contribution towards proton decay (Fig. 7). The decay products resulting from these processes would consist of $K$ mesons and neutrinos or muons, that is,

$$p \rightarrow K^+ \bar{\nu}_\mu \text{ or } K^0 \mu^+ , \qquad (11)$$

and so would differ from the *expected* decay products of $\pi$ mesons and positrons. This is very exciting because detection of the products of Eq. 11 not only may signal nucleon decay but also may provide the first signal of supersymmetry in nature. Experiments now running have all seen candidate events of this type. These events are, however, consistent with background. It may take several more years before a signal rises up above the background.

**Experiments.** An encouraging feature of the theory is that low-energy supersymmetry can be verified in the next ten years, possibly as early as next year with experiments now in progress at the CERN proton-antiproton collider.

Experimenters at CERN recently dis-

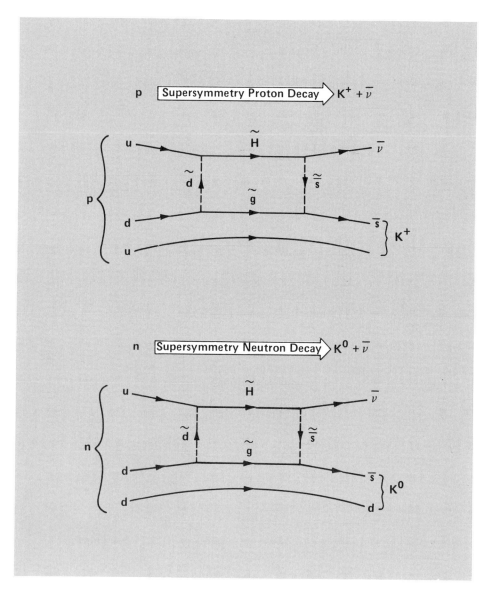

**Fig. 7. The dominant proton-decay and neutron-decay modes predicted by supersymmetry. The expected decay products are K mesons ($K^+$ and $K^0$) and neutrinos ($\bar{\nu}$).**

pected rate, then minimal SU(5) is in serious trouble.

The dominant decay modes predicted by minimal SU(5) for the nucleons are

$$p \rightarrow \pi^0 e^+$$

and

$$n \rightarrow \pi^- e^+ . \qquad (9)$$

These processes involve the exchange of a so-called $X$ or $Y$ boson with mass of order $M_G$ (Fig. 6), so that the predicted proton lifetime $\tau_p$ is

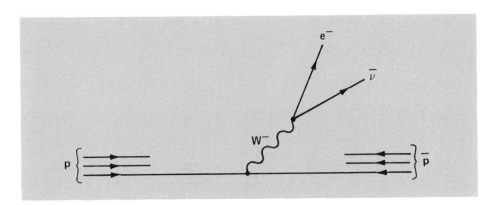

Fig. 8. The generation, in a high-energy proton-antiproton collision, of a W⁻ particle, which then decays into an electron (e⁻) and an antineutrino ($\bar{\nu}$).

covered the $W^\pm$ and $Z^0$ bosons, mediators of the weak interactions, and produced many of these bosons in high-energy collisions between protons and antiprotons (each with momentum $\sim 270$ GeV/$c$). For example, Fig. 8 shows the process for the generation of a $W^-$ boson, which then decays to a high-energy electron (detectable) and a high-energy neutrino (not detectable). A single electron with the characteristic energy of about 42 GeV was a clear signature for this process.

Let us now consider some of the signatures of supersymmetry for $p\bar{p}$ or $pp$ colliders. A clear signal for supersymmetry are multi-jet events with missing energy. For example, events containing one, two, three, or four hadronic jets and nothing more can be interpreted as a signal for either squark or gluino production (Figs. 5 and 9). A two- or four-jet signal is canonical, but these events can look like one- or three-jet events some fraction of the time.

There may also be events with two jets, a high-energy electron, and some missing energy. This is the characteristic signature of top quark production via $W$ decay (Fig. 10), and thus such events may be evidence for top quarks. But there is also an event predicted by supersymmetry with the same signature, namely, the production of a squark pair (Fig. 11). It would require many such events to disentangle these two possibilites.

The CERN proton-antiproton collider began taking more data in September 1984 with momentum increased to 320 GeV/$c$ per beam and with increased luminosity. No clear evidence for supersymmetric partners has been observed. As a result, the so-called UA-1 Collaboration at CERN has put lower limits on gluino and squark masses of approximately 60 and 80 GeV, respectively. As of this writing it is apparent that the discovery of supersymmetric partners, and perhaps also the top quark, must wait for the next generation of high-energy accelerators.

Hopefully, it will not be too long before we learn whether or not the underlying structure of the universe possesses this elegant, highly unifying type of symmetry. ■

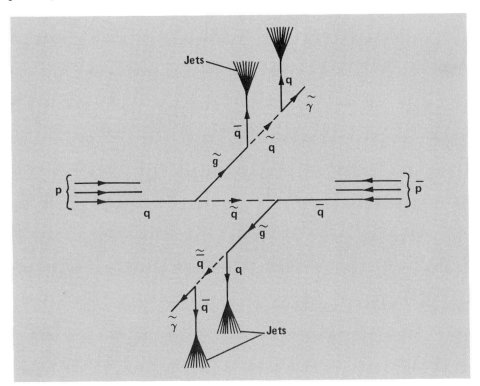

Fig. 9. A proton-antiproton collision involving supersymmetric particles (gluinos $\tilde{g}$, squarks $\tilde{q}$, antiquarks $\tilde{\bar{q}}$, and photinos $\tilde{\gamma}$) that generates four hadronic jets.

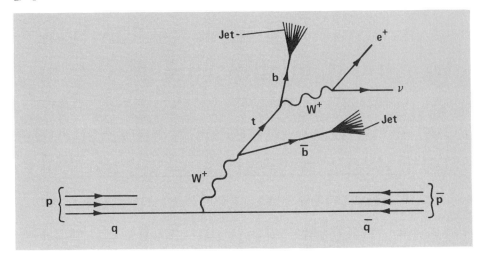

Fig. 10. Two-jet events observed by the UA-1 Collaboration at CERN can be interpreted, as shown here, as a process involving top quark t production.

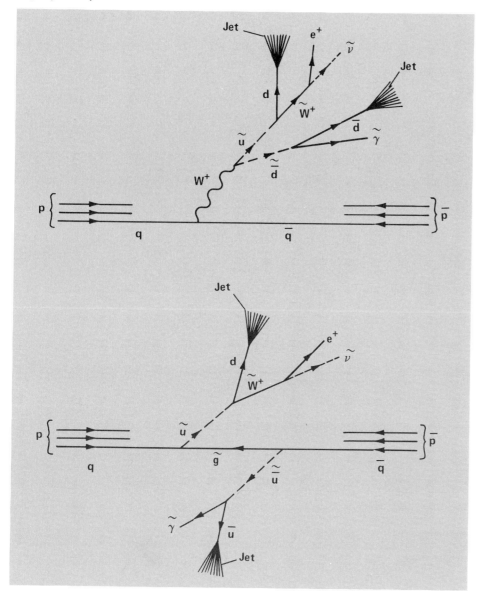

*Fig. 11. The same event discussed in Fig. 10, only here interpreted as a supersymmetric process involving squarks and antisquarks.*

## Further Reading

Daniel Z. Freedman and Peter van Nieuwenhuizen, "Supergravity and the Unification of the Laws of Physics." *Scientific American* (February 1978):126-143.

**Stuart A. Raby** did his undergraduate work at the University of Rochester, receiving his B.Sc. in physics in 1969. Stuart spent six years in Israel as a student/teacher, receiving a M.Sc. in physics from Tel Aviv University in 1973 and a Ph.D. in physics from the same institution in 1976. Upon graduating, he took a Research Associate position at Cornell University. From 1978 to 1980, Stuart was Acting Assistant Professor of physics at Stanford University and then moved over to a three-year assignment as Research Associate at the Stanford Linear Accelerator Center. He came to the Laboratory as a Temporary Staff Member in 1981, cutting short his SLAC position, and became a Staff Member of the Elementary Particles and Field Theory Group of Theoretical Division in 1982. He has recently served as Visiting Associate Research Scientist for the University of Michigan. He and his wife Michele have two children, Eric and Liat.

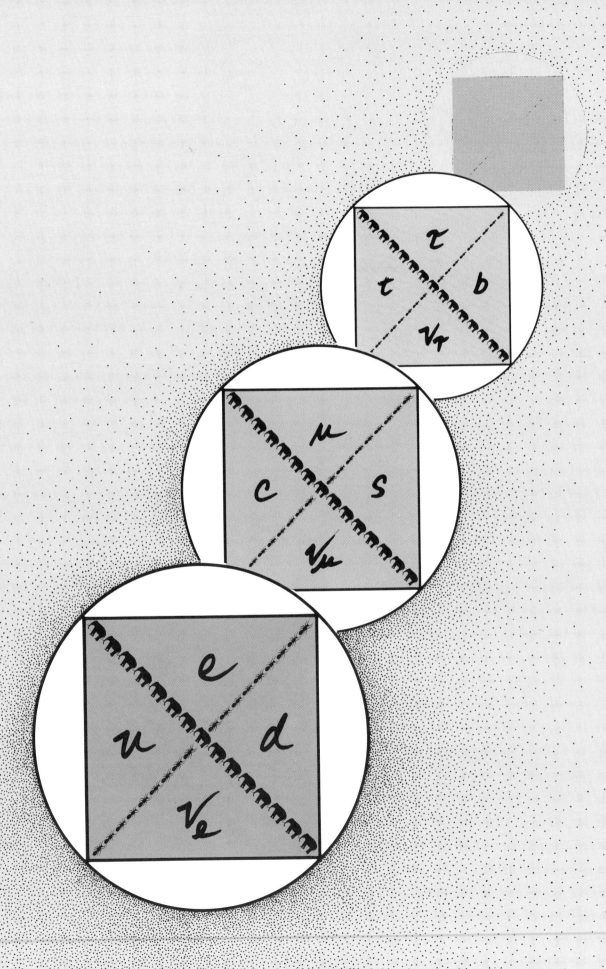

# The Family Problem

*by T. Goldman and Michael Martin Nieto*

*The roster of elementary particles includes replicas, exact in every detail but mass, of those that make up ordinary matter. More facts are needed to explain this seemingly unnecessary extravagance.*

The currently "standard" model of particle physics phenomenologically describes virtually all of our observations of the world at the level of elementary particles (see "Particle Physics and the Standard Model"). However, it does not explain them with any depth. Why is $SU(3)_C$ the gauge group of the strong force? Why is the symmetry of the electroweak force broken? Where does gravity fit in? How can all of these forces be unified? That is, from what viewpoint will they appear as aspects of a common, underlying principle? These questions lead us in the directions of supersymmetry and of grand unification, topics discussed in "Toward a Unified Theory."

Yet another feature of the standard model leaves particle physicists dissatisfied: the multiple repetitions of the representations* of the particles involved in the gauge interactions. By definition the adjoint representation† of the gauge fields must occur precisely once in a gauge theory. However, quantum chromodynamics includes no less than six occurrences of the color triplet representation of quarks: one for each of the *u*, *c*, *t*, *d*, *s*, and *b* quarks. The *u*, *c*, and *t* quarks have a common electric charge of ⅔ and so are distinguished from the *d*, *s*, and *b* quarks, which have a common electric charge of −⅓. But the quarks with a common charge are distinguished only by their dif-

ferent masses, as far as is now known. The electroweak theory presents an even worse situation, being burdened with nine left-chiral‡ quark doublets, three left-chiral lepton doublets, eighteen right-chiral quark singlets, and three right-chiral lepton singlets (Fig. 1).

Nonetheless, some organization can be discerned. The exact symmetry of the strong and electromagnetic gauge interactions, together with the nonzero masses of the quarks and charged leptons, implies that the right-chiral quarks and charged leptons and their left-chiral partners can be treated as single objects under these interactions. In addition, each neutral lepton is associated with a particular charged lepton, courtesy of the transformations induced by the weak interaction. Thus, it is natural to think in terms of three quark sets (*u* and *d*, *c* and *s*, and *t* and *b*) and three lepton sets ($e^-$ and $v_e$, $\mu^-$ and $v_\mu$, and $\tau^-$ and $v_\tau$) rather than thirty-three quite repetitive representations. Furthermore, the relative lightness of the *u* and *d* quark set and of the $e^-$ and $v_e$ lepton set long ago suggested to some that the quarks and leptons are also related (quark-lepton symmetry). Subtle mathematical properties of modern gauge field theories have provided new backing for this notion of three "quark-lepton families," each consisting of successively heavier quark and lepton sets (Table 1).

---

*We give a geometric definition of "representation," using as an example the $SU(3)_C$ triplet representation of, say, the up quark. (This triplet, the smallest non-singlet representation of $SU(3)_C$, is called the fundamental representation.) The members of this representation ($u_{red}$, $u_{blue}$, and $u_{green}$) correspond to the set of three vectors directed from the origin of a two-dimensional coordinate system to the vertices of an equilateral triangle centered at the origin. (The triangle is usually depicted as standing on a vertex.) The "conjugate" of the triplet representation, which contains the three anticolor varieties of the up quark with charge −⅔, can be defined similarly: it corresponds to the set of three vectors obtained by reflecting the vectors of the triplet representation through the origin. (The vectors of the conjugate representation are directed toward the vertices of an equilateral triangle standing on its side, like a pyramid.) The "group transformations" correspond to the set of operations by which any one of the quark or antiquark vectors is transformed into any other.*

---

†*The "adjoint" representation of $SU(3)_C$, which contains the eight vector bosons (the gluons), is found in the "product" of the triplet representation and its conjugate. This product corresponds to the set of nine vectors obtained by forming the vector sums of each member of the triplet representation with each member of its conjugate. This set can be decomposed into a singlet containing a null vector (a point at the origin) and an octet, the adjoint representation, containing two null vectors and six vectors directed from the origin to the vertices of a regular hexagon centered at the origin. Note that the adjoint representation is symmetric under reflection through the origin.*

---

‡*A massless particle is said to be left-handed (right-handed) if the direction of its spin vector is opposite (the same as) that of its momentum. Chirality is the Lorentz-invariant generalization of this handedness to massive particles and is equivalent to handedness for massless particles.*

If the underlying significance of this grouping by mass is not apparent to the reader, neither is it to particle physicists. No one has put forth any compelling reason for deciding which charge ⅔ quark and which charge −⅓ quark to combine into a quark set or for deciding which quark set and which charged and neutral lepton set should be combined in a quark-lepton family. Like Mendeleev, we are in possession of what appears to be an orderly grouping but without a clue as to its dynamical basis. This is one theme of "the family problem."

Still, we do refer to each quark and lepton set together as a family and thus reduce the problem to that of understanding only three families—unless, of course, there are more families as yet unobserved. This last is another question that a successful "theory of families" must answer. Grand unified theories, supersymmetry theories, and theories wherein quarks and leptons have a common substructure can all accommodate quark-lepton symmetry but as yet have not provided convincing predictions as to the number of families. (These predictions range from any even number to an infinite spectrum.)

Such concatenations of wild ideas (however intriguing) may not be the best approach to solving the family problem. A more conservative approach, emulating that leading to the standard model, is to attack the family problem as a separate question and to ask directly if the different families are dynamically related.

Here we face a formidable obstacle—a paucity of information. A fermion from one family has never been observed to change into a fermion from another family. Table 2 lists some family-changing decays that have been sought and the experimental limits on their occurrence. True, a $\mu^-$ may appear to decay into an $e^-$, but, as has been experimentally confirmed, it actually is transformed into a $\nu_\mu$, and simultaneously the $e^-$ and a $\bar{\nu}_e$ appear. Being an antiparticle, the $\bar{\nu}_e$ carries the opposite of whatever family quantum numbers distinguish an $e^-$ from any other charged lepton. Thus, no net "first-familiness" is created, and the "second-familiness"

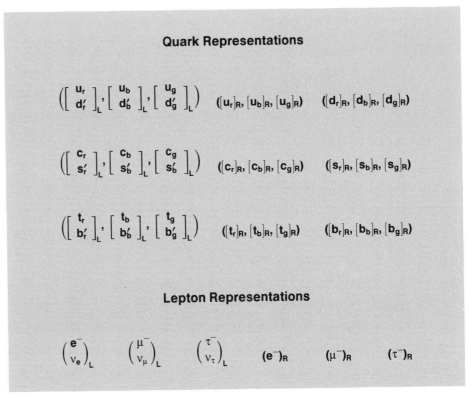

Fig. 1. The electroweak representations of the fermions of the standard model, which comprise nine left-chiral quark doublets, eighteen right-chiral quark singlets, three left-chiral lepton doublets, and three right-chiral lepton singlets. The subscripts r, b, and g denote the three color charges of the quarks, and the subscripts R and L denote right- and left-chiral projections. The symbols d', s', and b' indicate weak-interaction mass eigenstates, which, as discussed in the text, are mixtures of the strong-interaction mass eigenstates d, s, and b. Since quantum chromodynamics does not include the weak interaction, and hence is not concerned with chirality, the $SU(3)_C$ representations of the fermions are fewer in number: six triplets, each containing the three color-charge varieties of one of the quarks, and three singlets, each containing a charged lepton and its associated neutral lepton.

of the original $\mu^-$ is preserved in the $\nu_\mu$.

In spite of the lack of positive experimental results, current fashions (which are based on the successes of the standard model) make irresistible the temptation to assign a family symmetry group to the three known families. Some that have been considered include $SU(2)$, $SU(2) \times U(1)$, $SU(3)$, and $U(1) \times U(1) \times U(1)$. The impoverished level of our understanding is apparent from the $SU(2)$ case, in which we cannot even determine whether

the three families fall into a doublet and a singlet or simply form a triplet.

The clearest possible prediction from a family symmetry group, analogous to Mendeleev's prediction of new elements and their properties, would be the existence of one or more additional families necessary to complete a representation. Such a prediction can be obtained most naturally from either of two possibilities for the family symmetry: a spontaneously broken local gauge symmetry

## Table 1

Members of the three known quark-lepton families and their masses. Each family contains one particle from each of the four types of fermions: leptons with an electric charge of −1 (the electron, the muon, and the tau); neutral leptons (the electron neutrino, the muon neutrino, and the tau neutrino); quarks with an electric charge of ⅔ (the up, charmed, and top quarks); and quarks with an electric charge of −⅓ (the down, strange, and bottom quarks). Each family also contains the antiparticles of its members. (The antiparticles of the charged leptons are distinguished by opposite electric charge, those of the neutral leptons by opposite chirality, and those of the quarks by opposite electric and color charges. For historical reasons only the antielectron has a distinctive appellation, the positron.) Family membership is determined by mass, with the first family containing the least massive example of each type of fermion, the second containing the next most massive, and so on. What, if any, dynamical basis underlies this grouping by mass is not known, nor is it known whether other heavier families exist. The members of the first family dominate the ordinary world, whereas those of the second and third families are unstable and are found only among the debris of collisions between members of the first family.

| First Family | Second Family | Third Family |
|---|---|---|
| electron, $e^-$ <br> 0.5ll MeV/$c^2$ | muon, $\mu^-$ <br> 105.6 MeV/$c^2$ | tau, $\tau^-$ <br> 1782 MeV/$c^2$ |
| electron neutrino, $\nu_e$ <br> 0.00002 MeV/$c^2$ (?) | muon neutrino, $\nu_\mu$ <br> $\leq 0.5$ MeV/$c^2$ | tau neutrino, $\nu_\tau$ <br> $\leq 147$ MeV/$c^2$ |
| up quark, $u$ <br> $\simeq 5$ MeV/$c^2$ | charmed quark, $c$ <br> $\simeq 1500$ MeV/$c^2$ | top quark, $t$ <br> $\geq 40{,}000$ MeV/$c^2$ (?) |
| down quark, $d$ <br> $\simeq 10$ MeV/$c^2$ | strange quark, $s$ <br> $\simeq 170$ MeV/$c^2$ | bottom quark, $b$ <br> $\simeq 4500$ MeV/$c^2$ |

## Table 2

Experimental limits on the branching ratios for some family-changing decays. The branching ratio for a particular decay mode is defined as the ratio of the number of decays by that mode to the total number of decays by all modes. An experiment capable of determining a branching ratio for $\mu^+ \to e^+\gamma$ as low as $10^{-12}$ is currently in progress at Los Alamos (see "Experiments To Test Unification Schemes").

| Decay Mode | Branching Ratio (upper bound) | Dominant Decay Mode(s) |
|---|---|---|
| $\mu^+ \to e^+\gamma$ | $10^{-10}$ | $\mu^+ \to e^+\nu_e\bar{\nu}_\mu$ |
| $\mu^+ \to e^+e^+e^-$ | $10^{-12}$ | $\mu^+ \to e^+\nu_e\bar{\nu}_\mu$ |
| $\pi^0 \to \mu^\pm e^\mp$ | $10^{-7}$ | $\pi^0 \to \gamma\gamma$ |
| $K^+ \to \pi^+\mu^\pm e^\mp$ | $10^{-8}$ | $K^+ \to \pi^+\pi^0$ or $\mu^+\nu_\mu$ |
| $K_L \to \mu^\pm e^\mp$ | $10^{-8}$ | $K_L \to \pi^+\pi^-\pi^0$ or $\pi^0\pi^0\pi^0$ |
| $\Sigma^+ \to pu^\pm e^\mp$ | $10^{-5}$ | $\Sigma^+ \to p\pi^0$ |

or a spontaneously broken global symmetry.* What follows is a brief ramble (whose course depends little on detailed assumptions) through the salient features and implications of these two possibilities.

## Family Gauge Symmetry

All of the unseen decays listed in Table 2 would be strictly forbidden if the family gauge symmetry were an exact gauge symmetry as those of quantum electrodynamics and quantum chromodynamics are widely believed to be. Here, however, we do not expect exactness because that would imply the existence, contrary to experience, of an additional fundamental force mediated by a massless vector boson (such as a long-range force like that of the photon or a strong force like that of the gluons but extending to leptons as well as quarks). But we can, as in the standard model, assume a *broken* gauge symmetry.

We begin by placing one or more families in a representation of some family gauge symmetry group. (The correct group might be inferred from ideas such as grand unification or compositeness of fermions. However, it is much more likely that, as in the case of the standard model, this decision will best be guided by hints from experimental observations.) Together, the group and the representation determine currents that describe interactions between members of the representation. (These currents would be conserved if the family symmetry were exact.) For example, if the first and the second families are placed in the representation, an electrically neutral current describes the transformation $e^- \leftrightarrow \mu^-$, just as the charged weak current of the electroweak theory describes the transformation $e^- \leftrightarrow \nu_e$. Since the other family

*In principle, we should also consider the possibilities of a discrete symmetry or an explicit breaking of family symmetry (probably caused by some dynamics of a fermion substructure). However, these ideas would be radical departures from the gauge symmetries that have proved so successful to date. We will not pursue them here.

members necessarily fall into the same representation, the $e^- \leftrightarrow \mu^-$ current includes contributions from interactions between these other members ($d \leftrightarrow s$, for example), just as the charged weak current for $e^- \leftrightarrow \nu_e$ includes contributions from $\mu^- \leftrightarrow \nu_\mu$ and $\tau^- \leftrightarrow \nu_\tau$.

If we now allow the family symmetry to be a local gauge symmetry, we find a "family vector boson," $F$, that couples to these currents (Fig. 2) and mediates the family-changing interactions. As in the standard model, the coupled currents can be combined to yield dynamical predictions such as scattering amplitudes, decay rates, and relations between different processes.

**Scale of Family Gauge Symmetry Breaking.** Weak interactions occur relatively infrequently compared to electromagnetic and strong interactions because of the large dynamical scale (approximately 100 GeV) set by the masses of the $W^\pm$ and $Z^0$ bosons that break the electroweak symmetry. We can interpret the extremely low rate of family-changing interactions as being due to an analogous but even larger dynamical scale associated with the breaking of a local family gauge symmetry, that is, to a large value for the mass $M_F$ of the family vector boson. The branching-ratio limit listed in Table 2 for the reaction $K_L \rightarrow \mu^\pm + e^\mp$ allows us to estimate a lower bound for $M_F$ as follows.

Like the weak decay of muons, the $K_L \rightarrow \mu e$ decay proceeds through formation of a virtual family vector boson (Fig. 3). The rate for the decay, $\Gamma$, is given by

$$\Gamma \cong \frac{g_{\text{family}}^4}{M_F^4} \, m_K^5 . \tag{1}$$

Note that the fourth power of $M_F$ appears in Eq. 1 just as the fourth power of $M_W$ does (hiding in the square of the Fermi constant) in the rate equation for muon decay. (Certain chirality properties of the family interaction could require that two of the five powers of the kaon mass $m_K$ in Eq. 1 be replaced by the muon mass. However, since the inferred

value of $M_F$ varies as the fourth root of this term, the change would make little numerical difference.) It is usual to assume that $g_{\text{family}}$, the family coupling constant, is comparable in magnitude to those for the weak and electromagnetic interactions. This assumption reflects our prejudice that family-changing interactions may eventually be unified with those interactions. Using Eq. 1 and the branching-ratio limit from Table 2, we obtain

$$M_F \gtrsim 10^5 \, \text{GeV}/c^2 . \tag{2}$$

Such a large lower bound on $M_F$ implies that the breaking of a local family gauge symmetry produces interactions much weaker than the weak interactions.

Alternatively, processes like $K_L \rightarrow \mu e$ may be the result of family-conserving grand unified interactions in which quarks are turned into leptons. However, the experimental limit on the rate of proton decay implies that such interactions occur far less frequently than the family-violating interactions considered here.

Experiments with neutrinos, also, indicate a similarly large dynamical scale for the breaking of a local family gauge symmetry. A search for the radiative decay $\nu_\mu \rightarrow \nu_e + \gamma$ has yielded a lower bound on the $\nu_\mu$ lifetime of $10^5 \, (m_\nu/\text{MeV})$ seconds. If the mass of the muon neutrino is near its experimentally observed upper bound of 0.5 MeV/$c^2$, this lower bound on the lifetime is greater than the standard-model prediction of approximately $10^3 \, (\text{MeV}/m_\nu)^5$ seconds. Thus, some family-conservation principle may be suppressing the decay.

More definitive information is available from neutrino-scattering experiments. Positive pions decay overwhelmingly ($10^4$ to 1) into positive muons and muon neutrinos. In the absence of family-changing interactions, scattering of these neutrinos on nuclear targets should produce only negative muons. This has been accurately confirmed: neither positrons nor electrons appear more frequently than permitted by the present systematic experimental uncertainty of 0.1 per-

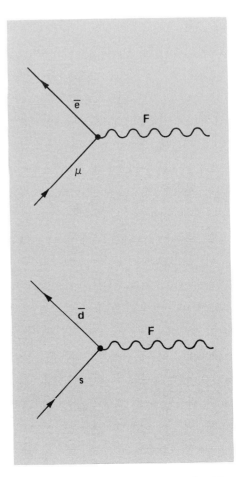

**Fig. 2. Examples of neutral family-changing currents coupled to a family vector boson (F). Such couplings follow from the assumption of a local gauge symmetry for the family symmetry.**

cent. An investigation of the neutrinos from muon decay has yielded similar results. The decay of a positive muon produces, in addition to a positron, an electron neutrino and a muon antineutrino. Again, in the absence of family-changing interactions, scattering of these neutrinos should produce only electrons and positive muons, respectively. A LAMPF experiment (E-31) has shown, with an uncertainty of about 5 percent, that no negative muons or positrons are produced.

The energy scale of Eq. 2 will not be directly accessible with accelerators in the

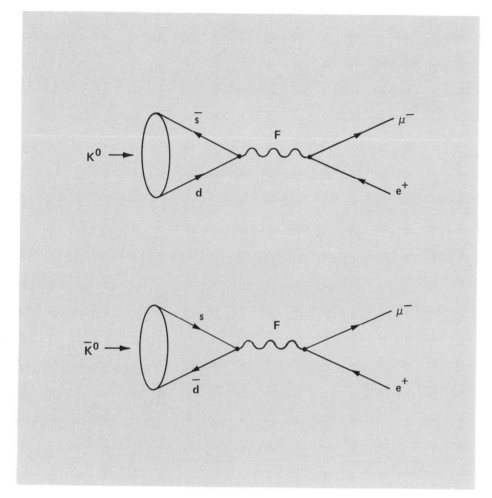

they are to be carried out by creatures with finite lifetimes!

For example, consider again the decay $K_L \to \mu e$. Since the rate of this decay varies inversely as the fourth power of the mass of the family vector boson, a value of $M_F$ in the million-GeV range implies a branching ratio lower by four orders of magnitude than the present limit. A search for so rare a decay would be quite feasible at a high-intensity, **medium-energy accelerator (such as the proposed LAMPF II) that is designed to produce kaon fluxes on the order of $10^8$ per second.** (Currently available kaon fluxes are on the order of $10^6$ per second.) A typical solid angle times efficiency factor for an in-flight decay experiment is on the order of 10 percent. Thus, $10^7$ kaons per second could be examined for the decay mode of interest. A branching ratio larger than $10^{-12}$ could be found in a one-day search, and a year-long experiment would be sensitive down to the $10^{-14}$ level. Of course, we do not know with absolute certainty whether a positive signal will be found at any level. Nonetheless, the need for such an observation to elucidate family dynamics impels us to make the attempt.

## Positive Evidence for Family Symmetry Breaking

Thus, despite expectations to the contrary, we have at present no positive evidence in any neutral process for nonconservation of a family quantum number, that is, for family-changing interactions mediated by exchange of an electrically neutral vector boson such as the $F$ of Figs. 2 and 3. Is it possible that our expectations are wrong—that this quantum number is exactly conserved as are electric charge and angular momentum? The answer is an unequivocal NO! We have—for quarks—positive evidence that family is a *broken* symmetry. To see this, we must examine the effect of the electroweak interaction on the quark mass eigenstates defined by the strong interaction.

We know, for instance, that a $K^+ (= u + \bar{s})$ decays by the weak interaction into a $\mu^+$ and

*Fig. 3. Feynman diagram for the family-changing decay $K_L \to \mu^- + e^+$, which is assumed to occur through formation of a virtual family vector boson (F). The $K_L$ meson is the longer lived of two possible mixtures of the neutral kaon ($K^0$) and its antiparticle ($\bar{K}^0$). Neither this decay nor the equally probable decay $K_L \to \mu^+ + e^-$ has been observed experimentally; the current upper bound on the branching ratio is $10^{-8}$.*

foreseeable future. The Superconducting Super Collider, which is currently being considered for construction next decade, is conceived of as reaching 40,000 GeV but is estimated to cost several billion dollars. We cannot expect something yet an order of magnitude more ambitious for a very long time. Thus, further information about the breaking of a local family gauge symmetry will not arise from a brute force approach but

rather, as it has till now, from discriminating searches for the needle of a rare event among a haystack of ordinary ones. Clearly, the larger the total number of events examined, the more definitive is the information obtained about the rate of the rare ones. For this reason the availability of high-intensity beams of the reacting particles is a very important factor in the experiments that need to be undertaken or refined, given that

a $\nu_\mu$ and also decays into a $\pi^+$ and a $\pi^0$ (Fig. 4). In quark terms this means that the $u$ quark and the $\bar{s}$ quark in the kaon are coupled through a $W^+$ boson. The two families (up-down and charmed-strange) defined by the quark mass eigenstates under the strong interaction are mixed by the weak interaction. Since the kaon decays occur in both purely leptonic and purely hadronic channels, they are not likely to be due to peculiar quark-lepton couplings. Similar evidence for family violation is found in the decays of $D$ mesons, which contain charmed quarks.

Weak-interaction eigenstates $d'$ and $s'$ may be defined in terms of the strong-interaction mass eigenstates $d$ and $s$ by

$$\begin{pmatrix} d' \\ s' \end{pmatrix} = \begin{pmatrix} \cos\theta_C & \sin\theta_C \\ -\sin\theta_C & \cos\theta_C \end{pmatrix} \begin{pmatrix} d \\ s \end{pmatrix}, \quad (3)$$

where $\theta_C$, the Cabibbo mixing angle, is experimentally found to be the angle whose sine is $0.23 \pm 0.01$. (The usual convention, which entails no loss of generality, is to assign all the mixing effects of the weak interaction to the down and strange quarks, leaving unchanged the up and charmed quarks.) The fact that the mass and weak-interaction eigenstates are different implies that a conserved family quantum number cannot be defined in the presence of both the strong and the weak interactions. We can easily show, however, that this conclusion does not contradict the observed absence of *neutral* family-violating interactions.

The weak charged-current interaction describing, say, the transformation of a $d'$ quark into a $u$ quark by absorption of a $W^+$ boson has the form

$$(\bar{u}d' + \bar{c}s')W^+ = (\bar{u}, \bar{c}) \begin{pmatrix} W^+ & 0 \\ 0 & W^+ \end{pmatrix} \begin{pmatrix} d' \\ s' \end{pmatrix}, \quad (4)$$

which, after substitution of Eq. 3, becomes

$$(\bar{u}d' + \bar{c}s')W^+ = \bar{u}(d\cos\theta_C + s\sin\theta_C)W^+ \\ + \bar{c}(-d\sin\theta_C + s\cos\theta_C)W^+. \quad (5)$$

(Here we suppress details of the Lorentz algebra.)

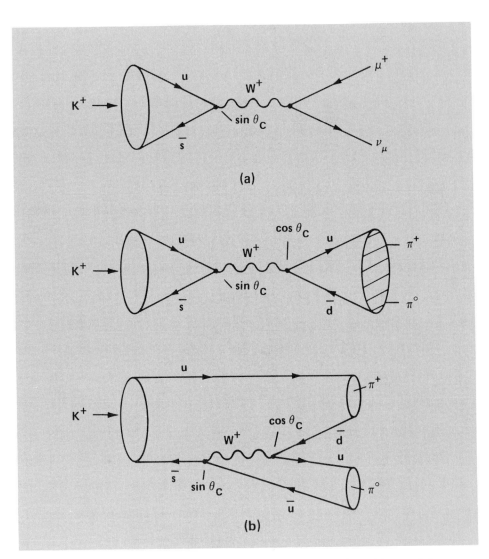

**Fig. 4. Feynman diagrams for the decays of a positive kaon into (a) a positive muon and a muon neutrino and (b) a positive and a neutral pion. The ellipse with diagonal lines represents any one of several possible pathways for production of a positive and a neutral pion from an up quark and an antidown quark. These decays, in which the up-down and charmed-strange quark families are mixed by the weak interaction (as indicated by $\sin\theta_C$ and $\cos\theta_C$), are evidence that the family symmetry of quarks is a broken symmmetry.**

Because of the mixing given by Eq. 3, the statement we made near the beginning of this article, that no family-changing decays have been observed, must be sharpened. True, no $s' \to u$ decay has been seen, but, of course, the $s \to u$ decay implied by Eq. 5 does occur.

Thus, "No family-changing decays of weak-interaction family eigenstates have been observed" is the more precise statement.

The weak neutral-current interaction describing the scattering of a $d'$ quark when it absorbs a $Z^0$ has a form like that of Eq. 4:

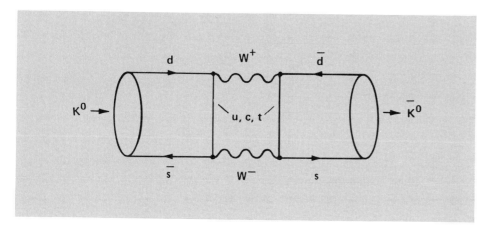

**Fig. 5. Feynman diagram for a CP-violating reaction that transforms the neutral kaon into its antiparticle. This second-order weak interaction occurs through formation of virtual intermediate states including either a u, c, or t quark.**

$$(\bar{d}'d' + \bar{s}'s')Z^0 = (\bar{d}', \bar{s}')\begin{pmatrix} Z^0 & 0 \\ 0 & Z^0 \end{pmatrix}\begin{pmatrix} d' \\ s' \end{pmatrix}. \tag{6}$$

Since the Cabibbo matrix in Eq. 3 is unitary, Eq. 6 is unchanged (except for the disappearance of primes on the quarks) by substitution of Eq. 3:

$$(\bar{d}'d' + \bar{s}'s')Z^0 = (\bar{d}d + \bar{s}s)Z^0 . \tag{7}$$

Thus, the weak neutral-current interaction does not change $d$ quarks into $s$ quarks anymore than it changes $d'$ quarks into $s'$ quarks. It is only the presumed family vector boson of mass greater than $10^5$ GeV that may effect such a change.

## Family Symmetry Violation and CP Violation

The combined operation of charge conjugation and parity reversal (CP) is, like parity reversal alone, now known not to be an exact symmetry of the world. An understanding of CP violation and proton decay would be of universal importance to explain "big-bang" cosmology and the observed excess of matter over antimatter.

The generalization by Kobayashi and Maskawa of Eq. 3 to the three-family case is introduced in "Particle Physics and the Standard Model"; it yields a relation between family symmetry violation and CP violation. Although other sources of CP violation may exist outside the standard model, this relation permits extraction of information about violation of family symmetry from studies of CP violation.

The phenomenon of CP violation has, so far, been observed only in the $K^0$-$\bar{K}^0$ system. The CP eigenstates of this system are the sum and the difference of the $K^0$ and $\bar{K}^0$ states. The violation is exhibited as a small tendency for the long-lived state, $K_L$, which normally decays into three pions, to decay into two pions (the normal decay mode of the short-lived state, $K_S$) with a branching ratio of approximately $10^{-3}$. This tendency can be described by saying that the $K_S$ and $K_L$ states differ from the sum and difference states by a mixing of order $\varepsilon$:

$$|K_S\rangle \cong |K^0\rangle + (1-\varepsilon)|\bar{K}^0\rangle$$

and

$$\tag{8}$$

$$|K_L\rangle \cong |K^0\rangle - (1-\varepsilon)|\bar{K}^0\rangle .$$

The quark-model analysis based on the work of Kobayashi and Maskawa and the second-order weak interaction shown in Fig. 5 predict an additional CP-violating effect not describable in terms of the mixing in Eq. 8; that is, it would occur even if $\varepsilon$ were zero. The effect, which is predicted to be of order $\varepsilon'$, where $\varepsilon'/\varepsilon$ is about $10^{-2}$, has not yet been observed, but experiments sufficiently sensitive are being mounted.

Both $\varepsilon$ and $\varepsilon'$ are related to the Kobayashi-Maskawa parameters that describe family symmetry violation. This guarantees that if the value of $\varepsilon'$ is found to be in the expected range, higher precision experiments will be needed to determine its exact value . If no positive result is obtained in the present round of experiments, it will be even more important to search for still smaller values. In either case intense kaon beams are highly desirable since the durations of such experiments are approaching the upper limit of reasonability.

Of course, in principle, CP violation can be studied in other quark systems involving the heavier $c$, $b$, and $t$ quarks. However, these are produced roughly $10^8$ times less copiously than are kaons, and the CP-violating effects are not expected to be as large as in the case of kaons.

## Global Family Symmetry

In our discussion of family-violating processes like $K \rightarrow \mu e$, we have, so far, assumed the existence of a massive gauge vector boson reflecting family dynamics. The general theorem, due to Goldstone, offers two mutually exclusive possibilities for the realization of a broken symmetry in field theory. One is the development of just such a massive vector boson from a massless one; the other is the absence of any vector boson and the appearance of a massless scalar boson, or Goldstone boson. The possible Goldstone boson associated with family symmetry has been called the familon and is denoted by $f$. As is generally true for such scalar bosons, the strength of its coupling falls inversely with the mass scale of the symmetry breaking. Cosmological argu-

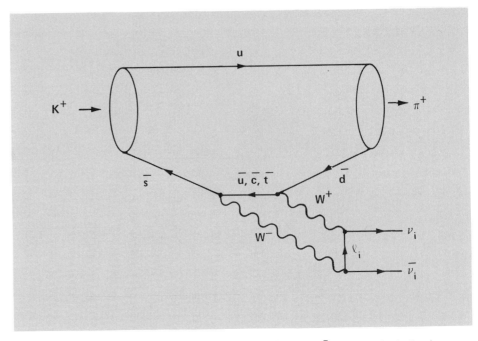

*Fig. 6. Feynman diagram for the decay* $K^+ \rightarrow \pi^+ + \nu_i + \bar{\nu}_i$, *where the index* i *covers all neutrino types light enough to appear in the reaction. The symbol* $\ell_i$ *stands for the charged lepton associated with* $\nu_i$ *and* $\bar{\nu}_i$.

ments suggest a lower bound on the coupling of approximately $10^{-12}$ GeV$^{-1}$, a value very near (within three orders of magnitude) the upper bound determined from particle-physics experiments.

The familon would appear in the two-body decays $\mu \rightarrow e + f$ and $s \rightarrow d + f$. The latter can be observed in the decay $K^+ (= u + \bar{s}) \rightarrow \pi^+ (= u + \bar{d}) + $ *nothing else seen*. The familon would not be seen because it is about as weakly interacting as a neutrino. The only signal that the decay had occurred would be the appearance of a positive pion at the kinematically determined momentum of 227 MeV/$c$.

Such a search for evidence of the familon would encounter an unavoidable background of positive pions from the reaction $K^+ \rightarrow \pi^+ + \nu_i + \bar{\nu}_i$, where the index $i$ covers all neutrino types light enough to appear in the reaction. This decay mode occurs through a one-loop quantum-field correction to the electroweak theory (Fig. 6) and is interesting in itself for two reasons. First, it depends on a different combination of the parameters involved in CP violation and on the number $N_\nu$ of light neutrino types. Since $N_\nu$ is expected to be determined in studies of $Z^0$ decay, an uncertainty in the value of a matrix element in the standard-model prediction of the $K^+ \rightarrow \pi^+ \nu_i \bar{\nu}_i$ branching ratio can be eliminated. Present estimates place the branching ratio in the range between $10^{-9}$ and $10^{-10}$ times $N_\nu$. Second, a discrepancy with the $N_\nu$ value determined from decay of the $Z^0$, which is heavier than the kaon, would be evidence for the existence of at least one neutrino with a mass greater than about 200 MeV/$c^2$.

## Fermion Masses and Family Symmetry Breaking

The mass spectrum of the fermions is itself unequivocal evidence that family symmetry is broken. These masses, which are listed in Table 1, should be compared to the $W^\pm$ and $Z^0$ masses of 83 and 92 GeV/$c^2$, respectively, which set the dynamical scale of electroweak interactions. (The masses quoted are the theoretical values, which agree well with the recently measured experimental values.) The very existence of the fermion masses violates electroweak symmetry by connecting doublet and singlet representations, and the variations in the pattern of mass splittings within each family show that family symmetry is broken. But since we neither know the mass scale nor understand the pattern of the family symmetry breaking, we do not really know the relation between the mass scale of electroweak symmetry breaking and the fermion mass spectrum. It is possible to devise models in which the first family is light because the family symmetry breaking suppresses the electroweak symmetry breaking. Thus, the "natural" scale of electroweak symmetry breaking among the fermions could remain approximately 100 GeV/$c^2$, despite the small masses (a few MeV/$c^2$) of some fermions.

Experiments to establish the masses of the neutrinos are of great interest to the family problem and to particle physics in general. Being electrically neutral, neutrinos are unique among the fermions in possibly being endowed with a so-called Majorana mass* in addition to the usual Dirac mass. One approach to determining these masses is by applying kinematics to suitable reactions. For example, one can measure the end-point energy of the electron in the beta decay $^3$H $\rightarrow$ $^3$He $+ e^- + \bar{\nu}_e$ or of the muon in the decay $\pi^+ \rightarrow \mu^+ + \nu_\mu$.

Another quite different approach is to search for "neutrino oscillations." If the neutrino masses are nonzero, weak interactions can be expected to mix neutrinos from different families just as they do the quarks. This mixing would cause a beam of, say, essentially muon neutrinos to be transformed into a mixture (varying in space and in time) of electron, muon, and tau neutrinos. Detection of these oscillations would not only settle the question of whether or not neutrinos have nonzero masses but would also provide information about the differences between the masses of neutrinos from different families. Experiments are in progress, but, since neutrino interactions are infamously rare, high-intensity beams are required to detect any neutrinos at all, let alone possible small oscillations in their family identity. (For details about the tritium beta decay and neutrino oscillation experiments in progress at Los Alamos, see "Experiments To Test Unification Schemes.")

## Conclusion

The family symmetry problem is a fundamental one in particle physics, apparently without sufficient information available at present to resolve it. Yet it is as crucial and important a problem as grand unification,

---

*\*Majorana mass terms are not allowed for electrically charged particles. Such terms induce transformations of particles into antiparticles and so would be inconsistent with conservation of electric charge.*

and it may well be a completely independent one. The known bound of $10^5$ GeV on the scale of family dynamics is an order of magnitude beyond the direct reach of any present or proposed accelerator, including the Superconducting Super Collider. These dynamics may, however, be accessible in studies of rare decays of kaons and other mesons, of CP violation, and of neutrino oscillations. To undertake these experiments at the necessary sensitivity requires intense fluxes of particles from the second or later families. A high-intensity, medium-energy accelerator could be a highly effective means of achieving these needs. Unlike many other experimental questions in particle physics, those on the high-intensity frontier are clearly defined. We await the answers expectantly. ■

## Further Reading

Howard Georgi. "A Unified Theory of Elementary Particles and Forces." Scientific American, April 1981, p. 48.

**T. (Terry) Goldman** received a B.Sc. in physics and mathematics in 1968 from the University of Manitoba and an A.M. and a Ph.D. in physics from Harvard in 1969 and 1973, respectively. He was a Woodrow Wilson Fellow from 1968 to 1969 and a National Research Council of Canada postdoctoral fellow at the Stanford Linear Accelerator Center from 1973 to 1975, when he joined the Laboratory's Theoretical Division as a postdoctoral fellow. In 1978 he became a staff member in the same division. From 1978 to 1980 he was on leave from Los Alamos as a Senior Research Fellow at California Institute of Technology, and during the academic year 1982-83 he was a Visiting Associate Professor at the University of California, Santa Cruz. His professional work has centered around weak interactions and grand unified theories. He is a member of the American Physical Society.

**Michael Martin Nieto** received a B.A. in physics from the University of California, Riverside, in 1961 and a Ph.D. in physics, with minors in mathematics and astrophysics, from Cornell University in 1966. He joined the Laboratory in 1972 after occupying research positions at the State University of New York at Stony Brook; the Niels Bohr Institute in Copenhagen; the University of California, Santa Barbara; Kyoto University; and Purdue University. His main interests are quantum mechanics, coherence phenomena, elementary particle physics, and astrophysics. He is a member of Phi Beta Kappa and the International Association of Mathematical Physicists and a Fellow of the American Physical Society.

Goldman and Nieto have co-authored several papers, including a recent one on tests of the gravitational properties of antimatter. The similarity of their other interests has also led them to work on related topics, although not always at the same time. For example, in 1972 Nieto authored a survey of important experiments in particle physics that could be done at the then new LAMPF. A decade later Goldman organized the Theoretical Symposium on Intense Medium-Energy Sources of Strangeness at the University of California, Santa Cruz, to study those new experiments that would be feasible at LAMPF II. The most recent manifestation of their similar interests is joint editorship of the proceedings of the annual meeting, held in Santa Fe, New Mexico from October 31 to November 3, 1984, of the Division of Particles and Fields of the American Physical Society.

# CP Violation in Heavy-Quark Systems

ere we extend the discussion of CP violation in "The Family Problem" to heavier quark systems. This requires generalizing the Cabibbo mixing matrix (Eq. 3 in the main text) to more than two families. The Cabibbo matrix relates the weak-interaction eigenstates of the $ud$ and $cs$ quark families to their strong-interaction mass eigenstates. Now, in general, the unitary transformation relating the weak and strong eigenstates among $n$ families will have $\frac{1}{2}n(n-1)$ rotations and $\frac{1}{2}(n-1)(n-2)$ physical phases.

We are interested in the generalization to three families since the third family, containing the $t$ and $b$ quarks, is known to exist. This

extension of the Cabibbo mixing matrix is called the Kobayashi-Maskawa (K-M) matrix after the two physicists who elucidated the problem. They realized that the mixing matrix for three families would naturally encompass a parameterization of CP violation. The K-M matrix can be written as a product of three rotations (which can be thought of as the Euler rotation angles of classical physics even though the convention is not the standard one) and a single physically meaningful phase (which can be identified as the CP-violating parameter). In particular, we define the K-M matrix $V$ for the three quark families $(ud, cs,$ and $tb)$ as follows:

$$\begin{pmatrix} d' \\ s' \\ b' \end{pmatrix} = V \begin{pmatrix} d \\ s \\ b \end{pmatrix} , \tag{A1}$$

where

$$V = \begin{pmatrix} 1 & 0 & 0 \\ 0 & c_2 & s_2 \\ 0 & -s_2 & c_2 \end{pmatrix} \begin{pmatrix} 1 & 0 & 0 \\ 0 & 1 & 0 \\ 0 & 0 & e^{i\delta} \end{pmatrix} \begin{pmatrix} c_1 & s_1 & 0 \\ -s_1 & c_1 & 0 \\ 0 & 0 & 1 \end{pmatrix} \begin{pmatrix} 1 & 0 & 0 \\ 0 & c_3 & s_3 \\ 0 & -s_3 & c_3 \end{pmatrix} \tag{A2}$$

$$= \begin{pmatrix} c_1 & s_1 c_3 & s_1 s_3 \\ -s_1 c_2 & c_1 c_2 c_3 - s_2 s_3 e^{i\delta} & c_1 c_2 s_3 + s_2 c_3 e^{i\delta} \\ s_1 s_2 & -c_1 s_2 c_3 - c_2 s_3 e^{i\delta} & -c_1 s_2 s_3 + c_2 c_3 e^{i\delta} \end{pmatrix} . \tag{A3}$$

Note the form of $V$ in Eq. A2. The first, third, and fourth matrices are rotations about particular axes. Except for the unusual convention, this is just a general orthogonal rotation in a three-dimensional Cartesian system. The $s_i$ and the $c_i$ are the sines and cosines of the three rotation angles $\theta_i$. Note that the $i = 1$ rotation is the Cabibbo rotation $\theta_C$ described in the text.

What is new is the second matrix factor in Eq. A2, which contains the complex amplitude with phase $\delta$ that parameterizes CP violation. Indeed, this is the factor that makes $V$ not an orthogonal transformation but a unitary transformation. $V$ is still norm-preserving, but contains phase information, something that quantum mechanics allows.

In principle, another matrix $U$ relates the weak and strong eigenstates of the $u$, $c$, and $t$ quarks, and the product $U^\dagger V$ describes the mixing of weak charged currents. However, we follow the standard convention and take $U = I$, thereby putting all of the physics of $U^\dagger V$ into $V$ itself. (Note that the unitarity of $V$ produces a result equivalent to that given by Eq. 7: there are still no *family-changing* neutral currents.) Because $V$ is "really" $U^\dagger V$, the rows of $V$ can be labeled by the $u$, $c$, and $t$ quarks. Thus, we can write $V$ as

$$V = \begin{pmatrix} V_{ud} & V_{us} & V_{ub} \\ V_{cd} & V_{cs} & V_{cb} \\ V_{td} & V_{ts} & V_{tb} \end{pmatrix} . \qquad (A4)$$

Physically, this means that the matrix elements $V_{ij}$ can be considered coupling constants or decay amplitudes between the quarks and the weak charged bosons $W^\pm$. For example, $V_{us} = \sin\theta_1 = \sin\theta_C$ is the left vertex in Fig. 4a of the main text, which can be considered a $u$ quark "decaying" into an $s$ quark.

We know from experiment that $\sin\theta_C = 0.23 \pm 0.01$. But further, from recent measurements of the lifetime of the $b$ quark and the branching ratio $\Gamma_{b\to u}/\Gamma_{b\to c}$, we know

that $\theta_2$ and $\theta_3$ are both small. That is, we have the information

$$|V_{cb}| = |c_1 c_2 c_3 + s_2 c_3 e^{i\delta}| = 0.044 \pm 0.005$$
$$(A5)$$

and

$$\left| \frac{V_{ub}}{V_{cs}} \right| = \left| \frac{s_1 s_3}{V_{cb}} \right| \le 0.12 . \qquad (A6)$$

These results imply that we can take $c_2$ and $c_3$ to be unity and obtain the approximation

$$V \approx \begin{pmatrix} c_1 & s_1 & s_1 s_3 \\ -s_1 & c_1 - s_2 s_3 e^{i\delta} & s_3 + s_2 e^{i\delta} \\ s_1 s_2 & -s_2 - s_3 e^{i\delta} & -s_2 s_3 + e^{i\delta} \end{pmatrix}$$
$$(A7)$$

In terms of quark mixing, CP violation in the $K^0$-$\bar{K}^0$ system is described by a second-order imaginary amplitude proportional to $s_2 s_3 \sin\delta$. In other words, the upper 2 by 2 piece of the matrix in Eq. A7 has this new imaginary contribution when compared with the Cabibbo matrix of Eq. 3. By using the Feynman diagram of Fig. 5 in the main text, the $K^0$-$\bar{K}^0$ transition-matrix element (traditionally called $M_{12}$) can be calculated in terms of the weak-interaction Hamiltonian and the entries of the mixing matrix $V$.

The older parameterization of CP violation, which involves the parameter $\varepsilon$, is model-independent. It focuses only on the properties of CP symmetry and the kaons themselves. It does not even need quarks. The value of $\varepsilon$ is determined by experiments (see below) and is directly related to $M_{12}$. It remains for a particular formalism (such as that described here) to successfully predict $M_{12}$ in a consistent manner. In particular, within the K-M formalism it is hard to obtain a large enough value for the CP-violating amplitude $\varepsilon$ even if one assumes $\delta = \pi/2$, because $s_2$ and $s_3$ are so small. In fact, agree-

ment with the measured value of $\varepsilon$ cannot be obtained *unless* the mass of the $t$ quark is equal to or greater than 60 GeV/$c^2$. Because the $t$ quark has not yet been found, this possibility remains open.

One way in which CP violation is observed in the $K^0$-$\bar{K}^0$ system was described in the main text. Another way is to detect an *anomalous* number of decays to leptons of the "wrong" sign. In the absence of mixing one ordinarily expects positively charged leptons from the $K^0$ parent and negatively charged leptons from the $\bar{K}^0$ parent; that is, $K^0 = d\bar{s}$ decays into $d(\bar{u}du)$ or $d(\bar{u}\ell^+\nu)$, and $\bar{K}^0 = \bar{d}s$ decays into $\bar{d}(ud\bar{u})$ or $\bar{d}(u\ell^-\bar{\nu})$, as shown in Fig. A1. However, to describe the propagation of a $K^0$ (or a $\bar{K}^0$), it must be decomposed into $K_L$ and $K_S$ states each of which is an approximate CP eigenstate containing approximately equal amplitudes of $K^0$ and $\bar{K}^0$. Since the $K_S$ lifetime is negligibly short, it is easy to design experiments to measure decays of the $K_L$ only. If CP were an exact symmetry, then the $K^0$ and $\bar{K}^0$ components of the $K_L$ would have equal amplitudes and would each provide exactly the same number of leptonic decays; that is, just as many "wrong"-sign leptons would come from decays of the $\bar{K}^0$ component (the antiparticles of Fig. A1) as "right"-sign leptons come from decays of the $K^0$ component (Fig. A1). The deviation from exact equality is another measure of CP violation.

What about CP violation in other neutral-boson systems? If one does the same type of anaylsis as is often done for the kaon system, one can *phenomenologically* describe CP violation by

$$|\varphi(t)\rangle = f_+(t)\,|\varphi^0\rangle + \frac{1 - \varepsilon_\varphi}{1 + \varepsilon_\varphi}\,f_-(t)\,|\bar\varphi^0\rangle ,$$
$$(A8)$$

where $\varphi^0$ is a neutral boson, $\bar\varphi^0$ is its conjugate under C, $\varepsilon_\varphi$ is the CP-violating parameter specific to that boson, and

$$f_\pm(t) = \tfrac{1}{2}\{\exp[-(im_1 + \Gamma_1/2)t] \\ \pm \exp[-(im_2 + \Gamma_2/2)t]\} .$$
$$(A9)$$

(Here the labels "1" and "2" refer to the approximate CP eigenstates.) The value of $|\varepsilon_\varphi| \ll 1$ gives the magnitude of the CP-violating amplitude relative to the CP-conserving amplitude.

For the kaon case, the decay widths $\Gamma_1$ and $\Gamma_2$ (or $\Gamma_L$ and $\Gamma_S$) are such that the mixing between $K^0$ and $\bar{K}^0$ is rapid. In particular, since $\Delta\Gamma/\Gamma \equiv 2(\Gamma_1 - \Gamma_2)/\Gamma_1 + \Gamma_2) \simeq$ (a number of order unity), starting either from $K^0$ or $\bar{K}^0$ the system quickly ends up in the $K_L$ state. This allows a detection of CP violation by observing the few $K_L \rightarrow 2\pi$ decays. However, for both the neutral $D$ and $T$ mesons ($\bar{D} = c\bar{u}$, $\bar{T} = t\bar{u}$), the values of $\Delta m \equiv (m_1 - m_2)$ and $\Delta\Gamma$ are both K-M suppressed (that is, small, given the values of $V_{ij}$ in Eq. A7), whereas the decay widths $\Gamma_i$ are not suppressed. Therefore both $\Delta m/\Gamma$ and $\Delta\Gamma/\Gamma$ are small, and so the time scale for mixing is long compared with that for decay. This situation can be thought of as "the mesons decaying before they have a chance to mix" into their approximate CP eigenstates. Since it is not possible to observe this mixing easily, it is naturally even harder to observe the deviation of the mixing from equal amplitudes of each component, which is the case for exact CP eigenstates. Thus, the observation of CP violation in the neutral $D$ and $T$ systems will be very difficult.

For the neutral $B$ mesons, however, the mixing can be large, as again both $\Gamma$ and $\Delta\Gamma$ are "Cabibbo"- (actually K-M-) suppressed by Eq. A5. Indeed, a large mixing of the neutral $B$ mesons containing a strange quark ($B_s^0 = \bar{b}s$ and $\bar{B}_s^0 = b\bar{s}$) has already been observed in the UA1 experiment at CERN. (Mixing of these mesons is shown in Fig. A2.) But the way this observation is done requires some explanation. The experiment looks for $b\bar{b}$ quark-pair production in proton-antiproton collisions. The signal for production of a $b$ quark is emission of a decay muon (from decay of the $b$ quark) with a large momentum component transverse to the axis defined by the proton and antiproton beams. According to QCD calculations, the overwhelming majority of observed back-to-back muon pairs with high transverse

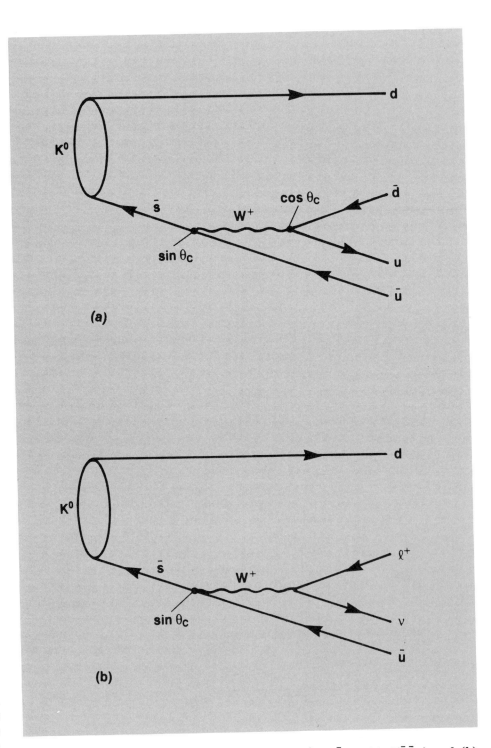

**Fig. A1. Feynman diagrams for the decays of $K^0 = d\bar{s}$ to (a) $d(\bar{u}du)$ and (b) $d(\bar{u}\ell^+ \nu)$. The analogous decays of $\bar{K}^0$ are ordinarily obtained from these simply by changing every particle into its antiparticle.**

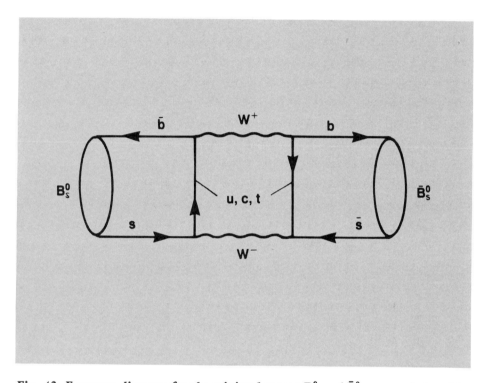

**Fig. A2. Feynman diagram for the mixing between $B_s^0$ and $\bar{B}_s^0$ mesons induced by second-order weak interactions. This diagram is analogous to that presented in the main text (Fig. 5) for mixing in the $K^0$-$\bar{K}^0$ system.**

momentum are the decay products of a common $b\bar{b}$ quark pair. Such parent quark pairs almost always appear as $B\bar{B}$ meson pairs.

Suppose there was little or no mixing between $B_s^0$ and $\bar{B}_s^0$. Then one would expect the observed ratio of the decays of $B_s^0\bar{B}_s^0$ pairs into back-to-back muon pairs with the same charge $[(++)$ or $(--)]$ to the decays of $B_s^0\bar{B}_s^0$ pairs with opposite charge $[(+-)]$ to be about 25 percent. This ratio is deduced by the following argument. Without mixing (a) the main contribution to unlike pairs comes from the direct decay of both quarks ($b \rightarrow c\mu^-\bar{\nu}$ and $\bar{b} \rightarrow \bar{c}\mu^+\nu$), and (b) the main contribution to like pairs comes from one primary decay and one secondary decay (for example, $b \rightarrow c\mu^-\bar{\nu}$ and $\bar{b} \rightarrow \bar{c} \rightarrow \bar{s}\mu^-\bar{\nu}$). The relative rates can be calculated from the known weak-decay parameters, and one obtains the value 0.24 for the ratio of like- to unlike-sign pairs.

However, with mixing (such as that shown in Fig. A2) one can sometimes have processes like $\bar{s}b \rightarrow \bar{s}c\mu^-\bar{\nu}$ and $s\bar{b} \rightarrow \bar{s}b \rightarrow \bar{s}c\mu^-\bar{\nu}$. This transforms some of the expected unlike-sign events into like-sign events. In fact, for a mixing of 10 percent, this changes the ratio of like- to unlike-sign events from about ¼ to about ½.

Indeed, the UA1 experiment at CERN sees a ratio of 50 percent. This result can be explained only by a large mixing between $B_s^0$ and $\bar{B}_s^0$, which overwhelms the tendency for the $b$ and $\bar{b}$ quarks to decay into opposite-sign pairs. Since one needs significant mixing to observe CP violation, there is hope of learning more about CP, depending on the (as yet undetermined) values of the mass-matrix parameters for $B_s^0$ and $\bar{B}_s^0$ (that is, $m_1$, $m_2$, $\Gamma_1$, and $\Gamma_2$).

For further details of this fascinating subject, we recommend the review "Quark Mixing in Weak Interactions" by Ling-Lie Chau (*Physics Reports* 95:1(1983)). ■

# Experiments to Test

*Flash chambers discharging like neon lights, giant spectrometers, stacks of crystals, tons of plastic scintillators, thousands of precisely strung wires—all employed to test the ideas of unified field theories.*

It has long been a dream of physicists to produce a unified field theory of the forces in nature. Much of the current experimental work designed to test such theories occurs at the highest energies capable of being produced by the latest accelerators. However, elegant experiments can be designed at lower energies that probe the details of the electroweak theory (in which the electromagnetic and weak interactions have been partially unified) and address key questions about the further unification of the electroweak and the strong interactions. (See "An Experimentalist's View of the Standard Model" for a brief look at the current status of the quest for a unified field theory.)

In this article we will describe four such experiments being conducted at Los Alamos, often with outside collaborators. The first, a careful study of the beta decay of tritium, is an attempt to determine whether or not the neutrino has a mass and thus whether or not there can be mixing between the three known lepton families (the electron, muon, and tau and their associated neutrinos).

Two other experiments examine the decay of the muon. The first is a search for *rare* decays that do not involve neutrinos, that is, the direct conversion across lepton families of the muon to an electron. The muon is a duplicate, except for a greater mass, of the electron, making such a decay seem almost mandatory. Detection of a rare decay, or even the lowering of the limits for its occurrence, would tell us once again more about the mixing between lepton families and about possible violation of lepton conservation laws. At the same time, precision studies of ordinary muon decay, in which neutrinos are generated (the muon is accompanied by its own neutrino and thereby preserves muon number), will help test the stucture of the present theory describing the weak interaction, for example, by setting limits on whether or not parity conservation is restored as a symmetry at high energies.

*The electron spectrometer for the tritium beta decay experiment under construction. The thin copper strips evident in the entrance cone region to the right and at the first narrow region toward the center are responsible for the greatly improved transmission of this spectrometer.*

# Unification Schemes

*by Gary H. Sanders*

The intent of the fourth experiment is to measure interference effects between the neutral and charged weak currents via scattering experiments with neutrinos and electrons. If destructive interference is detected, then the present electroweak theory should be applicable even at higher energies; if constructive interference is detected, then the theory will need to be expanded, say by including vector bosons beyond those (the $Z^0$ and the $W^{\pm}$) already in the standard model.

## Tritium Beta Decay

In 1930 Pauli argued that the continuous kinetic energy spectrum of electrons emitted in beta decay would be explained by a light, neutral particle. This particle, the neutrino, was used by Fermi in 1934 to account quantitatively for the kinematics of beta decay. In 1953, the elusive neutrino was observed directly by a Los Alamos team, Fred Reines and Clyde L. Cowan, using a reactor at Hanford.

Though the neutrino has generally been taken to be massless, no theory requires neutrinos to have zero mass. The current experimental upper limit on the electron neutrino mass is 55 electron volts (eV), and the Russian team responsible for this limit claims a lower limit of 20 eV. The mass of the neutrino is still generally taken to be zero, for historical reasons, because the experiments done by the Russian team are extremely complex, and because masslessness leads to a pleasing simplification of the theory.

A more careful look, however, shows that no respectable theory requires a mass that is identically zero. Since we have many neutrino flavors (electron, muon and tau neutrinos, at least), a nonzero mass would immediately open possibilities for mixing between these three known lepton families. Without regard to the minimal standard model or any unification schemes, the possible existence of massive neutrinos points out our basic ignorance of the origin of the known particle masses and the family structure of particles.

# An Experimentalist's View of the Standard Model

The dream of physicists to produce a unified field theory has, at different times in the history of physics, appeared in a different light. For example, one of the most astounding intellectual achievements in nineteenth century physics was the realization that electric forces and magnetic forces (and their corresponding fields) are different manifestations of a single electromagnetic field. Maxwell's construction of the differential equations relating these two fields paved the way for their later relation to special relativity.

**QED.** The most successful field theory to date, quantum electrodynamics (QED), appears to have provided us with a complete description of the electromagnetic force. This theory has withstood an extraordinary array of precision tests in atomic, nuclear, and particle physics, and at low and high energies. A generation of physicists has yearned for comparable field theories describing the remaining forces: the weak interaction, the strong interaction, and gravity.

An even more romantic goal has been the notion that a *single* field theory might describe all the known physical interactions.

**Electroweak Theory.** In the last two decades we have come a long way towards realizing this goal. The electromagnetic and weak interactions appear to be well described by the Weinberg-Salam-Glashow model that unifies the two fields in a gauge theory. (See "Particle Physics and the Standard Model" for a discussion of gauge theories and other details just briefly mentioned here.) This

electroweak theory appears to account for the apparent difference, at low energies, between the weak interaction and the electromagnetic interaction. As the energy of an interaction increases, a unification is achieved.

So far, at energies accessible to modern high-energy accelerators, the theory is supported by experiment. In fact, the discovery at CERN in 1983 of the heavy vector bosons $W^+$, $W^-$, and $Z^0$, whose large mass (compared to the photon) accounts for the relatively "weak" nature of the weak force, beautifully confirms and reinforces the new theory.

The electroweak theory has many experimental triumphs, but experimental physicists have been encouraged to press ever harder to test the theory, to explore its range of validity, and to search for new fundamental interactions and particles. The experience with QED, which has survived decades of precision tests, is the standard by which to judge tests of the newest field theories.

**QCD.** A recent, successful field theory that describes the strong force is quantum chromodynamics (QCD). In this theory the strong force is mediated by the exchange of color gluons and a coupling constant is determined analogous to the fine structure constant of the electroweak theory.

**Standard Model.** QCD and the electroweak theory are now embedded and united in the minimal standard model. This model organizes all three fields in a gauge

## Table

**The first three generations of elementary particles.**

| Family: | | I | II | III |
|---|---|---|---|---|
| Doublets | Quarks: | $\begin{pmatrix} u \\ d \end{pmatrix}_L$ | $\begin{pmatrix} c \\ s \end{pmatrix}_L$ | $\begin{pmatrix} t \\ b \end{pmatrix}_L$ |
| | Leptons: | $\begin{pmatrix} \nu_e \\ e \end{pmatrix}_L$ | $\begin{pmatrix} \nu_\mu \\ \mu \end{pmatrix}_L$ | $\begin{pmatrix} \nu_\tau \\ \tau \end{pmatrix}_L$ |
| Singlets: | | $u_R, d_R, e_R,$ | $c_R, s_R, \mu_R,$ | $t_R, b_R, \tau_R$ |

theory of electroweak and strong interactions. There are two classes of particles: spin-$\frac{1}{2}$ particles called fermions (quarks and leptons) that make up the particles of ordinary matter, and spin-1 particles called bosons that account for the interactions between the fermions.

In this theory the fermions are grouped asymmetrically according to the "handedness" of their spin to account for the experimentally observed violation of CP symmetry. Particles with right-handed spin are grouped in pairs or doublets; particles with left-handed spin are placed in singlets. The exchange of a charged vector boson can convert one particle in a given doublet to the other, whereas the singlet particles have no weak charge and so do not undergo such transitions.

The Table shows how the model, using this scheme, builds the first three generations of leptons and quarks. Since each quark ($u$, $d$, $c$, $s$, $t$, and $b$) comes in three colors and all fermions have antiparticles, the model includes 90 fundamental fermions.

The spin-1 boson mediating the electromagnetic force is a massless gauge boson,

that is, the photon $\gamma$. For the weak force, there are both neutral and charged currents that involve, respectively, the exchange of the neutral vector boson $Z^0$ and the charged vector bosons $W^+$ and $W^-$. The color force of QCD involves eight bosons called gluons that carry the color charge.

The coupling constants for the weak and electromagnetic interactions, $g_{wk}$ and $g_{em}$, are related by the Weinberg angle $\theta_W$, a mixing angle used in the theory to parametrize the combination of the weak and electromagnetic gauge fields. Specifically,

$$\sin \theta_W = g_{em}/g_{wk} .$$

Only objects required by experimental results are in the standard model, hence the term minimal. For example, no right-handed neutrinos are included. Other minimal assumptions are massless neutrinos and no requirement for conservation of total lepton number or of individual lepton flavor (that is, electron, muon, or tau number).

The theory, in fact, includes no mass for *any* of the elementary particles. Since the

vector bosons for the weak force and all the fermions (except perhaps the neutrinos) are known to be massive, the symmetry of the theory has to be broken. Such symmetry-breaking is accomplished by the Higgs mechanism in which another gauge field with its yet unseen Higgs particle is built into the theory. However, no other Higgs-type particles are included.

Many important features are built into the minimal standard model. For example, low-energy, charged-current weak interactions are dominated by $V - A$ (vector minus axial vector) currents; thus, only left-handed $W^\pm$ bosons have been included. Also, since neutrinos are taken to be massless, there are supposed to be no oscillations between neutrino flavors.

There are many possibilities for extensions to the standard model. New bosons, families of particles, or fundamental interactions may be discovered, or new substructures or symmetries may be required. The standard model, at this moment, has no demonstrated flaws, but there are many potential sources of trouble (or enlightenment).

**GUT.** One of the most dramatic notions that goes beyond the standard model is the grand unified theory (GUT). In such a theory, the coupling constants in the electroweak and strong sectors run together at extremely high energies ($10^{15}$ to $10^{19}$ giga-electron volts (GeV)). All the fields are unified under a single group structure, and a new object, the $X$, appears to generate this grand symmetry group. This very high-energy mass scale is not directly accessible at any conceivable accelerator. To explore the wilderness between present mass scales and the GUT scale, alas, all high-energy physicists will have to be content to work as low-energy physicists. Some seers believe the wilderness will be a desert, devoid of striking new physics. In the likely event that the desert is found blooming with unexplored phenomena, the journey through this terra incognita will be a long and fruitful one, even if we *are* restricted to feasible tools. ■

The reaction studied by all of the experiments mentioned is

$$^3\text{H} \rightarrow {}^3\text{He}^+ + e^- + \bar{\nu}_e .$$

This simple decay produces a spectrum of electrons with a definite *end point energy* (that is, conservation of energy in the reaction does not allow electrons to be emitted with energies higher than the end point energy). In the absence of neutrino mass, the spectrum, including this end point energy, can be calculated with considerable precision. Any experiment searching for a nonzero mass must measure the spectrum with sufficient resolution and control of systematic effects to determine if there is a deviation from the expected behavior.

Specifically, an end point energy lower than expected would be indicative of energy carried away as mass by the neutrino.

In 1972 Karl-Erik Bergkvist of the University of Stockholm reported that the mass of the electron antineutrino $\bar{\nu}_e$ was less than 55 eV. This experiment used tritium embedded in an aluminum oxide base and had a resolution of 50 eV. The Russian team set out to improve upon this result using a better spectrometer and tritium bound in valine molecules.

Valine is an organic compound, an amino acid. A molecular biologist in the Russian collaboration provided the expertise necessary to tag several of the hydrogen sites on the molecule with tritium. This knowledge is important since one of the effects limiting the accuracy of the result is the knowledge of the final molecular states after the decay.

Also important was the accurate determination of the *spectrometer resolution function*, which involved a measurement of the energy loss of the beta electrons in the valine. This was accomplished by placing an ytterbium-169 beta source in an identical source assembly and measuring the energy loss of these electrons as they passed through the valine.

The beta particles emitted from the source were analyzed magnetically in a toroidal beta spectrometer. This kind of spectrometer provides the largest acceptance for a given resolution of any known design, and the Russians made very significant advances. The Los Alamos research group, as we shall see, has improved the spectrometer design even further.

In 1980 the Russian group published a positive result for the electron antineutrino mass. After including corrections for the uncertainties in resolution and the final state spectrum, they quoted a 99 per cent confidence level value of

$$14 < m_{\bar{\nu}_e} < 46 \text{ eV} .$$

The result was received with great excitement, but two specific criticisms emerged. John J. Simpson of the University of Guelph pointed out that the spectrometer resolution was estimated neglecting the intrinsic linewidth of the spectrum of the ytterbium-169 calibration source. The experimenters then measured the source linewidth to be 6.3 eV; their revised analysis lowered the best value of the neutrino mass from 34.3 to 28 eV. The basic result of a finite mass survives this reanalysis, according to the authors, but it should be noted that the result is very sensitive to the calibration linewidth. Felix Boehm of the California Institute of Technology has observed that with an intrinsic linewidth of only 9 eV, the 99 per cent confidence level result would become consistent with zero.

The second criticism related to the assumption made about the energy of the final atomic states of helium-3. The valine molecule provides a complex environment, and the branching ratios into the $2s$ and $1s$ states of helium-3 are difficult to estimate. Thus the published result may prove to be false.

This discussion illustrates the difficulty of experiments of this kind. Each effort produces, in addition to the published measurement, a roadmap to the next generation experiment. The Russian team built upon its 1980 result and produced a substantially improved apparatus that yielded a new measurement in 1983.

The spectrometer was improved by adding an electrostatic field between the source and the magnetic spectrometer that could be used to accelerate the incoming electrons. The beta spectrum could then be measured, under conditions of constant magnetic field, by sweeping the electrostatic field to select different portions of the spectrum. This technique (originally suggested by the Los Alamos group) provides a number of advantages. The magnetic spectrometer always sees electrons in the same energy range, providing constant detection efficiency throughout the measured spectrum. The magnetic field can also be set above the beta spectrum end point with the electrostatic field accelerating electrons from decays in the source into the spectrometer acceptance. This reduces the background by a large factor by making the spectrometer insensitive to electrons from decays of tritium contamination in the spectrometer volume.

Also, finite source size, which produces a larger image at the spectrometer focal plane, was optically reduced by improved focusing at the source, yielding a higher count rate with better resolution.

The improved spectrometer had a resolution of 25 eV, compared to 45 eV in the 1980 experiment. Background was reduced by a factor of 20, and the region of the spectrum scanned was increased from 700 eV to 1750 eV.

The controversial spectrometer resolution function was determined using a different line of the ytterbium-169 source, and the Russians measured its intrinsic linewidth to be 14.7 eV. They also studied ionization losses by measuring the ytterbium-169 spectrum through varying thicknesses of valine, yielding a considerably more accurate resolution function.

The data were taken in 35 separate runs and the beta spectrum (Fig. 1) was fit by an expression that included the ideal spectral shape and the experimental corrections. The best fit gave

$$m_{\bar{\nu}_e} = 33.0 \pm 1.1 \text{ eV} ,$$

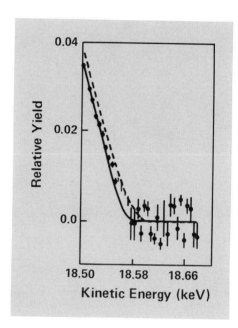

*Fig. 1. Electron energy spectrum for tritium decay. This figure shows the 1983 Russian data as the spectrum drops toward an end point energy of about 18.58 keV. The difference in the best fit to the data (solid line) and the fit for a zero neutrino mass (dashed line) is a shift to lower energies that corresponds to a mass of about 33.0 eV. (Figure adapted from Michael H. Shaevitz, "Experimental Results on Neutrino Masses and Neutrino Oscillations," page 140, in Proceedings of the 1983 International Symposium on Lepton and Photon Interactions at High Energies, edited by David G. Cassel and David L. Kreinick (Ithaca, New York:F.R. Newman Laboratory of Nuclear Studies, Cornell University, 1983).)*

with a 99 per cent confidence limit range of

$$20 < m_{\bar{\nu}_e} < 55 \, eV \, .$$

These results were derived by making particular choices for the final state spectra. Different assumptions for the valine molecu-

lar final states and the helium-3 molecular, atomic, and nuclear final states can produce widely varying results.

The physics community has been tantalized by the prospect that neutrinos have significant masses. Lepton flavor transitions, neutrino oscillations, and many other phenomena would be expected if the result is confirmed. The range of systematic effects, however, urges caution and enhanced efforts by experimenters to attack this problem in an independent manner. There are currently more than a dozen groups around the world engaged in improved experiments on tritium beta decay. A wide range of tritium sources, beta spectrometers, and analysis techniques are being employed.

**The Tritium Source.** In an ambitious attempt to use the simplest possible tritium source, a team from a broad array of technical fields at Los Alamos is attempting to develop a source that consists of a gas of free (unbound) tritium atoms. Combining diverse capabilities in experimental particle physics, nuclear physics, spectrometer design, cryogenics, tritium handling, ultraviolet laser technology, and materials science, this team has developed a nearly ideal source and has made numerous improvements in electrostatic-magnetic beta spectrometers.

The two most significant problems come from the scattering and energy loss of the electrons in the source and from the atomic and molecular final states of the helium-3 daughter. These effects are associated with any solid source. Thus the ideal source would appear to be free tritium nuclei, but this is ruled impractical by the repulsive effects of their charge.

The next best source is a gas of free tritium atoms. Detailed and accurate calculations of the atomic final states and electron energy losses can be performed. Molecular effects, including final state interactions, breakup, and energy loss in the substrate, are eliminated. Since the gas contains no inert atoms, the effect of energy loss and scattering in the source are reduced accordingly. Even the measurement of the beta spectrometer

resolution function is simplified.

The forbidding technical problem of such a design is building a source rich enough and compact enough to yield a useful count rate. Only one decay in $10^7$ produces an electron with energy in the interesting region near the end point where the spectrum is sensitive to neutrino mass.

The Los Alamos group was motivated by a 1979 talk given by Gerard Stephenson, of the Physics and Theoretical Divisions, on neutrino masses. They recognized quite early, in fact before the 1980 Russian result, that atomic tritium would be a nearly ideal source. In their first design, molecular tritium was to be passed through an extensive gas handling and purification system and atomic tritium prepared using a discharge in a radio-frequency dissociator. The pure jet of atomic tritium was then to be monitored for beta decays. It was clear, however, that the tritium atoms needed to be used more efficiently.

Key suggestions were made at this point by John Browne of the Physics Division and Daniel Kleppner of the Massachusetts Institute of Technology. Advances had been made in the production of dense gases of spin-polarized hydrogen. The new techniques—in which the atomic beam was cooled and then contained in a bottle made of carefully chosen materials observed to have a low probability for promoting recombination of the atoms—promised a possible intense source of free atomic tritium. The collaboration set out to develop and demonstrate this idea. Crucial to the effort was the participation of Laboratory cryogenics specialists.

The resulting tritium source (Fig. 2) circulates molecular tritium through a radio-frequency dissociator into a special tube of aluminum and aluminum oxide. Because the recombination rate for this material near 120 kelvins is very low, the system achieves 80 to 90 per cent purity of atomic tritium. The electrons from the beta decay of the atomic tritium are captured by a magnetic field, and then electrostatic acceleration, similar to that employed by the Russians, is used to trans-

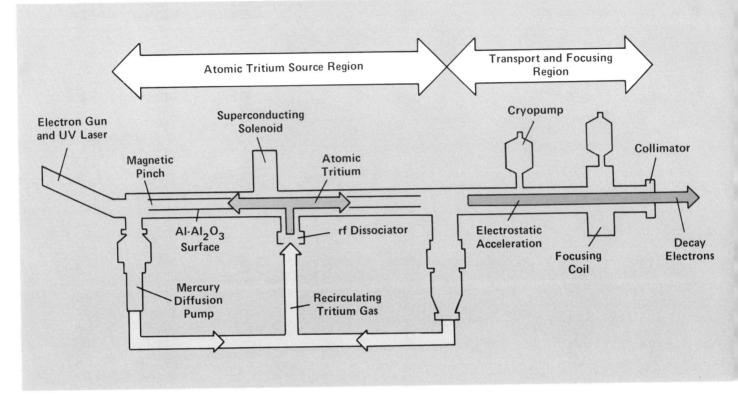

Fig. 2. The tritium source. Molecular tritium passes through the radio-frequency dissociator and then into a 4-meter-long tube as a gas of free atoms. The tube—aluminum with a surface layer of aluminum oxide—has a narrow range around a temperature of 120 kelvins at which the molecular recombination rate is very low, permitting an atom to experience approximately 50,000 collisions before a molecule is formed. The resulting diffuse atomic gas fills the tube, and mercury-diffusion pumps at the ends recirculate it through the dissociator. Typically, the system achieves 80 to 90 per cent purity of atomic tritium. By measuring the spectrum when the dissociator is off, the contribution from the 10 to 20 per cent contamination of molecular tritium can be determined and subtracted, resulting in a pure atomic tritium electron spectrum.

A superconducting coil surrounds the tube with a field of 1.5 kilogauss. At one end the winding has a reflecting field provided by a magnetic pinch. These fields capture electrons from beta decays with 95 per cent efficiency.

The other end of the tube connects to a vacuum region and has coils that transport and, importantly, focus an image of the electrons into the spectrometer (Fig 3). The tube is held at a selected voltage between −4 and −20 kilovolts, and electrons exit the source to ground potential. Thus, electrons from decays in the source tube are accelerated by a known amount to an energy above that of electrons from decays in

port the electrons toward the spectrometer. During this transport, focusing coils and a collimator are used to form a small image of the electron source in the spectrometer.

Development of this tritium source required solving an array of problems associated with a system that was to recirculate atomic tritium. Everything had to be extremely clean, and no organic materials were allowed; all surfaces are glass or metal. Conducting materials had to be used wherever insulators could collect charge and introduce a bias. The aluminum oxide coating in the tube is so thin that electrons simply tunnel through it, thus providing a conducting surface that does not encourage recombination. Special mercury-diffusion pumps and custom cryopumps, free of oil or other organic materials, had to be fabricated. Every part of the tritium source was an exercise in materials science.

The idea of using electrostatic acceleration at the output of the source was first proposed by the group at Los Alamos in 1980 and subsequently used in the measurement described in the 1983 Russian publication. Accelerating the electrons to an energy above that of electrons from tritium that decays *in* the spectrometer both strongly reduces the background and also improves the acceptance of electrons into the spectrometer. However, this technique necessitates a larger spectrometer.

There are two other important systematic effects that need to be dealt with: the source image seen by the spectrometer should be small, and electrons produced by decays in the tube that suffer scattering off the walls have an energy loss that distorts the measured spectrum. The focusing coil and the final collimator address both effects, providing a small image. The only energy loss mechanism remaining is in the tritium gas itself, where losses are less than 2 eV.

**The Spectrometer.** In addition to cryogenics, tritium handling, and laser technology, the Laboratory's powerful computing capabilities were employed in both the detailed optical design of the beta-electron spectrometer and in extensive Monte-Carlo modeling.

The spectrometer (Fig. 3) is an ambitious development of the Russian design. Electrons from the source pass through the entrance cone and are focused onto the spectrometer axis. One very significant improvement in the spectrometer is the design of the conductors running parallel to the spectrometer axis that do this focusing. In the Russian apparatus, the conductors were thick water-cooled tubes. Most electrons strike the tubes and, as a result of this loss,

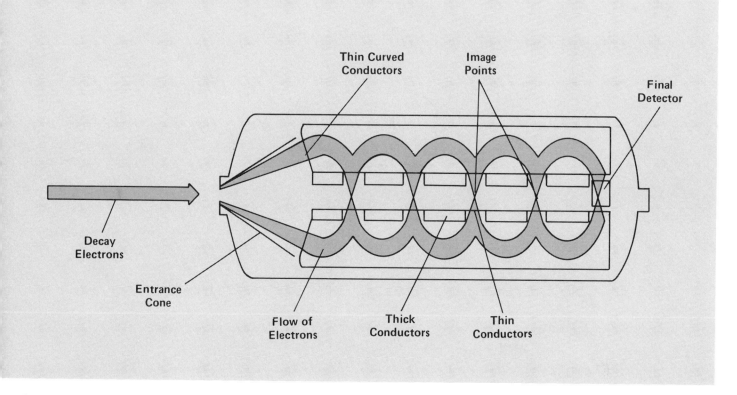

*Fig. 3. The spectrometer. Electrons from the source (Fig. 2) that pass through the collimator (with an approximate aperture of 1 centimeter) open into a cone shaped region in the spectrometer with a maximum half angle of 30 degrees. Electrons between 20 and 30 degrees pass between thin conducting strips into the spectrometer and are focused onto the spectrometer axis. This focus serves as a virtual image of the source. Transmission has been greatly improved over the Russian design through the use of thin conductors in all regions of electron flow (see opening photograph for a view of these conductors). The final focal plane detector is a position-sensitive, multi-wire proportional gas counter, also an improvement over previous detectors.*

*the spectrometer. Additional pumps also sharply reduce the amount of tritium escaping into the spectrometer.*

*Several sophisticated diagnostic systems monitor source output and stability. Beta detectors mounted in the focus region in front of the collimator measure the total decay rate from molecular and atomic tritium, whereas the fraction of tritium in molecular form is monitored by an ultraviolet (1027 angstroms wavelength) laser system developed by members of Chemistry Division that uses absorption lines of molecular tritium. A high-resolution electron gun is used to monitor energy loss in both the gas and the spectrometer. This gun is also used to measure the important spectrometer resolution function directly.*

their spectrometer has low transmission.

The Los Alamos spectrometer uses thin 20-mil strips for each of the conductors in the region within the transport aperture. This achieves an order of magnitude higher transmission, essential in yielding a useful count rate in an experiment with a dilute gas source.

Another benefit of the thin strips is that they can be formed easily. In fact, optical calculations accurate to third order dictate the curvature of the entrance and exit strips. The improved focusing properties of this arrangement yield an acceptance three times higher than the Russian device with no compromise in resolution.

The experimenters expect to be taking data throughout the latter part of 1984. They expect an order of magnitude less background and an order of magnitude larger geometric acceptance than the Russian ex-

periment. The design calls for a resolution between 20 and 30 eV, with a sensitivity to neutrino masses less than 10 eV. Even with their dilute gas source, they estimate a data rate in the region within 100 eV of the spectrum end point of about 1 hertz, fully competitive with rates obtained using solid sources.

Many groups around the world are vigorously pursuing this measurement. No other effort, however, will produce a result as free of systematic problems as the Los Alamos project. Other experiments are employing solid sources or, at best, molecular sources. Many have adopted an electrostatic grid system that introduces its own problems. To date, no design promises as clean a measurement. This year may well be the year in which the problem of neutrino mass is settled. The quantitative answer will be an important tool in uncovering the very poorly

understood relations between lepton families. No deep understanding of the models that unify the forces in nature can be expected without precise knowledge of the masses of neutrinos.

## Rare Decays of the Muon

The muon has been the source of one puzzle after another. It was discovered in 1937 in cosmic radiation by Anderson and Neddermeyer and by Street and Stevenson and was assumed to be the meson of Yukawa's theory of the nuclear force.

Yukawa postulated that the nuclear force, with its short range, should be mediated by the exchange of a massive particle, a meson. This differs from the massless photon of the infinite-range electromagnetic force. The muon mass, about 200 times the electron mass, fit Yukawa's theory well.

It was only after World War II ended that measurements of the muon's range in materials were found to be inconsistent with a particle interacting via a strong nuclear force. Discovery of the pion, or pi meson, settled the controversy. To this day, however, casual usage sometimes includes the erroneous phrase "mu meson".

With the resolution of the meson problem, however, the muon had no reason to be. It was simply not necessary. The muon appeared to be, in all known ways, a massive electron with no other distinguishing attributes. A famous quotation of I. I. Rabi summarized the mystery: "The muon, who ordered that?"

This question is none other than the family problem described earlier. Today, the mystery remains, but its complexity has grown. Three generations of fermions exist, and the mysterious relation of the muon to the electron is replicated in the existence of the tau, discovered in 1976 by Martin Perl and collaborators. The three generation scheme is built into the minimal standard model, but there is little insight to guide us to the ultimate number of generations.

Is there a conservation number associated with each family or generation? Are there selection rules or fundamental symmetries that account for the apparent absence of some transitions between these multiplets? Vertical and horizontal transitions between quark states do occur. Processes involving neutrinos connect the lepton generations. Can the pattern of these observed transitions give us a clue as to why we are blessed with this peculiar zoology? Should we look harder for the processes we have not observed? Rabi's question, in its most modern form, is a rich and bewildering one, and many experimental groups have taken up its challenge by pursuing high sensitivity studies of the rare and unobserved reactions that may connect the generations.

With the muon and electron virtual duplicates of each other, it was expected that the heavier muon would decay by simple, *neutrinoless* processes to the electron. Transitions such as $\mu^+ \to e^+ e^+ e^-$, $\mu^+ \to e^+ \gamma$, or

## Table 1

**The additive lepton numbers, their conservation laws, and some of the decays allowed or forbidden by those laws.**

| Family | Particles | Lepton Number |
|---|---|---|
| Electron | $e^-, \nu_e$ | $L_e = +1$ |
| | $e^+, \bar{\nu}_e$ | $L_e = -1$ |
| Muon | $\mu^-, \nu_\mu$ | $L_\mu = +1$ |
| | $\mu^+, \bar{\nu}_\mu$ | $L_\mu = -1$ |
| Tau | $\tau^-, \nu_\tau$ | $L_\tau = +1$ |
| | $\tau^+, \bar{\nu}_\tau$ | $L_\tau = -1$ |

Conservation Laws: $\Sigma L_e = $ Constant, $\Sigma L_\mu = $ Constant, $\Sigma L_\tau = $ Constant

Allowed Decay: $\mu^+ \to e^+ \nu_e \bar{\nu}_\mu$    Forbidden Decays: $\mu^+ \to e^+ \gamma$
$\mu^+ \to e^+ e^+ e^-$
$\mu^- Z \to e^- Z$
$\mu^- Z \to e^- (Z-2)$
$\mu^+ \to e^+ \bar{\nu}_e \nu_\mu$

$\mu^- Z \to e^- Z$ (where $Z$ signifies that the interaction is with a nucleus) were expected. Estimates of the rates for these processes using second-order, current-current weak interactions gave results too small to observe. In fact, the results were much smaller than the 1957 limit for the branching ratio for $\mu^+ \to e^+ \gamma$, which was $< 2 \times 10^{-5}$ (a branching ratio is the ratio of the probability a decay will occur to the probability of the most common decay).

A better early model appeared in 1957 when Schwinger proposed the intermediate vector boson (now called $W$ and observed directly in 1983) as the mediator of the charged-current weak interaction. With this model and under most assumptions, rates larger than the experimental limits were predicted for the three reactions. The failure to observe these decays required a dynamical suppression or a new conservation law. Despite the discussion to follow, the situation today has changed very little. The measured limits are more stringent, though, by many orders of magnitude.

The first proposal for lepton number con-

servation came in 1953. In fact, there have been three different schemes for conserving lepton number. The 1953 Konopinski-Mahmoud scheme cannot accommodate three lepton generations and has not survived. A scheme in which lepton number is conserved by a multiplicative law was proposed in 1961 by Feinberg and Weinberg, but this method is not the favored conservation law. An early experiment with a neutrino detector at the Clinton P. Anderson Meson Physics Facility in Los Alamos (LAMPF) has removed the multiplicative law from favor, and the current experiment to study neutrino-electron scattering, described later in this article, has set even more stringent limits on such a law.

The most favored scheme is additive lepton number conservation, proposed in 1957 by Schwinger, Nishijima, and Bludman. In this scheme, any process must separately conserve the sum of muon number and the sum of electron number. Table 1 shows the assignment of lepton numbers used. The extension to the third lepton flavor, tau, is obvious and natural.

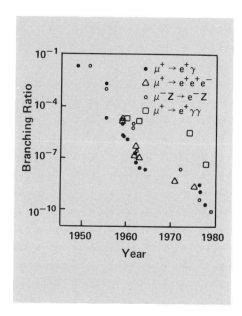

*Fig. 4. The progressive drop in the experimentally determined upper limit for the branching ratio of several muon-number violating processes shows a gap in the late 1960s. Essentially, this gap was the result of a belief by particle physicists in lepton number conservation.*

These schemes require, as the table hints, a distinct neutrino associated with each lepton. In a 1962 experiment the existence of separate muon and electron neutrinos was confirmed.

With a conservation law firmly entrenched in the minds of physicists, searches for decays that did not conserve lepton number seemed pointless. In a 1963 paper Sherman Frankel observed "Since it now appears that this decay is not lurking just beyond present experimental resolution, any further search . . . seems futile."

In retrospect it can be said that the particle physics community erred. The conclusion stated in the previous paragraph resulted in a nearly complete halt to efforts to detect processes that did not conserve lepton number—and this on the basis of a law postulated without any rigorous or fundamental basis!

It is easy to justify these assertions. Figure 4 shows that the experimental limits on rare decays were not aggressively addressed between 1964 and the late 1970s. This era of inattention ended abruptly when an experimental rumor circulated in 1977—an erroneous report terminated a decade of theoretical prejudice almost overnight! This could not have been the case if lepton conservation was required by fundamental ideas.

In 1977 a group searching for the process $\mu^+ \rightarrow e^+ \gamma$ at the Swiss Institute for Nuclear Research (SIN) became the inadvertent source of a report that the decay had been seen. The experiment, sometimes referred to as the "original SIN" experiment, was an order of magnitude more sensitive than any prior search for this decay and eventually set a limit on the branching ratio of $1.0 \times 10^{-9}$. A similar effort at the Canadian meson factory, TRIUMF, produced a limit of $3.6 \times 10^{-9}$ at about the same time.

**The Crystal Box.** The extraordinary controversy generated by the "original SIN" report motivated a Los Alamos group to attempt a search for $\mu^+ \rightarrow e^+ \gamma$ with a sensitivity to branching ratios of about $10^{-10}$. This experiment was carried out in 1978 and 1979, using several new technologies and a new type of muon beam at LAMPF, and yielded an upper limit of $1.7 \times 10^{-10}$ (90 per cent confidence level). That result stands as the most sensitive limit on the decay to date but should be surpassed this year by an experiment at LAMPF called the Crystal Box experiment.

This experiment was conceived as the earlier experiment came to an end. By searching for three rare muon decays simultaneously, the experiment would be a major advance in sensitivity and breadth. Several new technologies would be exploited as well as the capabilities of the LAMPF secondary beams.

In any search for a very rare decay, sensitivity is limited by two factors: the total number of candidate decays observed, and any other process that mimics the decay being searched for. The design of an experiment must allow the reliable estimate of the contribution of other processes to a false signal. This is generally done by a Monte-Carlo simulation of these decays that includes taking into account the detector properties.

In the absence of background or a positive signal for the process being studied, the number of seconds the experiment is run translates linearly into experimental sensitivity. However, when a background process is detected, sensitivity is gained only as the square root of the running time. This happens because one must subtract the number of background events from the number of observed events, and the statistical uncertainties in these numbers determine the limit. Generally, when an experiment reaches a level limited by background, it is time to think of an improved detector.

The Crystal Box detector is shown in Fig 5. A beam of muons from the LAMPF accelerator enters on the axis and is stopped in a thin polystyrene target. This beam consists of *surface* muons—a relatively new innovation developed during the 1970s and employed almost immediately at LAMPF and other meson factories.

Normal beams of muons are prepared in a three-step process: a proton beam from the accelerator strikes a target, generating pions; the pions decay in flight, producing muons; finally, the optics in the beam line are adjusted to transport the daughter muons to the experiment while rejecting any remaining pions. A more efficient way to collect low-momentum positive muons involves the use of a beam channel that collects muons from decays of positive pions generated in the target, but the muons collected are from pions that have only just enough momentum to travel from their production point in the target to its surface. Stopped in the surface, their decay produces positive muons of low momentum, near 29 MeV/$c$ (where $c$ is the speed of light). This technique enables experimenters to produce beams of surface muons that can be stopped in a thin experimental target with rates up to a hundred times more than conventional decay beams.

The muons stopped in the target decay virtually 100 per cent of the time by the mode

$$\mu^+ \rightarrow e^+ \, \nu_e \, \bar{\nu}_\mu \,,$$

with a characteristic muon lifetime of 2.2 microseconds. The Crystal Box detector accepts about 50 per cent of these decays and, therefore, must reject the positrons from several hundred thousand ordinary decays occurring each second. At the same time the detector must select those decays that appear to be generated by the processes of interest.

The Crystal Box was designed to simultaneously search for the decay modes

$$\mu^+ \rightarrow e^+ \, e^+ \, e^-$$
$$\rightarrow e^+ \, \gamma$$
$$\rightarrow e^+ \, \gamma \, \gamma \,.$$

(Since the Crystal Box does not measure the charge of the particles, we shall not generally distinquish between positrons and electrons in our discussion.)

The *detector properties* necessary for selecting final states from these reactions and rejecting events from ordinary muon decay are:

1. Energy resolution—The candidate decays produce two or three particles whose energies sum to the energy of a muon at rest. The ordinary muon decay and most background processes include particles from several decays or neutrinos that remain undetected but carry away some of the energy. These processes are extremely unlikely to yield the correct energy sum.

2. Momentum resolution— Given energy resolution adequate to accomplish the first requirement, vector momentum resolution requires a measurement of the directions of the particle trajectories. Since muons are stopped in the target, the decays being sought for will have vector momentum sums clustered, within experimental resolution, about zero. Particles from the leading background processes ($\mu^+ \rightarrow e^+ e^+ e^- \nu_e \bar{\nu}_\mu$, $\mu^+ \rightarrow e^+ \gamma \nu_e \bar{\nu}_\mu$, or coincidences of different ordinary muon decays) will tend to have non-

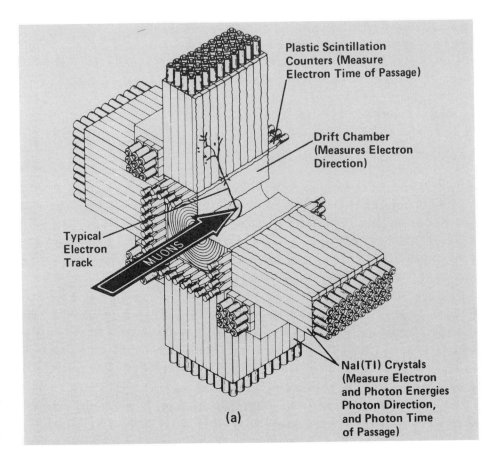

*Fig. 5. The Crystal Box detector. (a) A beam of muons enters the detector on axis. Because these are low-momenta surface muons, a thin polystyrene target is able to stop them at rates up to 100 times more than conventional muon beams. The beam intensity is generally chosen to be between 300,000 and 600,000 muons per second with pulses produced at a frequency of 120 hertz and a net duty factor between 6 and 10 per cent. Three kinds of detectors (drift chamber, plastic-scintillation counters, and NaI(Tl) crystals) surround the target. The detector elements are divided into four quadrants, each containing nine rows of crystals with a plastic scintillator in front of each row. This combination of detectors provides information on the energies, times of passage, and directions of the photons and electrons that result from muon decay in the target. The information is used to filter from several hundred thousand ordinary decays per second the perhaps several per second that may be of interest.*

*A sophisticated calibration and stabilization system was developed to achieve*

(c)

*and maintain the desired energy and time resolution for $4 \times 10^6$ seconds of data taking. Before a run starts, a plutonium-beryllium radioactive source is used for electron energy calibration. Also, a liquid hydrogen target is substituted periodically for the experimental target, and the photons emitted in the subsequent pion charge exchange are used for photon energy calibration. During data taking, energy calibration is monitored by a fiber optic flasher system that exposes each photomultiplier channel to a known light pulse. A small number of positrons are accepted from ordinary $\mu^+ \rightarrow e^+ \nu_e \bar{\nu}_\mu$ decays, and the muon decay spectrum cutoff at 52.8 MeV is used as a reference.*

*(b) The inner detector, the drift chamber, consists of 728 cells in 8 annular rings with about 5000 wires strung to provide the drift cell electrostatic geometry. A 5-axis, computer-controlled milling machine was used to accurately drill the array of 5000 holes in each end plate. These holes, many drilled at angles up to about 10 degrees, had to be located within 0.5 mil so that the chamber wires could be placed accurately enough to achieve a final resolution of about 1 millimeter in measuring the position of a muon decay in the target. The area of the stopping muon spot is about 100 cm². (Photo courtesy Richard Bolton.)*

*(c) The outer layer of the detector (here shown under construction) contains 396 thallium-doped sodium iodide crystals and achieves an electron and photon energy resolution of 5 to 6 per cent. This layer is highly segmented so that the electromagnetic shower produced by an event is spread among a cluster of crystals. A weighted average of the energy deposition can then be used to localize the interaction point of the photons with a position resolution of about 2 cm.*

zero vector sums.

3. Time resolution—Particles from the decay of a single muon are produced simultaneously. A leading source of background for, say $\mu^+ \rightarrow e^+ e^+ e^-$, is three electrons from the decay of three different muons. Such three-body final states are unlikely to occur simultaneously. Precision resolution in the time measurement, significantly better than 1 nanosecond, provides a powerful rejection of those random backgrounds.

4. Position resolution—Decays from a single muon will originate from a single point in the stopping target. Sometimes other processes will add extra particles to an event. The ability to accurately measure the trajectory of each particle in an event is crucial if experimental triggers that have extra tracks or that originate in separate vertices are to be rejected.

These parameters are used to filter measured events. In a sample of $10^{12}$ muons—the number required to reach sensitivities below the $10^{-11}$ level—most of this filtering must be done immediately, as the data is recorded. The Crystal Box experiment is exposed to approximately 500,000 muons stopping per second. The experimental "trigger" rate, the rate of decays that satisfy crude requirements, is about 1000 hertz. The detector has been designed with enough intelligence in its hardwired logic circuits to pass events to the data acquisition computer at a rate of less than 10 hertz. In turn, the program in the computer applies more refined filtering conditions so that events are written on magnetic tape at a rate of a few hertz.

Each condition used to narrow down the event sample to those that are real candidates provides a suppression factor. The combined suppression factors must permit the desired sensitivity. The design of the apparatus begins with the required suppressions and applies the necessary technology to achieve them.

A muon that stops in the target and decays by one of the subject decay modes produces only electrons, positrons or photons. The charged particles (hereafter referred to as electrons) are detected by an 8-layer wire drift chamber (Fig. 5 (b)) immediately surrounding the target. The drift chamber provides track information, pointing back at the origin of the event in the target and forward to the scintillators and crystals to follow. Its resolution and ability to operate in the high flux of electrons from ordinary muon decays in the target have pushed the performance limits of drift chambers; the chamber wires were placed accurately enough to achieve a final resolution of about 1 millimeter (mm) in measuring the position of a muon decay in the target.

Electrons are detected again in the next shell out from the target—a set of 36 plastic scintillation counters surrounding the drift chamber. These counters provide a measurement of the time of passage of the electrons with an accuracy of approximately 350 picoseconds. This accuracy is extraordinary for counters of the dimensions required (70 cm long by 6 cm wide by 1 cm thick) but is crucial to suppressing the random trigger background for the $\mu^+ \rightarrow e^+ e^+ e^-$ reaction. This performance is achieved by using two photomultiplier tubes, one at each end of the scintillator, and two special electronic timing circuits developed by the collaborators.

The electrons and photons that pass through the plastic scintillators deposit their energy in the next and outermost layer of the detector, a 396-crystal array of thallium-doped sodium iodide crystals. These crystals, acting as scintillators, provide fast precision measurement of both electron and photon energy (providing the energy and momentum filtering described earlier) and localize the interaction point of the photons with a position resolution of about 2 cm. The use of such large, highly segmented arrays of inorganic scintillator crystals was pioneered in high-energy physics in the late 1970's by the Crystal Ball detector at the Stanford Linear Accelerator Center. This technology is now widespread in particle physics research, with detectors planned that involve as many as 12,000 crystals.

The sodium iodide array also provides accurate time measurements on the photons.

A fast photomultiplier tube and electronics with special pulse shaping, amplification, and a custom-tailored, constant-fraction timing discriminator were melded into a system that gives subnanosecond accuracy.

The major detector elements—the drift chamber, plastic scintillators and sodium iodide crystals—are used in logical combinations to select events that may be of interest. A $\mu^+ \rightarrow e^+ e^+ e^-$ event is selected when three or more non-adjacent plastic scintillators are triggered and energy deposit occurs in the sodium iodide rows behind them. The special circuits developed for the scintillators are used for this selection: one high-speed circuit insures that the three or more triggers are coincident within a very tight time interval (approximately 5 nanoseconds), the second circuit requires the three or more hits to be in non-adjacent counters. The last requirement suppresses events in which low momentum radiative daughters trigger adjacent counters or when an electron crosses the crack between two counters.

An even more sophisticated trigger processor was constructed to insure that the three particles triggering the apparatus conform to a topology consistent with a three-body decay of a particle at rest. Thus, a pattern of tracks that, say, necessarily has net momentum in one direction (Fig. 6 (a)) is rejected, but a pattern with the requisite symmetry (Fig. 6 (b)) is accepted. This "geometry box" is an array of programmable read-only-memory circuits loaded with all legal hit patterns as determined by a Monte-Carlo simulation of the $\mu^+ \rightarrow e^+ e^+ e^-$ experiment.

Finally, the total energy deposited in the sodium iodide must be, within the real-time energy resolution, consistent with the rest energy of a muon.

The $\mu^+ \rightarrow e^+ \gamma$ and $\mu^+ \rightarrow e^+ \gamma \gamma$ reactions are selected by combining an identified electron (a plastic scintillator counter triggered coincident with sodium iodide signals) and one or more photons (a sodium iodide signal triggered with no count in the plastic scintillator in front of it). Also, these events must be in the appropriate geometric pattern (for example, directly opposite each other for $\mu^+$

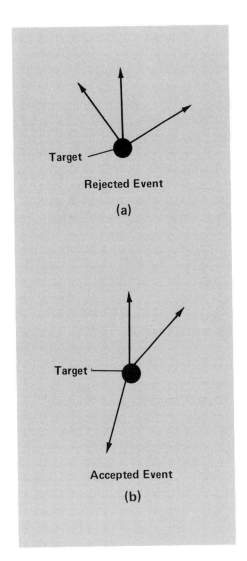

**Fig. 6. (a) A pattern of tracks with net momentum is not consistent with the neutrinoless decay of a muon at rest, and such an event will be rejected, whereas an event with a pattern such as the one in (b) will be accepted.**

$\rightarrow e^+ \gamma$) and have the correct energy balance.

The Crystal Box should report limits in the $10^{-11}$ range on the three reactions of interest this calendar year. It will also be used during the next year in a search for the $\pi^0 \rightarrow \gamma \gamma \gamma$ decay, which violates charge conjugation invariance. A search for only the $\mu^+ \rightarrow e^+ e^+ e^-$

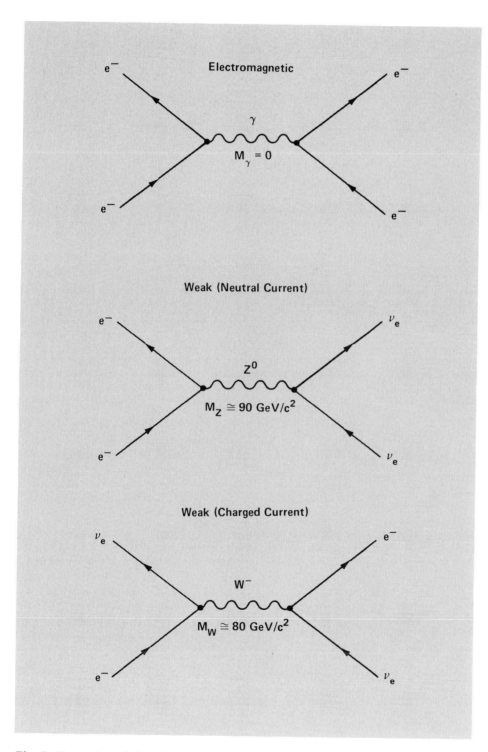

**Electromagnetic**

$M_\gamma = 0$

**Weak (Neutral Current)**

$M_Z \cong 90 \text{ GeV/c}^2$

**Weak (Charged Current)**

$M_W \cong 80 \text{ GeV/c}^2$

*Fig. 7. Examples of the electromagnetic and weak interactions in quantum field theory.*

process is being carried out at the Swiss Institute for Nuclear Research with an ultimate sensitivity of $10^{-12}$ available in the next year.

A third LAMPF $\mu^+ \rightarrow e^+ \gamma$ experiment is planned after the Crystal Box experiment. With present meson factory beams and foreseeable detector technology, this next generation experiment may well be the final round.

## Neutrino-Electron Scattering

The unification of the electromagnetic and weak interactions is a treatment of physical processes described by the exchange of three fundamental bosons. The exchange of a photon yields an electromagnetic current, and the $W^\pm$ and $Z^0$ bosons are exchanged in interactions classified as charged and neutral weak currents, respectively. Figure 7 illustrates how quantum field theory represents these processes.

A traditional method of probing electroweak unification in the standard model has been to determine the precise onset of weak effects in an interaction that is otherwise electromagnetic. Especially important are experiments—with polarized electron scattering at fixed target accelerators and more recent studies at electron-positron colliders—that probe the *interference* between the amplitudes of the electromagnetic and neutral-current weak interactions. Interference effects may be easier to observe than direct measurement of the small amplitudes of the weak interaction.

An Irvine-Los Alamos-Maryland team is conducting a unique and novel search for another interference. They have set out to probe the *purely* weak interference between the amplitudes of the charged and neutral currents. In the same way that electron scattering experiments search for interference between photon and $Z^0$ boson interactions, the Los Alamos based experiment is searching for the interference between charged-current $W$ interactions and neutral-current $Z^0$ interactions.

This experiment is attempting a unique

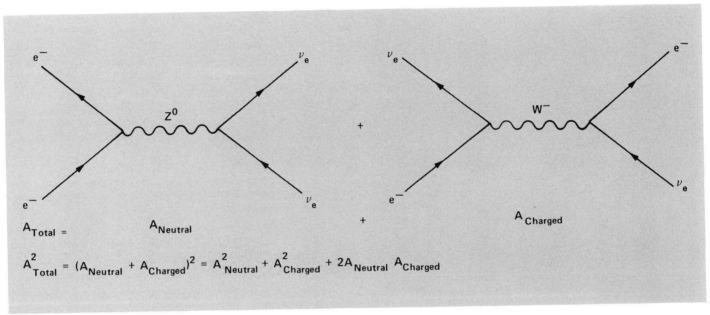

$$A_{Total} = \quad A_{Neutral}$$

$$A^2_{Total} = (A_{Neutral} + A_{Charged})^2 = A^2_{Neutral} + A^2_{Charged} + 2A_{Neutral}\,A_{Charged}$$

**Fig. 8. The interaction between an electron and its neutrino can take place via either the neutral current (with a $Z^0$) or the charged current (with a $W^-$), which results in an interference term $(2A_{Neutral}A_{Charged})$ in the expression for the** square of the total amplitude $A_{Total}$. An experiment at LAMPF will probe this purely weak interference by studying $\nu_e$-electron scattering.

measurement because Los Alamos is currently the only laboratory in the world with the requisite source of electron neutrinos. Moreover, the experiment gains importance from the fact that comparatively little is known about the physics of the $Z^0$ relative to that of the $W$.

The measurement is a simple variation on the electron-electron scattering experiments. To substitute the $W$ current for the electromagnetic current, the experimenters substitute the electron neutrino $\nu_e$ as the projectile and set out to measure the frequency of electron-neutrino elastic scattering from electrons. While this is conceptually simple, it is, in fact, technically quite difficult. The experiment must yield a sufficiently precise measure of the frequency of these scatters to separate out theoretical predictions made with different assumptions. To illustrate how the experiment tests the standard model, we must examine the nature of the model's predictions for $\nu_e$-$e$ scattering.

Electroweak theory obeys the group structure SU(2) $\times$ U(1). The SU(2) group has three generators, $W^+$, $W^-$, and $W^3$, which are the charged and neutral vector bosons identified with the gauge fields. The U(1) group has a single neutral boson generator $B$. The familiar phenomenological neutral photon field is constructed from the linear combination

$$A_\mu = W^3 \sin\theta_W + B\cos\theta_W ,$$

(where $\theta_W$ is the Weinberg angle, a measure of the ratio of the contributions of the weak and the electromagnetic forces to the total interaction). The phenomenological neutral current carried by the $Z^0$ is similarly constructed from

$$Z^0 = W^3 \cos\theta_W - B\sin\theta_W .$$

In the standard model the process

$$\nu_e + e^- \rightarrow \nu_e + e^-$$

can take place by the exchange of either the neutral-current boson $Z^0$ or the charged-current boson $W^-$ (Fig. 8), resulting in the usual interference term for the probablity of a process that can take place in either of two ways. The question then is what form will this interference take.

All models of the weak interaction that are currently considered viable predict a *negative, or destructive,* interference term. A model that can produce *constructive* interference is one that includes additional neutral gauge bosons beyond the $Z^0$. Thus, the observation of a $\nu_e$-$e$ scattering cross section consistent with constructive interference would indicate a phenomenal change in our picture of electroweak physics. Since the common $Z^0$ with about the predicted mass was directly observed only last year, and since higher mass regions will be accessible during this decade, such a result would set off a vigorous search by the particle physics community.

How will the traditional low-energy theory

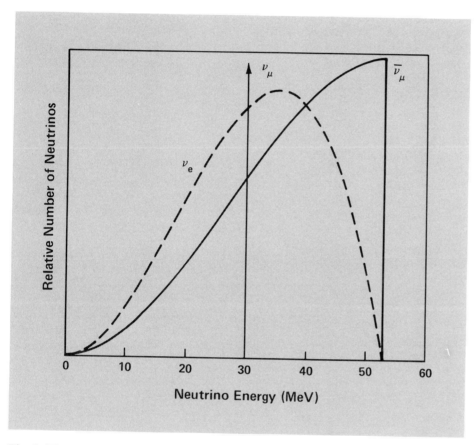

*Fig. 9. The energy spectra for the three types of neutrinos that result from the decay of a positive pion ($\pi^+ \rightarrow \mu^+ + \nu_\mu$, $\mu^+ \rightarrow e^+ + \nu_e + \bar{\nu}_\mu$).*

result of a two-body decay, is monoenergetic with an energy at about 30 MeV. The $\bar{\nu}_\mu$ spectrum has a cutoff energy at about 53 MeV, and the $\nu_e$ spectrum peaks around 35 or 40 MeV then falls off, also at about 53 MeV. These three particles are the source of many possible measurements.

The primary goal is the study of the $\nu_e$-$e$ elastic scattering already discussed. The detector, which we shall describe in more detail shortly, must detect electrons characteristic of the elastic scattering, that is, they should have energies between 0 and 53 MeV and lie within about 15 degrees of the forward direction (the tracks must point back to the neutrino source).

Also, by selecting events with electrons below 35 MeV, the group will search for the first observation of an exclusive neutrino-induced nuclear transition. The process

$$\nu_e + {}^{12}C \rightarrow e^- + {}^{12}N$$

would produce electrons with less than 35 MeV energy that lie predominantly outside the angular region for the elastic scattering events.

Another important physics goal, neutrino oscillations, can be addressed simultaneously. A process, called an "appearance," in which the $\bar{\nu}_\mu$ species disappears from the beam and $\bar{\nu}_e$ appears, can be probed by searching for the presence of $\bar{\nu}_e$ in the beam. This type of neutrino does not exist in the original neutrino source, so its presence downstream could be evidence for the $\bar{\nu}_\mu$-$\bar{\nu}_e$ oscillation. The experimental signature for such a process is the presence of isotropic single positrons produced by the reaction

$$\bar{\nu}_e + p \rightarrow n + e^+ ,$$

combined with a selection in energy of more than 35 MeV, which can be used to isolate these candidate events from the nuclear transition process discussed above.

In all three of the processes studied, the technical problem to be solved is the separation of the desired events from competing

of weak interactions (apparently governed by $V - A$ currents) mesh with future observations at higher energies? The standard model prediction, which contains negative interference, is that the cross section for $\nu_e$-$e$ elastic scattering should be about 60 per cent of the cross section in the traditional $V - A$ theory. The LAMPF experiment must measure the cross section with an accuracy of about 15 per cent to be able to detect the lower rate that would occur in the presence of interference and thus be able to determine whether interference effects are present or not.

In addition, the magnitude of the interference is a function of $\sin^2\theta_W$, and a precise measurement of the interference constitutes a measurement of this factor. In fact, it is

statistically more efficient to do this with a neutral current process because the charged current contains $\sin^2\theta_W$ ($\approx 0.25$) summed with unity, whereas for the neutral current the leading term is $\sin^2\theta_W$.

**The Experiments.** The LAMPF proton beam ends in a thick beam stop where pions ($\pi^+$) are produced. These pions decay by the process

$$\pi^+ \rightarrow \mu^+ + \nu_\mu$$
$$\phantom{\pi^+ \rightarrow} \hookrightarrow e^+ + \nu_e + \bar{\nu}_\mu ,$$

yielding three types of neutrinos exiting the beam stop. The $\nu_e$ and $\bar{\nu}_\mu$ are each produced with a continuous spectrum (Fig. 9) typical of muon decay, whereas the $\nu_\mu$ spectrum, the

143

background processes. The properties of the detector (Fig. 10) needed to do this include:

1. Passive shielding—Lead, iron, and concrete are used to absorb charged and neutral cosmic ray particles entering the detector volume. However, the shield is not thick enough to insure that events seen in the inner detector come only from neutrinos entering the detector and not from residual cosmic ray backgrounds. The outer shield merely reduces the flux, consisting mainly of muons and hadrons from cosmic rays and of neutrons from the LAMPF beam stop.

The LAMPF beam is on between 6 and 10 per cent of each second so that the periods between pulses will provide an important normalizing measurement indicating how well the passive shielding works.

2. Active anti-coincidence shield—This multilayer device is an active detector that surrounds the inner detector and serves many purposes. For example, muons from cosmic rays that penetrate the passive shield are detected here by being coincident in time with an inner detector trigger. This allows the rejection of these "prompt" muons, with less than one muon in $10^4$ surviving the rejection. Data acquisition electronics that store the history of the anti-coincidence shield for 32 microseconds prior to an inner detector trigger serve an even more complex purpose. This information is used to reject any inner detector electrons coming from a muon that stopped in the outer shield and that took up to 32 microseconds to decay. The mean muon lifetime is only 2.2 microseconds, so this is a very satisfactory way to reject such events.

3. Inner converter—Photons penetrating the anti-coincidence layer, produced perhaps by cosmic rays or particles associated with the beam, strike an additional layer of steel and are either absorbed or converted into electronic showers that are seen as tracks connected to the edge of the inner detector. Such events are discarded in the data analysis.

4. Inner detector—This module's primary role is to measure the trajectory and energy deposition of electrons and other charged

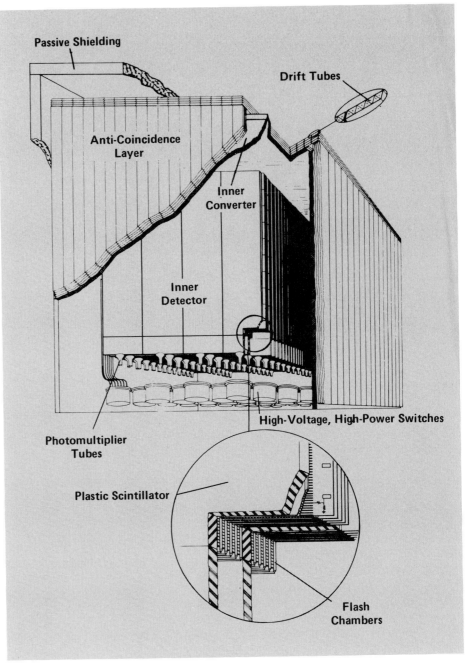

Fig. 10. *The detector for the neutrino-electron scattering experiments. The outer layer of passive shielding (mainly steel) cuts down the flux of neutral solar particles.*

*The anti-coincidence shield rejects muons from cosmic rays and electrons coming from the decay of muons stopped in the outer shield. It consists of four layers of drift tubes, totaling 603 counters, each 6 meters long. A total of 4824 wires provides a fine-grained, highly effective screen, with an inefficiency (and therefore a suppression) of $2 \times 10^{-5}$.*

*Another steel layer, the inner converter, is used to reject photons from cosmic rays or other particles associated with the beam.*

*The inner detector consists of 10 tons of plastic scintillators interleaved with 4.5 tons of tracking chambers. The plastic scintillators sample the electron energy every 10 layers of track chamber. There are 160 counters, each 75 cm by 300 cm by 2.5 cm thick, and they measure the energy to about 10 per cent accuracy. The track chambers are a classic technology: they are flash chambers that behave like neon lights when struck by an ionizing particle, discharging in a luminous and climactic way. There are a phenomenal 208,000 flash tubes in the detector, and they measure the electron tracks and sort them into angular bins about 7 degrees wide.*

particles. Electron tracks are the signature of the desired neutrino reactions, but recoil protons generated by neutrons from the beam stop and from cosmic rays must also be detected and filtered out in the data analysis. The inner detector contains layers of plastic scintillators that sample the particle energy deposited along its path for particle identification and also provide a calorimetric measurement of the total energy. Trajectory measurement is provided by a compact system of flash chambers interleaved with the plastic scintillators.

When this detector is turned on, it counts about $10^8$ raw events per day, mostly from cosmic rays. To illustrate the selectivity required of this experiment, a recent data run of a few months is expected to produce somewhat less than 50 $v_e$-$e$ elastic scattering events.

This highly segmented detector is necessarily extremely compact. The neutrino flux produced in the beam stop is emitted in all directions and therefore has an intensity that falls off inversely with the square of the distance. Thus there was a strong design premium for developing a compact, dense detector and placing it as close to the source as feasible.

The detector is now running around the clock, even when the LAMPF beam is off (to pin down background processes). The data already taken include many $v_e$-$e$ events that are being reported, as are preliminary results on lepton number conservation and neutrino oscillations. Data taken with additional neutron shielding during the next year or two are expected to provide the precision test of the standard model that the experimenters seek.

Beyond this effort, the beginnings of a much larger and ambitious neutrino program at Los Alamos are evident. A group (Los Alamos; University of New Mexico; Temple University; University of California, Los Angeles and Riverside; Valparaiso University; University of Texas) working in a new LAMPF beam line are mounting the prototype for a much larger fine-grained neutrino detector. Currently, a focused beam

source of neutrinos is being developed that will eventually employ a rapidly pulsed "horn" to focus pions that decay to neutrinos. This development will be used to provide neutrinos for a major new detector. The group is not content to work merely on developing the facility but is using a preliminary detector to measure some key cross sections and set new limits on neutrino oscillations as well.

Another group (Ohio State, Louisiana State, Argonne, California Institute of Technology, Los Alamos) is assembling the first components of an aggressive effort to search for the $\bar{v}_e$ appearance mode. Other physicists at the laboratory are preparing a solar neutrino initiative.

The exciting field of neutrino research, begun by Los Alamos scientists, is clearly entering a golden period.

## Precision Studies of Normal Muon Decay

The measurement of the electron energy spectrum and angular distribution from ordinary muon decay,

$$\mu \rightarrow e + \bar{v}_e + v_\mu ,$$

is one of the most fundamental in particle physics in that it is the best way to determine the constants of the weak interaction. These studies have led to limits on the $V - A$ character of the theory.

The spectrum of ordinary muon decay may be precisely calculated from the standard model. Built into the minimal standard model—consistent with the idea that everything in the model must be required by measurements—are the assumptions that neutrinos are massless and the only interactions that enter are of vector and axial vector form (that is, $V - A$, or equal magnitude and opposite sign). Lepton flavor conservation is also taken to be exact.

This $V - A$ structure of the weak interaction can be tested by precise measurements of the electron spectrum from ordinary

muon decay. The spectrum is characterized (to first order in $m_e/m_\mu$ and integrated over the electron polarization) by

$$\frac{dN}{x^2 \, dx \, d(\cos \theta)} \propto (3 - 2x)$$

$$+ \left( \frac{4}{3} \rho - 1 \right)(4x - 3)$$

$$+ 12 \frac{m_e}{m_\mu} \frac{(1-x)}{x^2} \eta + \left[ (2x - 1) + \right.$$

$$\left. \left( \frac{4}{3} \delta - 1 \right)(4x - 3) \right] \xi \, \mathbf{P}_\mu \cos \theta ,$$

where $m_e$ is the electron mass, $\theta$ is the angle of emission of the electron with respect to the muon polarization vector $\mathbf{P}_\mu$, $m_\mu$ is the muon mass, and $x$ is the reduced electron energy ($x = 2E/m_\mu$ where $E$ is the electron energy). The Michel parameters $\rho$, $\eta$, $\xi$, $\delta$ characterize the spectrum.

The standard model predicts that

$$\rho = \delta = \frac{3}{4}, \quad \xi = 1, \quad \text{and} \quad \eta = 0 .$$

One can also measure several parameters characterizing the longitudinal polarization of the electron and its two transverse components. Table 2 gives the current world average values for the Michel parameters. These data have been used to place limits on the weak interaction coupling constants, as shown in Table 3. As can be seen, the current limits allow up to a 30 per cent admixture of something other than a pure $V - A$ structure. Other analyses, with other model-dependent assumptions, set the limit below 10 per cent.

One of the extensions of the minimal standard model is a theory with left-right symmetry. The gauge symmetry group that embodies the left-handed symmetry would be joined by one for right-handed symmetry, and the charged-current bosons $W^+$ and $W^-$ would be expanded in terms of a symmetric combination of fields $W_L$ and $W_R$. Such an extension is important from a theoretical standpoint for several reasons. First, it restores parity conservation as a high-energy symmetry of the weak interaction. The well-known observation of parity violation in weak processes would then be relegated to the status of a low-energy phenomenon due to the fact that the mass of the right-handed $W$ is much larger than that of the left-handed $W$. Each lepton generation would probably require two neutrinos, a light left-handed one and a very heavy right-handed member.

The dominance of the left-handed charged current at presently accessible energies would be due to a very large mass for $W_R$, but the $W_L - W_R$ mass splitting would still be small on the scale of the grand unification mass $M_X$. Thus the precision study of a weak decay such as ordinary muon decay or nucleon beta decay can be used to set a limit on the left-right symmetry of the weak interaction.

With such plums as the $V - A$ nature of the weak interaction and the existence of right-handed $W$ bosons accessible to such precision studies, it is not surprising that several experimental teams at meson factories are carrying out a variety of studies of ordinary muon decay. One team working at the Canadian facility TRIUMF has already collected data and set a lower limit of 380 GeV on the mass of the right-handed $W$. This was done with a muon beam of only a few MeV!

**The Time Projection Chamber.** A Los Alamos - University of Chicago-NRC Canada collaboration is carrying out a particularly comprehensive and sensitive study of the muon decay spectrum using a novel and elaborate device known as a time projection chamber (TPC).

The TPC (Fig. 11) is a very large volume drift chamber. In a conventional drift chamber, an array of wires at carefully determined potentials collects the ionization left in a gas by a passing charged particle. The time of arrival of the packet of ionization in the cell near each wire is used to calculate the path of the particle through the cell. The gas

## Table 2

**Theoretical and experimental values for the weak-interaction Michel parameters.**

| Michel Parameter | $V - A$ Prediction | Current Value | Expected Los Alamos Accuracy |
|---|---|---|---|
| $\rho$ | ¾ | $0.752 \pm 0.003$ | $\pm 0.00023$ |
| $\eta$ | 0 | $-0.12 \pm 0.21$ | $\pm 0.0061$ |
| $\xi$ | 1 | $0.972 \pm 0.014$ | $\pm 0.001$ |
| $\delta$ | ¾ | $0.755 \pm 0.008$ | $\pm 0.00064$ |

## Table 3

**Experimental limits on the weak-interaction coupling constants, including the expected limit for the Los Alamos Experiment.**

| Constant | Present Limit | Expected Limit |
|---|---|---|
| Axial Vector | $0.76 < g_A < 1.20$ | $0.988 < g_A < 1.052$ |
| Tensor | $g_T < 0.28$ | $g_T < 0.027$ |
| Scalar | $g_S < 0.33$ | $g_S < 0.048$ |
| Pseudo Scalar | $g_P < 0.33$ | $g_T < 0.048$ |
| Vecto-axial Vector Phase | $\varphi_{VA} = 180° \pm 15°$ | $\varphi_{VA} = 180° \pm 2.6°$ |

and the field in the cell are chosen so that the ionization drifts at a constant terminal velocity. Thus the calculation of the position from the drift time can be done accurately. Many drift chambers provide coordinate measurements accurate to less than 100 micrometers.

On the other hand, a TPC uses the same drift velocity phenomenon but employs it in a large volume with no wires in the sensitive region. The path of ionization drifts en masse under the influence of an electric field along the axis of the chamber. The ionization is collected on a series of electrodes, called

pads, on the chamber endcaps, providing precision measurement of trajectory charge and energy. The pad signal also gives a time measurement, relative to the event trigger, that can be used to reconstruct the spatial coordinate of each point on the trajectory.

The TPC in the Los Alamos experiment is placed in a magnetic field sufficiently strong that the decay electrons, whose energies range up to about 53 MeV, follow helical paths. The magnetic field is accurate enough to make absolute momentum measurements of the decay electrons.

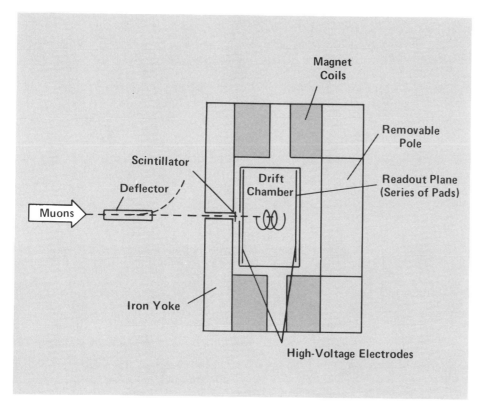

*Fig. 11. The time projection chamber (TPC), a device to study the muon decay spectrum. A beam of muons from LAMPF enters the TPC via a 2-inch beam pipe that extends through the magnet pole parallel to the magnetic field direction. Before entering the chamber, the muons pass through a 10-mil thick scintillator that serves as a muon detector. The scintillator is viewed, via fiber optic light guides, by two photomultiplier tubes located outside the magnet. The thresholds for the discriminators on these photomultiplier channels are adjusted to produce a coincidence for the more heavily ionizing muons while the minimum-ionizing beam electrons are ignored. A deflector located in the beam line 2 meters upstream of the magnet produces a region of crossed electric and magnetic fields through which the beam passes. This device acts first as a beam separator, purifying the muon flux—in particular, reducing the number of electrons in the beam from about 200 to about 1.5 for every muon. The device also acts as a deflector, keeping additional particles out of the chamber by switching off the electric field once a muon has been observed entering the detector. The magnetic field in this detector is provided by an iron-enclosed solenoid, with the maximum field in the current arrangement being 6.6 kilogauss. The field has been carefully measured and found to be uniform to better then 0.6 per cent within the entire TPC-sensitive volume of 52 cm in length by 122 cm in diameter. The TPC readout, on the chamber endcaps, consists of 21 identical modules, each of which has 15 sense wires and 255 pads arranged under the sense wires in rows of 17 pads each. The sense wires provide the high field gradient necessary for gas amplification of the track ionization. The 21 modules are arranged to cover most of the 122-centimeter diameter of the chamber.*

A beam of muons from LAMPF passes first through a device that acts as a beam separator, purifying the muon flux (especially of electrons, which are reduced by this device from an electron-to-muon ratio of 200:1 to about 3:2). The device also acts as a deflector, keeping additional particles from entering the chamber once a muon is inside. With a proper choice of beam intensity, only one muon is allowed in the TPC at a time. Next the beam passes through a 10-mil thick scintillator (serving both as a muon detector and a device used to reject events caused by the remaining beam electrons) and continues into the TPC along a line parallel to the magnetic field direction.

The requirement for an event to be triggered is that one muon enters the TPC during the LAMPF beam pulse and stops in the central 10 cm of the drift region. The entering muon is detected by a signal coincidence from photomultipliers attached to the 10-mil scintillator (this signal operates the deflector that keeps other muons out). The scintillator signal must also be coincident—including a delay that corresponds to the drift time from the central 10 cm of the TPC—with a high level signal from any of the central wires of the TPC. If no delayed coincidence occurs, indicating that the muon did not penetrate far enough into the TPC, or a high level output is detected before the selected time window, indicating that the muon penetrated too far, the event is rejected and all electronics are reset. Then 250 microseconds later (to allow for complete clearing of all tracks in the TPC) the beam is allowed to re-enter for another attempt. The event is also rejected if a second muon enters the TPC during the 200-nanosecond period required to turn off the deflector electric field.

If the event is accepted, the computer reads 20 microseconds of stored data. This corresponds to five muon decay lifetimes plus the 9 microseconds it takes for a track to drift the full length of the TPC.

The experiment is expected to collect about $10^8$ muon decay events, at a trigger rate of 120 events per second, during the next year. Preliminary data have already been taken, showing that the key resolution for electron momentum falls in the target range, namely $\Delta p/p$ is 0.7 per cent averaged over the entire spectrum. Figure 12 shows one of the elegant helical tracks obtained in these early runs.

Ultimately, this experiment will be able to improve upon the four parameters shown in Table 2, although the initial emphasis will be on $\rho$. In the context of left-right symmetric models, an improved measurement of $\rho$ will place a new limit on the allowed mixing angle between $W_R$ and $W_L$ that is almost independent of the mass of the $W_R$.

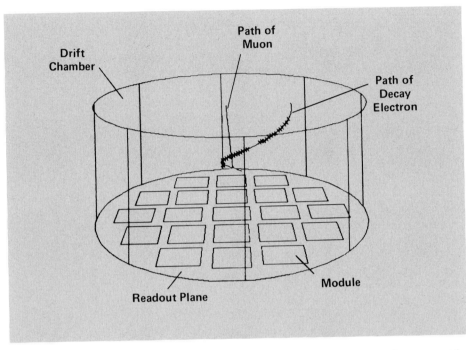

## Summary

The particle physics community is aggressively pursuing research that will lead to verification or elaboration of the minimal standard model. Most of the world-wide activity is centered at the high-energy colliding beam facilities, and the last few years have yielded a bountiful harvest of new results, including the direct observation of the $W^\pm$ and $Z^0$ bosons. Many of the key measurements of the 1980s are likely to be made at the medium-energy facilities, such as LAMPF, or in experiments far from accelerators, deep underground and at reactors, where studies of proton decay, solar neutrino physics, neutrino oscillations, tritium beta decay, and other bellwether research is being carried out. ∎

*Fig. 12. An example of the typical helical track observed for a muon-decay event in an early run with the TPC. (The detector here is shown on end compared to Fig. 11.)*

**Gary H. Sanders** learned his physics on the east coast, starting at Stuyvesant High School in New York City, then Columbia and an A.B. in physics in 1967, and finally a Ph.D. from the Massachusetts Institute of Technology in 1971. The work for his doctoral thesis, which dealt with the photoproduction of neutral rho mesons on complex nuclei, was performed at DESY's electron synchrotron in Hamburg, West Germany under the guidance of Sam Ting. After seven years at Princeton University, during which time he used the beams at Brookhaven National Laboratory and Fermi National Accelerator Laboratory, he came west to join the Laboratory's Medium Energy Physics Division and use the beams at LAMPF. A great deal of his research has dealt with the study of muons and with the design of the beams, detectors, and signal processing equipment needed for these experiments.

# An Experimental Update

In the two years since "Experiments to Test Unification Schemes" was written, each of the four experiments described has completed a substantial program of measurements and published its first results. In one case, the entire program is complete with final results submitted for publication. So far, all results are fully consistent with the minimal standard model. Opportunities for theories with new physics have been substantially constrained.

**Tritium Beta Decay.** Although dissociation into atomic tritium has not yet been employed to make a physics measurement, the study of the tritium beta decay spectrum using *molelcular* tritium has been completed. An upper limit of 26.8 eV (95 per cent confidence level) has been placed on the mass of the electron antineutrino, with a best fit value of zero mass. This result is inconsistent with the best fit value most recently reported by the Russians (30 ± 2 eV) and excludes a large fraction of their latest mass range of 17 to 40 eV. Several other experiments, including those done by teams from Lawrence Livermore National Laboratory, the Swiss Institute for Nuclear Research, and a Japanese group, have also begun to erode the Russian claim of a nonzero neutrino mass. Improvements in these limits, including the Los Alamos atomic tritium measurement, are expected soon.

**Rare Decays of the Muon.** The Crystal Box detector has completed its search for rare decays of the muon and, for the three processes sought, has published (at the 90 per cent confidence level) the following new limits:

$$B(\mu^+ \rightarrow e^+ e^+ e^-) \ < 3.1 \times 10^{-11}$$

$$B(\mu^+ \rightarrow e^+ \gamma) < 4.9 \times 10^{-11}$$

$$B(\mu^+ \rightarrow e^+ \gamma \gamma) < 7.2 \times 10^{-11} \ .$$

An experiment at the Swiss Institute for Nuclear Research has also obtained a limit on the first process of about $2.4 \times 10^{-12}$. These four results place severe lower limits on the masses of new objects that could produce nonconservation of lepton number. For example, in one analysis of deviations from the standard model, the Crystal Box limit on $\mu^+ \rightarrow e^+ \gamma$ sets the scale for new interactions to $10^4$ TeV or higher.

A new and far more ambitious search for $\mu^+ \rightarrow e^+ \gamma$ has been undertaken by some members of the Crystal Box group together with a large group of new collaborators. Their design sensitivity is set at less than $1 \times 10^{-13}$, and their new detector is under construction. In addition, four other groups, using rare decays of the kaon, are searching for processes that violate lepton number conservation, such as $K_L \rightarrow \mu e$ and $K_L \rightarrow \pi \mu e$.

**Neutrino-Electron Scattering.** This experiment has now collected $121 \pm 25 \ \nu_x e^-$ scattering events, of which $99 \pm 25$ are identified as $\nu_e e^-$ scatterings. The resulting cross section agrees with the standard electroweak theory, rules out constructive interference between weak charged-current and neutral-current interactions, and favors the existence of destructive interference between these two interactions. Additional data are being collected, and this result will be made considerably more precise.

**Normal Muon Decay.** After preliminary studies, the team using the time projection chamber to study normal muon decay decided to concentrate on the measurement of the $\rho$ parameter (Table 2). Since this parameter is measured by averaging over the muon spin, it was not necessary to preserve the spin direction of the muon stopping in the chamber. The researchers used an improved entrance separator to rotate the spin perpendicular to the beam direction, and precession in the chamber magnetic field then averaged the polarization. They also took advantage of a small entrance scintillator to trigger the apparatus on muon stops. This technique purified the experimental sample but perturbed the muon spin, which, however, is acceptable for the measurement of $\rho$. A higher event rate was possible because the entrance scintillator signal eliminated the need for the beam to be pulsed. The scintillator, by itself, effectively ensured a single stopping muon. In this mode, the group collected $5 \times 10^7$ events, which are now being analyzed, and this sample is expected to sharpen the knowledge of $\rho$ by a factor of 5. Limits on charged right-handed currents from a related measurement have now been reported by the TRIUMF group.

The minimal standard model has, to date, survived these demanding tests. Where is the edge of its validity? We shall have to wait a little longer to find the answer. Experimentalists are already mounting the next round of detectors in this inquiry. ∎

# the march toward

I n the Book of Genesis, we are told that ... *unto Enoch was born Irad: and Irad begat Mehujael: and Mehujael begat Methusael: and Methusael begat Lamech. And Lamech* ... And so it is with particle accelerators! Each generation of these machines answers a set of important questions, makes some fundamental discoveries, and gives rise to new questions that can be answered only by a new generation of accelerators, usually of higher energy than the previous one. For example, in the decade of the 1950s, the Berkeley Bevatron was built to confirm the existence of the antiproton, and it was subsequently used to discover an unexpected array of new "particles." These were our earliest clues about the existence of quarks but were not recognized as such until 1964, when the $\Omega^-$ particle was discovered at the Brookhaven AGS, a much more powerful proton accelerator than the Bevatron. In more recent times the brilliant discovery of the $W^\pm$ and $Z^0$ bosons at the CERN $Sp\bar{p}S$, a proton-antiproton collider that imparts ten times more energy to particle beams than the AGS, has confirmed the Nobel-prize-winning gauge theory of Glashow, Weinberg, and Salam. And now we are faced with understanding the physics behind the masses of these bosons, which will require an accelerator at least ten times more powerful than the $Sp\bar{p}S$!

# higher energies

*by S. Peter Rosen*

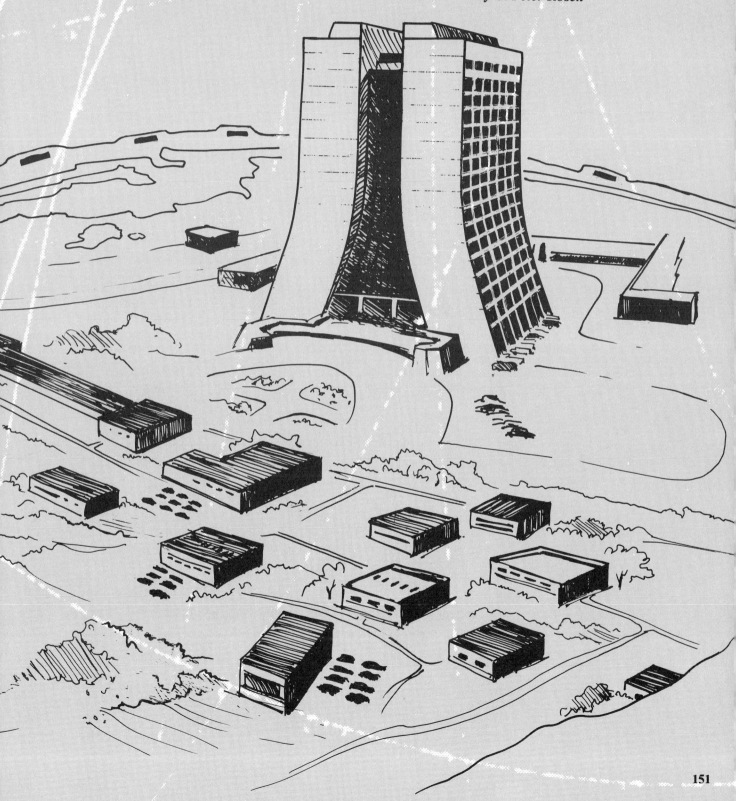

When these questions have been answered, we may expect the cycle to repeat itself until we run out of resources—or out of space. So far the field of particle physics has been fortunate: every time it seems to have reached the end of the energy line, some new technical development has come along to extend it into new realms. Synchrotrons such as the Bevatron and the Cosmotron, its sister and rival at Brookhaven, both represented an order-of-magnitude improvement over synchrocyclotrons, which in their time overcame relativistic problems to extend the energy of cyclotrons from tens of MeV into the hundreds. What allowed these developments was the synchronous principle invented independently by E. McMillan at Berkeley and V. Veksler in the Soviet Union.

In a cyclotron a proton travels in a circular orbit under the influence of a constant magnetic field. Every time it crosses a particular diameter, it receives an accelerating kick from an rf electric field oscillating at a constant frequency equal to the orbital frequency of the proton at some (low) kinetic energy. Increasing the kinetic energy of the proton increases the radius of its orbit but does not change its orbital frequency until the effects of the relativistic mass increase become significant. For this reason a cyclotron cannot efficiently accelerate protons to energies above about 20 MeV. The solution introduced by McMillan and Veksler was to vary the frequency of the rf field so that the proton and the field remained in synchronization. With such synchrocyclotrons proton energies of hundreds of MeV became accessible.

In a synchrotron the protons are confined to a narrow range of orbits during the entire acceleration cycle by varying also the magnetic field, and the magnetic field can then be supplied by a ring of magnets rather than by the solid circular magnet of a cyclotron. Nevertheless, the magnets in early synchrotrons were still very large, requiring 10,000 tons of iron in the case of the Bevatron, and for all practical purposes the synchrotron appeared to have reached its economic limit with this 6-GeV machine. Just at the right

time a group of accelerator physicists at Brookhaven invented the principle of "strong focusing," and Ernest Courant, in May 1953, looked forward to the day when protons could be accelerated to 100 GeV—fifty times the energy available from the Cosmotron—with much smaller magnets! In the meantime Courant and his colleagues contented themselves with building a machine ten times more energetic, namely, the AGS (Alternating Gradient Synchrotron).

Courant proved to be most farsighted, but even his optimistic goal was far surpassed in the twenty years following the invention of strong focusing. The accelerator at Fermilab (Fermi National Accelerator Laboratory) achieved proton energies of 400 GeV in 1972, and at CERN (Organisation Europeene pour Recherche Nucléaire) the SPS (Super Proton Synchrotron) followed suit in 1976. Size is the most striking feature of these machines. Whereas the Bevatron had a circumference of 0.1 kilometer and could easily fit into a single building, the CERN and Fermilab accelerators have circumferences between 6 and 7 kilometers and are themselves hosts to large buildings.

Both the Fermilab accelerator and the SPS are capable of accelerating protons to 500 GeV, but prolonged operation at that energy is prohibited by excessive power costs. This economic hurdle has recently been overcome by the successful development of superconducting magnets. Fermilab has now installed a ring of superconducting magnets in the same tunnel that houses the original main ring and has achieved proton energies of 800 GeV, or close to 1 TeV. The success of the Tevatron, as it is called, has convinced the high-energy physics community that a 20-TeV proton accelerator is now within our technological grasp, and studies are under way to develop a proposal for such an accelerator, which would be between 90 and 160 kilometers in circumference. Whether this machine, known as the SSC (Superconducting Super Collider), will be the terminus of the energy line, only time will tell; but if the past is any guide, we can expect some-

thing to turn up. (See "The SSC—An Engineering Challenge.")

Paralleling the higher and higher energy proton accelerators has been the development of electron accelerators. In the 1950s the emphasis was on linear accelerators, or linacs, in order to avoid the problem of energy loss by synchrotron radiation, which is much more serious for the electron than for the more massive proton. The development of linacs culminated in the two-mile-long accelerator at SLAC (Stanford Linear Accelerator Center), which today accelerates electrons to 40 GeV. This machine has had an enormous impact upon particle physics, both direct and indirect.

The direct impact includes the discovery of the "scaling" phenomenon in the late 1960s and of parity-violating electromagnetic forces in the late 1970s. By the scaling phenomenon is meant the behavior of electrons scattered off nucleons through very large angles: they appear to have been deflected by very hard, pointlike objects inside the nucleons. In exactly the same way that the experiments of Rutherford revealed the existence of an almost pointlike nucleus inside the atom, so the scaling experiments provided a major new piece of evidence for the existence of quarks. This evidence was further explored and extended in the '70s by neutrino experiments at Fermilab and CERN.

Whereas the scaling phenomenon opened a new vista on the physics of nucleons, the 1978 discovery of parity violation in the scattering of polarized electrons by deuterons and protons closed a chapter in the history of weak interactions. In 1973 the phenomenon of weak neutral currents had been discovered in neutrino reactions at the CERN PS (Proton Synchrotron), an accelerator very similar in energy to the AGS. This discovery constituted strong evidence in favor of the Glashow-Weinberg-Salam theory unifying electromagnetic and weak interactions. During the next five years more and more favorable evidence accumulated until only one vital piece was missing—the demonstration of parity violation in electron-nucleon

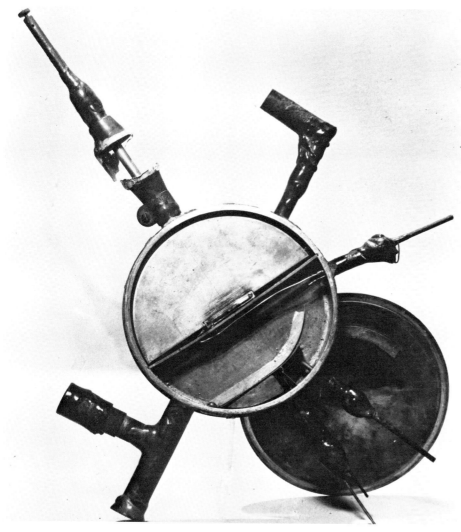

*The "string and sealing wax" version of a cyclotron. With this 4-inch device E. O. Lawrence and graduate student M. S. Livingston successfully demonstrated the feasibility of the cyclotron principle on January 2, 1931. The device accelerated protons to 80 keV. (Photo courtesy of Lawrence Berkeley Laboratory.)*

GeV per beam. In the fall of 1974, the ψ particle, which provided the first evidence for the fourth, or charmed, quark was found among the products of electron-positron collisions at SPEAR; at the same time the *J* particle, exactly the same object as ψ, was discovered in proton collisions at the AGS. With the advent of *J*/ψ, the point of view that all hadrons are made of quarks gained universal acceptance. (The up, down, and strange quarks had been "found" experimentally; the existence of the charmed quark had been postulated in 1964 by Glashow and J. Bjorken to equalize the number of quarks and leptons and again in 1970 by Glashow, J. Iliopoulos, and L. Maiani to explain the apparent nonoccurrence of strangeness-changing neutral currents.

The discovery of *J*/ψ, together with the discovery of neutral currents the year before, revitalized the entire field of high-energy physics. In particular, it set the building of electron-positron storage rings going with a vengeance! Plans were immediately laid at SLAC for PEP (Positron Electron Project), a larger storage ring capable of producing 18-GeV beams of electrons and positrons, and in Hamburg, home of DORIS (Doppel-Ring-Speicher), the European counterpart of SPEAR, a 19-GeV storage ring named PETRA (Positron Electron Tandem Ring Accelerator) was designed. Subsequently a third storage ring producing 8-GeV beams of positrons and electrons was built at Cornell; it goes by the name of CESR (Cornell Electron Storage Ring).

Although the gluon, the gauge boson of quantum chromodynamics, was discovered at PETRA, and the surprisingly long lifetime of the *b* quark was established at PEP, the most interesting energy range turned out to be occupied by CESR. Very shortly before this machine became operative, L. Lederman and his coworkers, in an experiment at Fermilab similar to the *J* experiment at Brookhaven, discovered the ϒ particle at 9.4 GeV; it is the *b*-quark analogue of *J*/ψ at 3.1 GeV. By good fortune CESR is in just the right energy range to explore the properties of the ϒ system, just as SPEAR was able to

reactions at a very small, but precisely predicted, level. In a brilliant experiment C. Prescott and R. Taylor and their colleagues found the missing link and thereby set the seal on the unification of weak and electromagnetic interactions.

A less direct but equally significant impact of the two-mile linac arose from the electron-positron storage ring known as SPEAR

(Stanford Positron Electron Accelerating Ring). Electrons and positrons from the linac are accumulated in two counterrotating beams in a circular ring of magnets and shielding, which, from the outside, looks like a reconstruction of Stonehenge. Inside, enough rf power is supplied to overcome synchrotron radiation losses and to allow some modest acceleration from about 1 to 4

elucidate the ψ system. Many interesting results about Υ, its excited states, and mesons containing the *b* quark are emerging from this unique facility at Cornell.

The next round for positrons and electrons includes two new machines, one a CERN storage ring called LEP (Large Electron-Positron) and the other a novel facility at SLAC called SLC (Stanford Linear Collider). LEP will be located about 800 meters under the Jura Mountains and will have a circumference of 30 kilometers. Providing 86-GeV electron and positron beams initially and later 130-GeV beams, this machine will be an excellent tool for exploring the properties of the $W^\pm$ bosons. SLC is an attempt to overcome the problem of synchrotron radiation losses by causing two *linear* beams to collide head on. If successful, this scheme could well establish the basic design for future machines of extremely high energy. At present SLC is expected to operate at 50 GeV per beam, an ideal energy with which to study the $Z^0$ boson.

High energy is not the only frontier against which accelerators are pushing. Here at Los Alamos LAMPF (Los Alamos Meson Physics Facility) has been the scene of pioneering work on the frontier of high intensity for more than ten years. At present this 800-MeV proton linac carries an average current of 1 milliampere. To emphasize just how great an intensity that is, we note that most of the accelerators mentioned above hardly ever attain an average current of 10 microamperes. LAMPF is one of three so-called meson factories in the world; the other two are highly advanced synchrocyclotrons at TRIUMF (Tri-University Meson Facility) in Vancouver, Canada, and at SIN (Schweizerisches Institut für Nuklearforschung) near Zurich, Switzerland.

The high intensity available at LAMPF has given rise to fundamental contributions in nuclear physics, including confirmation of the recently developed Dirac formulation of nucleon-nucleus interactions and discovery of giant collective excitations in nuclei. In addition, its copious muon and neutrino

*A state-of-the art version of a proton synchrotron. Here at Fermilab protons will be accelerated to an energy close to 1 TeV in a 6562-foot-diameter ring of superconducting magnets. Wilson Hall, headquarters of the laboratory and a fitting monument to a master accelerator builder, appears at the lower left. (Photo courtesy of Fermi National Accelerator Laboratory.)*

beams have been applied to advantage in particle physics, especially in the areas of rare modes of particle decay and neutrino physics.

The search for rare decay modes (such as $\mu^+ \rightarrow e^+ + \gamma$) remains high on the agenda of particle physics because our present failure to see them indicates that certain conservation laws seem to be valid. Grand unified theories of strong and electroweak interactions tell us that, apart from energy and momentum, the only strictly conserved quantity is electric charge. According to these theories, the conservation of all other quantities, including lepton number and baryon number, is only approximate, and violations of these conservation laws must occur, although perhaps at levels the minutest of the minute.

Meson factories are ideally suited to the search for rare processes, and here at Los Alamos, at TRIUMF, and at SIN plans are being drawn up to extend the range of present machines from pions to kaons. (See "LAMPF II and the High-Intensity Frontier.") Several rare decays of kaons can provide important insights into grand uni-

fied theories, as well as into theories that address the question of $W^\pm$ and $Z^0$ masses, and so the search for them can be expected to warm up in the next few years.

Another reason for studying kaon decays is CP violation, a phenomenon discovered twenty years ago at the AGS and still today not well understood. Because the effects of CP violation have been detected only in kaon decays and nowhere else, extremely precise measurements of the relevant parameters are needed to help determine the underlying cause. In this case too, kaon factories are very well suited to attack a fundamental problem of particle physics.

In the area of neutrino physics, LAMPF has made important studies of the identity of neutrinos emitted in muon decay and is now engaged in a pioneering study of neutrino-electron scattering. High-precision measurements of the cross section are needed as a test of the Glashow-Weinberg-Salam theory and are likely to be a major part of the experimental program at kaon factories.

While the main thrust of particle physics has always been carried by accelerator-based experiments, there are, and there have always been, important experiments performed without accelerators. The first evidence for strange particles was found in the late 1940s in photographic emulsions exposed to cosmic rays, and in 1956 the neutrino was first detected in an experiment at a nuclear reactor. In both cases accelerators took up these discoveries to explore and extend them as far as possible.

Another example is the discovery of parity nonconservation in late 1956. The original impetus came from the famous $\tau$-$\theta$ puzzle concerning the decay of $K$ mesons into two and three pions, and it had its origins in accelerator-based experiments. But the definitive experiment that demonstrated the nonconservation of parity involved the beta decay of cobalt-60. Further studies of nuclear beta decay led to a beautiful clarification of the Fermi theory of weak interactions and laid the foundations for modern gauge theories. The history of this era reveals a remarkable interplay between accelerator and non-accelerator experiments.

In more recent times the solar neutrino experiment carried out by R. Davis and his colleagues deep in a gold mine provided the original motivation for the idea of neutrino oscillations. Other experiments deep underground have set lower limits of order $10^{32}$ years on the lifetime of the proton and may yet reveal that "diamonds are not forever."

And the limits set at reactors on the electric dipole moment of the neutron have proved to be a most rigorous test for the many models of CP violation that have been proposed.

In 1958, a time of much expansion and optimism for the future, Robert R. Wilson, the master accelerator builder, compared the building of particle accelerators in this century with the building of great cathedrals in 12th and 13th century France. And just as the cathedral builders thrust upward toward Heaven with all the technical prowess at their command, so the accelerator builders strive to extract ever more energy from their mighty machines. Just as the cathedral builders sought to be among the Heavenly Hosts, bathed in the radiance of Eternal Light, so the accelerator builders seek to unlock the deepest secrets of Nature and live in a state of Perpetual Enlightenment:

> *Ah, but a man's reach should exceed*
> *his grasp,*
> *Or what's a heaven for?*
>
> Robert Browning

Wilson went on to build his great accelerator, and his cathedral too, at Fermilab near Batavia, Illinois. In its time, the early to mid 1970s, the main ring at Fermilab was the most powerful accelerator in the world, and it will soon regain that honor as the Tevatron begins to operate. The central laboratory building, Wilson Hall, rises up to sixteen stories like a pair of hands joined in prayer, and it stands upon the plain of northcentral Illinois much as York Minster stands upon the plain of York in England, visible for miles around. Some wag once dubbed the laboratory building "Minster Wilson, or the Cathedral of St. Robert," and he observed that the quadrupole logo of Fermilab should be called "the Cross of Batavia." But Wilson Hall serves to remind the citizens of northern Illinois that science is ever present in their lives, just as York Minster reassured the peasants of medieval Yorkshire that God was always nearby.

The times we live in are much less optimistic than those when Wilson first made his comparison, and our resources are no longer as plentiful for our needs. But we may draw comfort from the search for a few nuggets of truth in an uncertain world.

> *To gaze up from the ruins of the*
> *oppressive present towards the stars is*
> *to recognise the indestructible world of*
> *laws, to strengthen faith in reason, to*
> *realise the "harmonia mundi" that*
> *transfuses all phenomena, and that*
> *never has been, nor will be, disturbed.*
>
> Hermann Weyl, 1919

**S. Peter Rosen,** a native of London, was educated at Merton College, Oxford, receiving from that institution a B.S. in mathematics in 1954 and both an M.A. and D.Phil. in theoretical physics in 1957. Peter first came to this country as a Research Associate at Washington University and then worked as Scientist for the Midwestern Universities Research Association at Madison, Wisconsin. A NATO Fellowship took him to the Clarendon Laboratory at Oxford in 1961. He then returned to the United States to Purdue University, where he retains a professorship in physics. Peter has served as Senior Theoretical Physicist for the U.S. Energy Research and Development Administration's High Energy Physics Program, as Program Associate for Theoretical Physics with the National Science Foundation, and as Chairman of the U.S. Department of Energy's Technical Assessment Panel for Proton Decay and on the Governing Board for the Lewes Center for Physics in Delaware. His association with the Laboratory extends back to 1977 when he came as Visiting Staff Member. He has served as Consultant with the Theoretical Division and as a member of the Program Advisory Committee and Chairman of the Neutrino Subcommittee of LAMPF. He is currently Associate Division Leader for Nuclear and Particle Physics of the Theoretical Division. Peter's research specialties are symmetries of elementary particles and the theory of weak interactions.

# The Next Step in Energy

Two years have passed since this article was written, and the high-energy physics commuinity is now poised to take the next major step forward in energy. Between now and 1990 a progression of new accelerators (see table) will raise the center-of-mass energy of proton-antiproton collisions to 2 TeV and that of electron-positron collisions to 100 GeV—more than enough to produce $W^\pm$ and $Z^0$ bosons in large quantities. In addition, the HERA accelerator at DESY will enable us to collide an electron beam with a proton beam, producing over 300 GeV in the center of mass. Thus over the next five years we can look forward to a wealth of new data and much new physics.

If the past is any guide, we can anticipate many surprises and discoveries of new phenomena as the energy of accelerators marches upward. But even if we are surprised by a lack of surprises, there is still much important physics to be explored in this new domain. As explained in the article by Raby, Slansky, and West, we have a "standard model" of particle physics that does a beautiful job of describing all known phenomena but has the unsatisfactory feature of requiring far too many arbitrary parameters to be put in by hand. It is therefore important to look for physics beyond the standard model, the discovery of which could lead us to a more general, more highly unified model with far fewer arbitrary parameters.

One avenue for searching beyond the standard model is the precise measurement of properties of the particles it includes. For example, the model provides a well-defined relationship between the masses of the $W^\pm$ and $Z^0$ bosons on the one hand and the weaᵗ

neutral-current mixing angle θ on the other. The theoretical corrections to this relationship due to virtual quantum mechanical processes (the so-called radiative corrections) can be reliably calculated in perturbation theory. To achieve the experimental precision of 1 percent required to test these calculations, we must observe many thousands of events.

Because of their higher energies, both the Tevatron and SLC are expected to be much more copious sources of electroweak bosons than the $S\bar{p}pS$ facility at which they were discovered. Whereas the $S\bar{p}pS$ has produced approximately 200 $Z^0 \to e^+e^-$ and 2000 $W^\pm \to e^\pm \nu$ events in a period of three to four

years, the design luminosity of the Tevatron is such that it should yield 1500 $Z^0 \to \mu^+\mu^-$ and 15,000 $W^\pm \to \ell^\pm \nu_\ell$ in a good year. Even more impressive is SLC, which will produce 3,000,000 $Z^0$ bosons per year at its design luminosity! So we look forward to precise determinations of the properties of the $W^\pm$ bosons from the Tevatron and of the $Z^0$ boson from SLC.

Another fundamental test for the standard model is the existence of a neutral scalar boson, a component of the Higgs boson multiplet responsible for generating the masses of the $W^\pm$ and $Z^0$ gauge bosons. While theory imposes a lower limit of a few GeV on the mass of this Higgs particle, it

---

**Information about the new colliding-beam accelerators that are planned to be in operation by 1990.**

| Accelerator | Location | Types and Energies of Colliding Beams | Operational Date |
|---|---|---|---|
| Tevatron | Fermilab, Illinois | 1-TeV proton 1-TeV antiproton | December 1986 |
| Tristan | KEK, Japan | 30-GeV electron 30-GeV positron | December 1986 |
| SLC | SLAC, California | 50-GeV electron 50-GeV positron | Spring 1987 |
| LEP | CERN, Geneva | 50-GeV electron 50-GeV positron | Fall 1988 |
| HERA | DESY, Hamburg | 30-GeV electron 820-GeV proton | Spring 1990 |

gives no firm prediction for its magnitude, nor even an upper bound, and so we have to conduct a systematic search over a wide band of energies. Should the mass of the Higgs particle be less than that of the $Z^0$, then we have a good chance of finding it at SLC through such processes as $Z^0 \to H^0\gamma$, $H^0 e^+ e^-$, and $H^0\mu^+\mu^-$. If the Higgs particle is more massive than the $Z^0$, then it may show up at the Tevatron through such decays as $H^0 \to Z^0\gamma$ and $Z^0\mu^+\mu^-$. Should it prove to be beyond the range of the Tevatron, then we shall have to wait until the SSC comes on line in the mid 1990s.

Another avenue for exploration beyond the standard model is the search for particles it does not include. For example, are there more than three families of fermions? Precisions studies of the width for $Z^0 \to \Sigma_\ell \bar{\nu}_\ell \nu_\ell$ will enable us to count the number of neutrino species, while apparently anomalous decays of the $W^+$ will enable us to detect new charged leptons, provided of course that the lepton mass is less than that of the $W^+$. In a similar way, decays of $W^\pm$ into jets of hadrons may reveal the existence of new heavy quarks, including the top quark required to fill out the existing three families of elementary fermions. A hint of the top quark was found by the UA1 detector at the $S\bar{p}pS$, but there were too few events for it to be convincing. The much higher event rates of the new accelerators will be extremely useful in these searches.

Besides the Higgs scalar boson and further replications of the known fermion families, there are hosts of new particles predicted by theories that unify the strong and electroweak interactions with one another and with gravity. Some of these ideas are discussed in "Toward a Unified Theory" and "Supersymmetry at 100 GeV." Perhaps the most prevalent of such ideas is that of supersymmetry, which predicts that for every particle there exists a sypersymmetric partner, or sparticle, differing in spin by ½ and obeying opposite statistics. Thus for each fermion there exists a scalar boson, an *s* fermion, and for each gauge boson there exists a fermion called a gaugino. Some of these sparticles

could be sufficiently light, between a few GeV and a few tens of GeV, to be in the mass range that can be explored with the new accelerators.

There may also be new interactions that appear only at the higher mass scales open to the new accelerators. Right-handed currents are absent from known low-energy weak interactions, possibly because the corresponding gauge bosons are much heavier than the $W^\pm$ and $Z^0$ of the standard model. Depending upon their masses, these gauge bosons could be produced at LEP II (an energy upgrade of LEP) or at the Tevatron; the tails of their propagators might even show up at SLC and at LEP itself.

Another way of searching for new interactions is provided by the electron-proton collider HERA, located at DESY in Hamburg. There collision of 820-GeV protons with 30-GeV electrons provides 314 GeV in the center of mass and momentum transfers as large as $10^5$ GeV$^2$. Furthermore the electrons are naturally polarized perpendicular to the plane of the ring around which they rotate, and this polarization can easily be converted to a longitudinal direction, left-handed or right-handed. We know how the left-handed state will interact through the known $W^\pm$ and $Z^0$; with the right-handed state we can see if a new type of weak interaction comes into play at these high energies.

As the energy of accelerators increases, so the resulting collisions become less like interactions of the complicated hadronic structures that make up protons and antiprotons and more like collisions between the elementary constituents of these structures, namely quarks and gluons. In electron-positron collisions we begin with what we believe are two elementary and point-like objects. (Think of them as the ideal mass points of mechanics rather than the heavy balls found upon the billiard table.) The hadrons produced in low-energy collisions tend to emerge over large angles, but, as the incident energy increases, so does the tendency for the hadrons to become highly correlated in a small number of directions. These correlated groupings of hadrons are called jets, and we believe that

they signify, as closely as is physically possible, the production of quarks and gluons. These elementary objects cannot emerge from the collision regions as free particles because of the confinement properties and the color neutrality of the strong force of quantum chromodynamics. Instead they "hadronize" into jets of highly correlated groupings of color-neutral hadrons. In practice gluon jets, which were first discovered at the Doris accelerator in Hamburg, are slightly fatter than quark jets, which were originally found at PEP.

From such a point of view, the Tevatron becomes a quark-quark, quark-gluon, and gluon-gluon collider, while HERA is an electron-quark collider. The reduction to elementary fermions and bosons enables us to interpret events much more simply than might otherwise be possible, but it does reduce the effective energy available for collisions. At the $S\bar{p}pS$, for example, gluons take up half of the energy of the proton, and so each quark has, on the average, one-sixth of the energy of the parent proton. At the SSC the fraction will be somewhat lower. We can therefore anticipate that in the decades to come there will be a strong impetus to push available energies well beyond that of the SSC.

Theoretical motivation for the continued thrust toward higher energies may come from the notion of compositeness. Fifty years ago the electron and proton were thought to be elementary objects, but we know today that the proton is far from elementary. It is possible that in the coming period of experimentation we will discover that electrons also are not elementary, but are made up of other, more fundamental entities. Indeed there are theories in which leptons and quarks are all composite objects, made from things called rishons, or preons. Should this be the case, we will need energies much higher than that of the SSC to explore their properties. The lesson is very simple: at whatever energy scale we may be located, there is always much more to learn. Today's elementary particles may be tomorrow's atoms. ∎

# LAMPF II and the High-Intensity Frontier

*by Henry A. Thiessen*

A small Los Alamos group has spent the past two years planning an addition to LAMPF, the 800-MeV, 1-milliampere proton linac on Mesita de Los Alamos. Dubbed LAMPF II and consisting of two high-current synchrotrons fed by LAMPF, the addition will provide beams of protons with a maximum energy of 45 GeV and a maximum current of 200 microamperes. Compared to its best existing competitor, the AGS at Brookhaven National Laboratory, LAMPF II will produce approximately 90 times more neutrinos, 300 times more kaons, and 1000 times more antiprotons. Figure 1 shows a layout of the proposed facility.

## Why Do We Need LAMPF II?

The new accelerator will continue the tradition set by LAMPF of operating in the intersection region between nuclear physics and particle physics. Other articles in this issue ("The Family Problem" and "Experiments To Test Unification Schemes") have discussed crucial experiments in particle physics that require high-intensity beams of secondary particles. For example, the large mass estimated for a "family vector boson" implies that, now and for the foreseeable future, the possibility of family-changing interactions can be in-

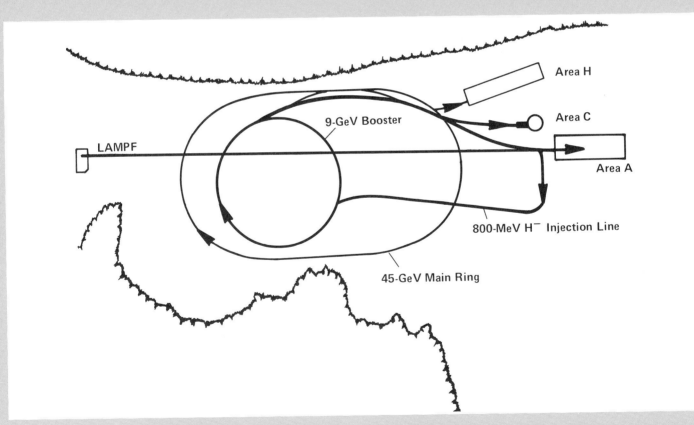

*Fig. 1. LAMPF II, the proposed addition to LAMPF, is designed to produce protons beams with a maximum energy of 45 GeV and a maximum current of 200 microamperes. These proton beams will provide intense beams of antiprotons, kaons, muons, and neutrinos for use in experiments important to both particle and nuclear physics. The addition consists of two synchrotrons, both located 20 meters below the existing LAMPF linac. The booster is a 9-GeV, 60- hertz, 200-microampere machine fed by LAMPF, and the main ring is a 45-GeV, 6-hertz, 40-microampere machine. Proton beams will be delivered to the main experimental area of LAMPF (Area A) and to an area for experiments with neutrino beams and short, pulsed beams of other secondary particles (Area C). A new area for experiments with high-energy secondary beams (Area H) will be constructed to make full use of the 45-GeV proton beam.*

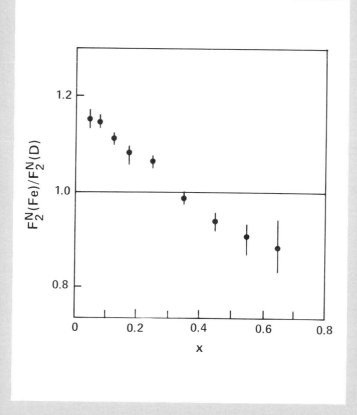

**Fig. 2. The "EMC effect" was first observed in data on the scattering of muons from deuterium and iron nuclei at high momentum transfer. The ratio of the two nucleon structure functions ($F_2^N(Fe)$ and $F_2^N(D)$) deduced from these data by regarding a nucleus as simply a collection of nucleons is shown above as a function of x, a parameter representing the fraction of the momentum carried by the nucleon struck in the collision. The observed variation of the ratio from unity is quite contrary to expectations; it can be interpreted as a manifestation of the quark substructure of the nucleons within a nucleus. (Adapted from J. J. Aubert et al. (The European Muon Collaboration), Physics Letters 123B(1983):175.)**

vestigated only with high-intensity beams of kaons and muons. And studies of neutrino masses and neutrino-electron scattering, which are among the most important tests of possible extensions of the standard model, demand high-intensity beams of neutrinos to compensate for the notorious infrequency of their interactions.

Here I take the opportunity to discuss some of the experiments in *nuclear* physics that can be addressed at LAMPF II. The examples will include the search for quark effects with the Drell-Yan process, the production of quark-gluon plasma by annihilation of antiprotons in nuclei, the extraction of nuclear properties from hypernuclei, and low-energy tests of quantum chromodynamics.

**Quark Effects.** A major problem facing today's generation of nuclear physicists is to develop a model of the nucleus in terms of its fundamental constituents—quarks and gluons. In terms of nucleons the venerable nuclear shell model has been as successful at interpreting nuclear phenomena as its analogue, the atomic shell model, has been at interpreting the structure and chemistry of atoms. But nucleons are known to be made of quarks and gluons and thus must possess some additional internal degrees of freedom. Can we see some of the effects of these additional degrees of freedom? And then can we use these observations to construct a theory of nuclei based on quarks and gluons?

Defining an experiment to answer the first question is difficult for two reasons. First, we know from the success of the shell model that nucleons dominate the observable properties of nuclei, and when this model fails, the facts can still be explained in terms of the exchange of pions or other mesons between the nucleons. Second, the current theory of quarks and gluons (quantum chromodynamics, or QCD) is simple only in the limit of extremely high energy and extremely high momentum transfer, the domain of "asymptotic QCD." But the world of nuclear physics is very far from that domain. Thus, theoretical guidance from the more complicated domain of low-energy QCD is sparse.

To date no phenomenon has been observed that can be interpreted unambiguously as an effect of the quark-gluon substructure of nucleons. However, the results of an experiment at CERN by the "European Muon Collaboration"[1] are a good candidate for a quark effect, although other explanations are possible. This group determined the nuclear structure functions for iron and deuterium from data on the inelastic scattering of muons at high momentum transfers. (A nuclear structure function is a multiplicative correction to the Mott cross section; it is indicative of the momentum distribution of the quarks within the nucleus.) From these structure functions they then inferred values for the nucleon structure function by assuming that the nucleus is simply a collection of nucleons. (If this assumption were true, the inferred nucleon structure function would not vary from nucleus to nucleus.) Their results (Fig. 2) imply that an iron nucleus contains more high-momentum quarks and fewer low-momentum quarks than does deuterium. This was quite unexpected but was quickly corroborated by a re-analysis[2] of some ten-year-old electron-scattering data from SLAC and has now been confirmed in great detail by several new experiments.[3,4] The facts are clear, but how are they to be interpreted?

The larger number of low-momentum quarks in iron than in deuterium may mean that the quarks in iron are sharing their momenta, perhaps with other quarks through formation of, say, six-quark states. Another interpretation, that iron contains many more pions acting as nuclear "glue" than does deuterium, has already been discounted by the results of a LAMPF experiment on the scattering of polarized protons from hydrogen and lead.[5] Whatever the final interpretation of the "EMC effect" may be, it clearly indicates that the internal structure of the nucleon changes in the nucleus.

Interpretation of the EMC effect is complicated by the fact that the contribution of the "valence" quarks (the three quarks that predominantly make up a nucleon) to the lepton-scattering amplitude is not distinguishable from the contribution of the "sea" quarks (the virtual quark-antiquark pairs that can exist within the nucleon for short times). One way to sort out these contributions is to measure the amplitude for production of lepton-antilepton pairs in high-energy hadron-hadron collisions.[6] When the momentum of the lepton-antilepton pair transverse to the hadron beam is small, the dominant amplitude for this Drell-Yan process arises from the annihilation of a quark and an antiquark into a photon, which then decays into the lepton-antilepton pair (Fig. 3). Since valence and sea quarks from different hadronic probes make different contributions to the amplitude, measurement of these differences with the 45-GeV proton beam of LAMPF II and its secondary beams of pions, kaons, and antiprotons can help to decide among the possible explanations of the EMC effect.

**Quark-Gluon Plasma.** Quantum chromodynamics predicts that at a sufficiently high temperature or density the vacuum can turn into a state of quarks, antiquarks, and gluons called quark-gluon plasma. (Such a plasma is expected to have been formed in the first few microseconds after the creation of the universe.) The present generation of relativistic heavy-ion experiments is designed to produce this plasma by achieving high density. However, since the predicted uncertainty in the transition temperature is much smaller than the predicted uncertainty in the transition density, achieving high temperature is regarded as the better approach to producing such a plasma.

D. Strottman and W. Gibbs of Los Alamos have investigated the possibility of heating a nucleus to the required high temperature by annihilation of high-energy antiprotons within the nucleus.[7] The results of a calculation by Strottman (Fig. 4), which were based on a hydrodynamic model, indicate that in a nearly head-on collision between a 10-GeV antiproton and a uranium nucleus, most of the available energy is deposited within the nucleus, raising its temperature to that necessary for formation of the quark-gluon plasma. Gibbs has performed such a calculation with the intranuclear cascade model and obtained very similar results.

Like relativistic heavy-ion experiments, such antiproton experiments pose two problems: isolating from among many events the rare head-on collisions and finding a signature of the transition to plasma. The high intensity of antiprotons to be available at LAMPF II will

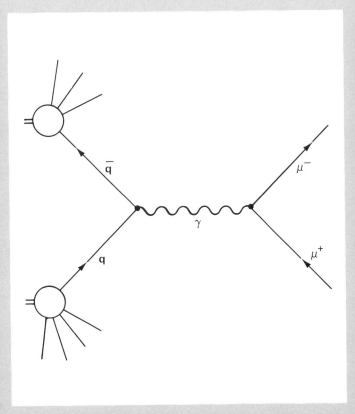

**Fig. 3. The Drell-Yan process is the name given to the production of a lepton-antilepton pair in a collision between two hadrons. When the momentum of the lepton pair transverse to the projectile hadron is small, the dominant amplitude for the Drell-Yan process arises from the interaction pictured above: a quark and an antiquark from the two hadrons annihilate to form a photon, which then decays into the lepton-antilepton pair (here shown as a muon-antimuon pair).**

help solve these problems by providing large numbers of events for study.

**Nuclear Properties from Hypernuclei.** A "hypernucleus" is a nucleus in which a neutron is replaced by a strange heavy baryon, the Lambda ($\Lambda$). (The valence-quark composition of a neutron is *udd*, and that of a $\Lambda$ is *uds*.) Such hypernuclei are produced in collisions of kaons with ordinary nuclei. The properties of hypernuclei are accessible to measurement because their lifetimes are relatively long (similar to that of the free $\Lambda$, about $10^{-10}$ second). These properties provide information about the forces among the nucleons with the nucleus. In fact, the $\Lambda$ plays a role in studies of the nuclear environment similar

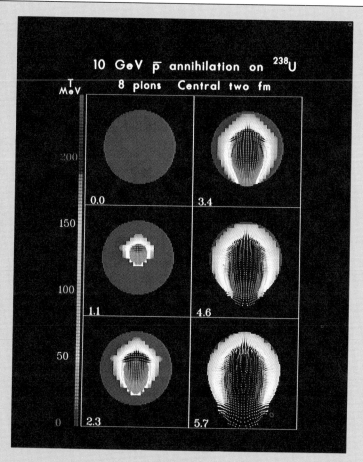

*Fig. 4. A color-coded computer-graphic display of the temperature (in MeV) within a uranium-238 nucleus at various times (in $10^{-23}$ second) after annihilation of a 10-GeV antiproton with a nucleon. (The temperatures were calculated by D. Strottman on the basis of a hydrodynamic model.) Annihilation of the antiproton produces approximately eight pions with a mean momentum of 1.2 GeV/c. Interaction of these pions with the nucleus significantly increases the temperature of the central region of the nucleus (third frame). This hot region expands, and finally energy begins to escape from the nucleus (sixth frame). The temperatures achieved are sufficiently high for formation of a predicted state of matter known as quark-gluon plasma.*

Pauli exclusion principle would not be applicable.) The energy levels of these hypernuclei would be indicative of the nuclear potential in the interior of the nucleus, a property that is is otherwise difficult to measure.

A particularly interesting feature of the light hypernuclei is the nearly zero value of the spin-orbit interaction between the $\Lambda$ and the nucleus.[8,9,10,11] Although this result was completely unexpected, it has since been explained in terms of both a valence-quark model of the baryons and a conventional meson-exchange model of nuclear forces. However, these two "orthogonal" descriptions of nuclear matter yield very different predictions for the spin-orbit interaction between the $\Sigma$ (another strange baryon) and the nucleus. Data that might distinguish between the two models has yet to be taken.

Most experimentalists working in the field of hypernuclei are hampered by the low intensity and poor energy definition of the kaon beams available at existing accelerators. The much higher intensity and better energy definition of the kaon beams to be provided by LAMPF II will greatly benefit this field.

**Low-Energy Tests of QCD.** A striking prediction of QCD is the existence of "glueballs," bound states containing only gluons. Also predicted are bound states containing mixtures of quarks and gluons, known as meiktons or hermaphrodites. These objects, if they exist, should be produced in hadron-nucleon collisions. However, since they are predicted to occur in a region already populated by a large number of hadrons, finding them will be a difficult job, requiring detailed phase-shift analyses of exclusive few-body channels in the predicted region. The high-intensity beams of LAMPF II, especially the pure kaon beams, will be extremely useful in searches for glueballs and meiktons.

Another expectation based on QCD is the near absence of polarization effects in inelastic hadron-nucleon scattering. But the few experiments on the exclusive channels at high momentum transfer have revealed strong polarization effects.[12] In contrast, the quark counting rules of QCD for the energy dependence of the elastic scattering cross section have been observed to be valid, even though the theory is not applicable in this energy regime. The challenge to both theory and experiment is to find out why some facets of QCD agree with experiment when they are not expected to, and vice versa. Obviously, more data are needed.

Also needed are more data on hadron spectroscopy, particularly in the area of kaon-nucleon scattering, which has received little attention for more than a decade. Such data are needed to help guide the development of quark-confinement theories.

## LAMPF II Design

LAMPF II was designed with two goals in mind: production of a 45-GeV, 40-microampere proton beam as economically as possible,

to that played by, say, a carbon-13 nucleus in NMR studies of the electronic environment within a molecule. For example, consider those hypernuclei in which a low neutron energy level is occupied by a $\Lambda$ in addition to the maximum allowable number of neutrons. (Such hypernuclei should exist since it is widely thought that the

**161**

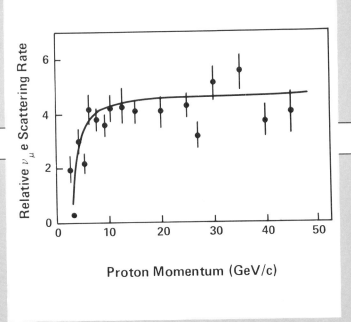

*Fig. 5. Monte-Carlo calculation of the rate of scattering between muon neutrinos and electrons (in an unbiased 4-meter by 4-meter detector located 90 meters from a beryllium neutrino-production target) as a function of the momentum of the protons producing the neutrinos. (The solid curve is simply a guide to the eye.) The calculations are based on various experimental values of the pion-production rate. The scattering rate plotted is the rate per unit power in the proton beam. The momentum of the protons to be produced by the LAMPF II booster (9.9 GeV/c) is well above the knee of the yield curve.*

and minimum disruption to the ongoing experimental programs at LAMPF. The designs of both of the new synchrotons reflect these goals.

The booster, or first stage, will be fed by the world's best H⁻ injector, LAMPF. This booster will provide a 9-GeV, 200-microampere beam of protons at 60 hertz. The 200-microampere current is the maximum consistent with continued use of the 800-MeV LAMPF beam by the Weapons Neutron Research Facility and the Proton Storage Ring. The 9-GeV energy is ideal not only for injection into the second stage but also for production of neutrinos to be used in scattering experiments (Fig. 5). Eighty percent of the booster current will be dedicated to the neutrino program. In contrast, the booster stage at other accelerators usually sits idle between pulses in the main ring. Since the phase space of the LAMPF beam is smaller in all six dimensions than the injection requirements of LAMPF II, lossless injection at a correct phase space is straightforward.

The 45-GeV main ring is shaped like a racetrack for two reasons: it fits nicely on the long, narrow mesa site and it provides the long straight sections necessary for efficient slow extraction. The main ring is basically a 12-hertz machine but will be operated at 6 hertz to permit slow extraction of a beam at a duty factor of 50 percent. This compromise minimizes the initial cost yet preserves the option of doubling the current and increasing the duty factor by adding a stretcher at a later date. The 45-GeV proton energy will provide kaons and antiprotons with energies up to 25 GeV. Such high energies should prove especially useful for the experiments mentioned above on the Drell-Yan process and exclusive hadron interactions.

The booster has a second operating mode: 12 GeV at 30 hertz and 100 microamperes with a duty factor of 30 percent. This 12-GeV mode will be useful for producing kaons in the early years if the main ring is delayed for financial reasons.

The most difficult technical problem posed by LAMPF II is the rf system, which must provide up to 10 megavolts at a peak power of 10 megawatts and be tunable from 50 to 60 megahertz. Furthermore, tuning must be rapid; that is, the bandpass of the tuning circuit must be on the order of 30 kilohertz. The ferrite-tuned rf systems used in the past are typically capable of providing only 5 to 10 kilovolts per gap at up to 50 kilowatts and, in addition, are limited by power dissipation in the ferrite tuners and plagued by strong, uncontrollable nonlinear effects. We have chosen to concentrate the modest development funds available at present on the rf system. A teststand is being built, and various ferrites are being studied to gain a better understanding of their behavior.

Following a lead from the microwave industry (one recently applied in a buncher cavity developed by the Laboratory's Accelerator Technology Division for the Proton Storage Ring), we have chosen a bias magnetic field perpendicular to the rf magnetic field. (All other proton accelerators employ parallel bias.) The advantage of perpendicular bias is a reduction in the ferrite losses by as much as

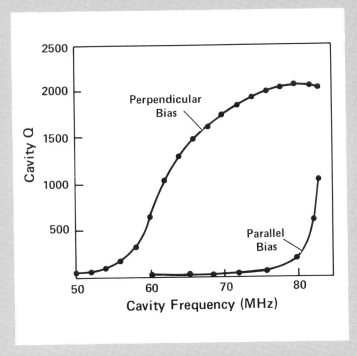

*Fig. 6. Performance of ferrite-tuned test cavities with parallel and perpendicular bias magnetic fields. The data shown are for a Ni-Zn ferrite; other types of ferrites give similar results.*

two orders of magnitude (Fig. 6). Since the loss in the ferrite is proportional to the square of the voltage on each gap, reducing these losses is essential to achieving the performance required of the LAMPF II system.

A collaboration led by R. Carlini and including the Medium Energy and Accelerator Technology divisions and the University of Colorado has made a number of tests of the perpendicular bias idea. Their results indicate that in certain ferrites the low losses persist at power levels greater than that needed for the LAMPF II cavities. A full-scale cavity is now being constructed to demonstrate that 100 kilovolts per gap at 300 kilowatts is possible. This prototype will also help us make a choice of ferrite based on both rf performance and cost of the bias system. A full-scale, full-power prototype of the rf system is less than a year away.

## Conclusion

This presentation of interesting experiments that could be carried out at LAMPF II is of necessity incomplete. In fact, the range of possibilities offered by LAMPF II is greater than that offered by any other facility being considered by the nuclear science community. Its funding would yield an extraordinary return. ■

## References

1. J. J. Aubert et al. (The European Muon Collaboration). "The Ratio of the Nucleon Structure Functions $F_2^N$ for Iron and Deuterium." *Physics Letters* 123B(1983):275.

2. A. Bodek et al. "Electron Scattering from Nuclear Targets and Quark Distributions in Nuclei." *Physical Review Letters* 50(1983):1431.

3. A. Bodek et al. "Comparison of the Deep-Inelastic Structure Functions of Deuterium and Aluminum Nuclei." *Physical Review Letters* 51(1983):534.

4. R. G. Arnold et al. "Measurements of the $A$ Dependence of Deep-Inelastic Electron Scattering from Nuclei." *Physical Review Letters* 52(1984):727.

5. T. A. Carey, K. W. Jones, J. B. McClelland, J. M. Moss, L. B. Rees, N. Tanaka, and A. D. Bacher. "Inclusive Scattering of 500-MeV Protons and Pionic Enhancement of the Nuclear Sea-Quark Distribution." *Physical Review Letters* 53(1984):144.

6. I. R. Kenyon. "The Drell-Yan Process." *Reports on Progress in Physics* 45(1982):1261.

7. D. Strottman and W. R. Gibbs. "High Nuclear Temperatures by Antimatter-Matter Annihilation." Accepted for publication in *Physics Letters*.

8. W. Brückner et al. "Spin-Orbit Interaction of Lambda Particles in Nuclei." *Physics Letters* 79B(1978):157.

9. R. Bertini et al. "A Full Set of Nuclear Shell Orbitals for the $\Lambda$ Particle Observed in $^{32}_{\Lambda}$S and $^{40}_{\Lambda}$Ca."

10. A. Bouyssy. "Strangeness Exchange Reactions and Hypernuclear Spectroscopy." *Physics Letters* 84B(1979):41.

11. M. May et al. "Observation of Levels in $^{13}_{\Lambda}$C, $^{14}_{\Lambda}$N, and $^{18}_{\Lambda}$O Hypernuclei." *Physical Review Letters* 47(1981):1106.

12. A review of these experiments was presented by G. Bunce at the Conference on the Intersections between Particle and Nuclear Physics, Steamboat Springs, Colorado, May 23-30, 1984.and will be published in the conference proceedings.

# The SSC—
# An Engineering Challenge

*by Mahlon T. Wilson*

The accelerator known as the SSC (Superconducting Super Collider) is a bold idea that will enable a giant step forward in high-energy physics. Within a circular ring fifty to one hundred miles around, two proton beams will collide and liberate enough energy to create new particles up to fifty times heavier than the weak bosons. These energies are necessary to go beyond the plateau of understanding summarized by the standard model. Specific issues to be addressed include the mechanism for breaking the symmetry between electromagnetic and weak interactions, the possibility that quarks and leptons are composite particles, and the existence of quark-lepton families heavier than those now known. In addition, exploration of this higher energy region is quite likely to uncover entirely new phenomena.

To bring some order to the multitude of suggestions put forth for what should be attempted with this machine and how it should be built, the high-energy physics community has held a series of workshops both here and abroad. The workshops resulted in a decision to study in detail the technical feasibility and estimated cost of achieving one particular set of beam parameters. Over 150 representatives from a number of national laboratories and universities and a few commercial firms contributed to this Reference Designs Study, which was headquartered at Lawrence Berkeley National Laboratory and directed by Maury Tigner of Cornell University. This heroic effort occupied the first four months of this year and produced many thousands of pages of text and cost estimates. From these has been extracted a summary document of about two thousand pages, which will serve as a point of reference for continued discussion and development of a proposal to the Department of Energy for funding.

The objective addressed in the Reference Designs Study was provision of two 20-TeV proton beams capable of being collided head-on at up to six locations. The maximum luminosity of each beam was set at $10^{33}$ per square centimeter per second. Three design concepts for the magnetic field were considered, all incorporating superconducting magnets of niobium-titanium cooled by liquid helium to 4.5 kelvins. The accompanying table lists some features of the three designs worthy of the adjective "super." Much care was taken to include in the reference designs components whose performance and cost were based on those of existing equipment. When this was not possible, advocates of a proposed component were required to break the component down into items of known cost and to defend their estimate of total cost. A disagreement of even a few dollars in the estimated cost of any one item can be significant, since thousands of each of hundreds of items are needed for the accelerator. The similarity of the estimated total costs for the three reference designs reflects a similarity between the greater costs associated with higher magnetic fields (more superconducting material) and those associated with physically larger accelerators (more cryogenic equipment, more excavation, more piping and cables, and so on).

The Reference Designs Study brought to light several engineering challenges that can be characterized as interesting, to say the least. A good first question is how to lay out an 18- to 33-mile-diameter circle with the required dimensional accuracy. The sheer size of the facilities being considered—the circumferences of which range from the highway distance between Los Alamos and Cochiti Pueblo to that between Los Alamos and Albuquerque—create unusual problems in communications.

The long magnets present challenges in fabrication, transportation, field testing, and alignment. For example, the 3-tesla magnets, which are about one and one-half football fields long but only a bit over one foot in diameter, will behave like wet noodles if improperly lifted. And although such long magnets can be bent sufficiently to conform to the topography of specially tailored roads, they must be supported during transport at intervals of about every ten feet. All the magnet versions raise other issues. The numerous plumbing and wiring connections must be of the highest quality. Several inches

Features of the three SSC designs considered in the Reference Designs Study. The 6.5-tesla design involves a conductor-dominated field with both beam tubes in a common cold-iron yoke that contributes slightly to shaping the field. In this design the dipole magnet, beam tubes, and yoke are supported within a single cryostat. The 5-tesla design involves a conductor-dominated dipole field with a heavy-walled iron cryostat to attenuate the fringe field. This single-bore design requires two separate rings of dipole magnets. The 3-tesla design is similar to the 6.5 tesla design except that the field is shaped predominantly by the cold-iron yoke rather than by the conductor.

| Dipole Field (T) | Dipole Magnet Length (ft) | Accelerator Diameter (mi) | Total Estimated Cost ($) |
|---|---|---|---|
| 6.5 | 57 | 18 | 2.72 billion |
| 5 | 46 | 23 | 3.05 billion |
| 3 | 460 | 33 | 2.70 billion |

of thermal contraction of the components within the cryostats must be accommodated. Heat leaks from power and instrumentation leads must be minimized, as must those from the magnet supports. (What is needed are supports with the strength of an ox yoke but the substance of a spider web.) Alignment will require some means for knowing the exact location of the magnets within their cryostats. And if a leak should develop in any of the piping within a magnet's cryostat, there needs to be a method for locating the "sick" magnet and determining where within it the problem exists.

Questions of safety, also, must be addressed. For example, the refrigerator locations every 2 to 5 miles around the ring are logical sites for personnel access, but is this often enough? What happens if a helium line should rupture? After all, a person can run only a few feet breathing helium. Will it be necessary to exclude personnel from the tunnel when the system is cold, or can this problem be solved with, say, supplied-air suits or vehicles?

Achieving head-on collisions of the beams presents further challenges. Each beam must be focused down to 10 microns and, more taxing, be positioned to within an accuracy of about 1 micron. It takes a reasonably good microscope even to see something that small! Will a truck rumbling by shake the beams out of a collision course? What will be the effect of earth tides or earthquakes? Does the ground heave due to annual changes in temperature or water-table level? How stable is the ground in the first place? That is, does part of the accelerator move relative to the remainder? Will it be desirable, or necessary, to have a robot system constantly moving around the ring tweaking the positions of the magnets? What would the robot, or any surveyor, use as a reference for alignment?

These are but a few of the many issues that have been raised about construction and operation of the SSC. Resolving them will require considerable technology and ingenuity.

In April of this year, the Department of Energy assigned authority over the SSC effort to Universities Research Association (URA), the consortium of fifty-four universities that runs Fermilab. URA, in turn, assigned management responsibilities to a separate board of overseers under Boyce McDaniel of Cornell University. This board selected Maury Tigner as director and Stanley Wojcicki of Stanford University as deputy director for SSC research and development. A headquarters is being established at Lawrence Berkeley National Laboratory, and a team will be drawn together to define what the SSC must do and how best that can be done. Secretary of Energy Donald Hodel has approved the release of funds to support the first year of research and development. Since the $20 million provided was about half the amount felt necessary for progress at the desired rate, shortcuts must be taken in reaching a decision on magnet type so that site selection can begin soon.

Los Alamos has been involved in the efforts on the SSC since the beginning. We have participated in numerous workshops, collated siting information and published a Site Atlas, and contributed to the portions of the Reference Designs Study on beam dynamics and the injector. We may be called upon to provide the injector linac, kicker magnets, accelerating cavities, and numerous other accelerator components. Our research on magnetic refrigeration has the potential of halving the operating cost of the cryogenic system for the SSC. Although the results of this research may be too late to be incorporated in the initial design, magnetic refrigerator replacements for conventional units would quickly repay the investment. ∎

# SCIENCE UNDERGROUND

**R**emarkable though it may seem, some of our most direct information about processes involving energies far beyond those available at any conceivable particle accelerator and far beyond those ever observed in cosmic rays may come from patiently watching a large quantity of water, located deep underground, for indications of improbable behavior of its constituents. Equally remarkable, our most direct information about the energy-producing processes deep in the cores of stars comes not from telescopes or satellites but from carefully sifting a large volume of cleaning fluid, again located deep underground, for indications of rare interactions with messengers from the sun. In what follows we will explore some of the science behind these statements, learn a bit about how such experiments are carried out, and venture into what the future may hold.

The experiments that we will discuss, which can be characterized as searches for exceedingly rare processes, have two features in common: they are carried out deep below the surface of the earth, and they involve a large mass of material capable of undergoing or participating in the rare process in question. The latter feature arises from the desire to increase the probability of observing the process within a reasonable length of time. The underground site is necessary to shield the experiment from secondary cosmic rays. These products of the interactions of primary cosmic rays within our atmosphere would create an overwhelming background of confusing, misleading "noise." Since about 75 percent of the secondary cosmic rays are extremely penetrating muons (resulting from the decays of pions and kaons), effective shielding requires overburdens on the order of a kilometer or so of solid rock (Fig. 1).

What are the goals of the experiments that make worthwhile these journeys into the hazardous depths of mines and tunnels with complex, sensitive equipment? The largest and in many ways the most spectacular experiments—the searches for decay of protons or

# *the search for rare events*
*by L. M. Simmons, Jr.*

neutrons—are aimed at understanding the basic interactions of nature. The oldest seeks to verify the postulated mechanism of stellar energy production by detecting solar neutrinos—the lone truthful witnesses to the nuclear reactions in our star's core. Smaller experiments investigate double beta decay, the rarest process yet observed in nature, to elucidate properties of the neutrino. Muon "telescopes" will observe the numbers, energies, and directions of cosmic-ray muons to obtain information about the composition and energy spectra of primary cosmic rays. Large neutrino detectors will measure the upward and downward flux of neutrinos through the earth and hence search for neutrino oscillations with the diameter of the earth as a baseline. These detectors can also serve as monitors for signals of rare galactic events, such as the intense burst of neutrinos that is expected to accompany the gravitational collapse of a stellar core.

A site that can accommodate the increasingly sophisticated technology required will encourage the mounting of underground experiments to probe these and other processes in ever greater detail.

## The Search for Nucleon Instability

The universe is thought to be about ten billion ($10^{10}$) years old, and of this unimaginable span of time, the life of mankind has occupied but a tiny fraction. The lifetime of the universe, while immense on the scale of the lifetime of the human species, which is itself huge on the scale of our own lives, is totally insignificant when compared to the time scale on which matter is known to be stable. It is now certain that protons and (bound) neutrons have lifetimes on the order of $10^{31}$ years or more. Thus for all practical purposes these particles are totally stable. Why examine the issue any further?

The incentive is one of principle. The mass of a proton or neutron, about 940 MeV/$c^2$, is considerably greater than that of many other particles: the photon (zero mass), the neutrinos (very small, perhaps zero mass), the electron (0.5ll MeV/$c^2$), the muon (106 MeV/$c^2$), and

the charged and neutral pions (140 MeV/$c^2$ and 135 MeV/$c^2$), to name only the most familiar. Therefore, energy conservation alone does not preclude the possibility of nucleon decay. Bearing in mind Murray Gell-Mann's famous dictum that "Everything not compulsory is forbidden," we are obligated to search for nucleon decay unless we know of something that forbids it.

Conservation laws forbidding nucleon decay had been independently postulated by Weyl in 1929, Stueckelberg in 1938, and Wigner in 1949 and 1952. But Lee and Yang argued in 1955 that such laws would imply the existence of a long-range force coupled to a conserved quantum number known as baryon number. (The baryon number of a particle is the sum of the baryon numbers of its quark constituents, $+\frac{1}{3}$ for each quark and $-\frac{1}{3}$ for each antiquark. The proton and the neutron thus have baryon numbers of $+1$.) Lee and Yang's reasoning followed the lines that lead to the derivation of the Coulomb force from the law of conservation of electric charge. However, no such long-range force is observed, or, more accurately, the strength of such a force, if it exists, must be many orders of magnitude weaker than that of the weakest force known, the gravitational force. Thus, although no information was available as to just how unstable nucleons might be, no theoretical argument demanded exact conservation of baryon number.

Los Alamos has the distinction of being the site of the first searches for evidence of nucleon decay. In 1954 F. Reines, C. Cowan, and M. Goldhaber placed a scintillation detector in an underground room at a depth of about 100 feet and set a lower limit on the nucleon lifetime of $10^{22}$ years. In 1957 Reines, Cowan, and H. Kruse deduced a greater limit of $4 \times 10^{23}$ years from an improved version of the experiment located at a depth of about 200 feet (in "the icehouse," an area excavated in the north wall of Los Alamos Canyon). Since these early Los Alamos experiments, the limit on the lifetime of the proton has been increased by many orders of magnitude.

Nonconservation of baryon number is also favored as an explanation for a difficulty with the big-bang theory of creation of the universe. The difficulty is that the big bang supposedly created baryons and antibaryons in equal numbers, whereas today we observe a dramatic excess of matter over antimatter (and an equally dramatic excess of photons over matter). In 1967 A. Sakharov pointed out that this asymmetry must be due to the occurrence of processes that do not conserve baryon number; his original argument has since been elaborated in terms of grand unified theories by several authors. The very existence of physicists engaged in searches for nucleon decay is mute testimony to the baryon asymmetry of the universe and, by inference, to the decay of nucleons at some level.

The recent resurgence of interest in the stability of nucleons arises in part from the success of the unified theory of electromagnetic and weak interactions by Glashow, Salam, and Weinberg. This non-Abelian gauge theory, which is consistent with all available data and correctly predicts the existence and strength of the neutral-current weak interaction and the masses of the $Z^0$ and $W^\pm$ gauge bosons, involves essentially only one parameter (apart from the masses of the elementary particles). The measured value of this parameter (the Weinberg angle) is given by $\sin^2\theta_W = 0.22 \pm 0.01$. The success of the electroweak model gave considerable legitimacy to the idea that gauge theories may be the key to unifying all the interactions of nature.

The simplest gauge theory to be applied to unifying the electroweak and strong interactions (minimal SU(5)) gave rise to two exciting predictions. One, that $\sin^2\theta_W = 0.215$, agreed dramatically with experiment, and the other, that the lifetime of the proton against decay into a positron and a neutral pion (the predicted dominant decay mode) lay between $1.6 \times 10^{28}$ and $6.4 \times 10^{30}$ years, implied that experiments to detect nucleon decay were technically feasible.

Experimentalists responded with a series of increasingly sensitive experiments to test

this prediction of grand unification. What approach is followed in these experiments? Out of the question is the direct production of the gauge bosons assumed to mediate the interactions that lead to nucleon decay. (This was the approach followed recently and successfully to test the electroweak theory.) The grand unified theory based on minimal SU(5) predicts that the masses of these bosons are on the order of $10^{14}$ GeV/$c^2$, in contrast to the approximately $10^2$-GeV/$c^2$ masses of the electroweak bosons and many orders of magnitude greater than the masses of particles that can be produced by any existing or conceivable accelerator or by the highest energy cosmic ray. Thus, the only feasible approach is to observe a huge number of nucleons with the hope of catching a few of them in the quantum-mechanically possible but highly unlikely act of decay.

The largest of these experiments (the IMB experiment) is that of a collaboration including the University of California, Irvine, the University of Michigan, and Brookhaven National Laboratory. In this experiment (Fig. 2) an array of 2048 photomultipliers views 7000 tons of water at a depth of 1570 meters of water equivalent (mwe) in the Morton-Thiokol salt mine near Cleveland, Ohio. The water serves as both the source of (possibly) decadent nucleons and as the medium in which the signal of a decay is generated. The energy released by nucleon decay would produce a number of charged particles with so much energy that their speed in the water exceeds that of light in the water (about $0.75c$, where $c$ is the speed of light in vacuum). These particles then emit cones of Cerenkov radiation at directions characteristic of their velocities. The photomultipliers arrayed on the periphery of the water detect this light as it nears the surfaces. From the arrival times of the light pulses and the patterns of their intersections with the planes of the photomultipliers, the directions of the parent charged particles can be inferred. Their energies can be estimated from the amount of light observed, in conjunction with calibration studies based on the vertical passage of muons through the detector. (The

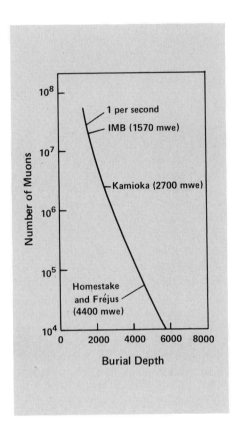

Fig. 1. For some experiments the only practical way to sufficiently reduce the background caused by cosmic-ray muons is to locate the experiments deep underground. Shown above is the number of cosmic-ray muons incident per year upon a cube 10 meters on an edge as a function of depth of burial. By convention depths of burial in rocks of various densities are normalized to meters of water equivalent (mwe). The depths of some of the experiments discussed in the text are indicated.

impressive sensitivity of such an experiment is well illustrated by the information that the light from a charged particle at a distance of 10 meters in water is less than that on the earth from a photoflash on the moon.)

This "water Cerenkov" detection scheme was chosen in part for its simplicity, in part for its relatively low cost, and in part for its

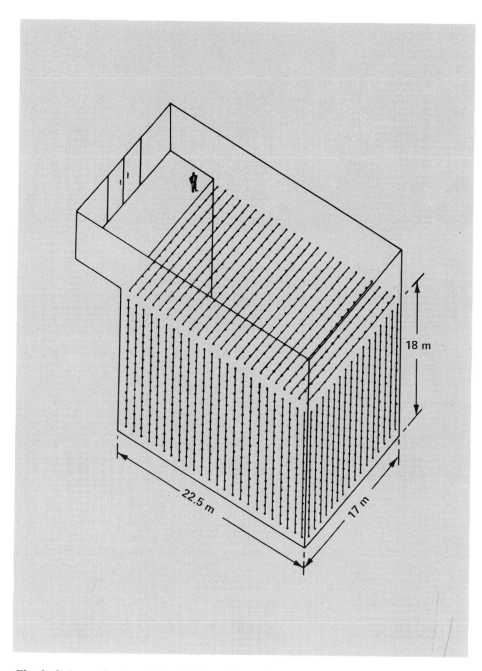

*Fig. 2. Schematic view of the IMB nucleon-decay detector. A total of 2048 5-inch photomultipliers are arrayed about the periphery of 7000 tons of water contained within a plastic-lined excavation at a depth of 1570 mwe in a salt mine near Cleveland, Ohio. The photomultipliers monitor the water for pulses of Cerenkov radiation, some of which may signal the decay of a proton or a neutron. (From R. M. Bionta et al., "IMB Detector—The First 30 Days," in* Science Underground *(Los Alamos, 1982) (American Institute of Physics, New York, 1982)).*

high efficiency at detecting the electrons that are the ultimate result of the $p \to e^+ + \pi^0$ decay. (The neutral pion immediately decays to two photons, which produce showers of electrons in the water.) Note, however, that although this two-body decay is especially easy to detect because of the back-to-back orientation of the decay products, it must be distinguished, at the relatively shallow depth of the IMB experiment, among a background of about $2 \times 10^5$ muon-induced events per day. (The lower limit on the proton lifetime predicted by minimal SU(5) implies a maximum rate for $p \to e^+ \pi^0$ of several events per day.)

Another experiment employing the water Cerenkov detection scheme is being carried out at a depth of 2700 mwe by a collaboration including the University of Tokyo, KEK (National Laboratory for High-Energy Physics), Niigata University, and the University of Tsukuba. The experiment is located under Mt. Ikenayama in the deepest active mine in Japan, the Kamioka lead-zinc mine of the Mitsui Mining and Smelting Co. Although the mass of the water viewed in this experiment (3000 tons) is substantiallly less than that in the IMB experiment, its greater depth of burial results in lower background rates. More important, 1000 20-inch photomultipliers are deployed at Kamioka (Fig. 3), in contrast to the 2048 5-inch photomultipliers at IMB. As a result, a ten times greater fraction of the water surface at Kamioka is covered by photocathode material, and the light-collection efficiency is greater by a factor of about 12. Thus the track detection and identification capabilities of the Kamioka experiment are considerably better.

To date neither the IMB experiment nor the Kamioka experiment has seen any candidate for $p \to e^+\pi^0$. These negative results yield a proton lifetime greater than $3 \times 10^{32}$ years for this decay mode, well outside the range predicted by the grand unified theory based on minimal SU(5). Since this theory has a number of other deficiencies (it fails to predict the correct ratio for the masses of the light quarks and predicts a drastically incorrect ratio for the number of baryons and photons produced by the big bang), it is therefore now thought to be the wrong unification model. Other models, at the current stage of their development, have too little predictive power to yield decay rates that can be unambiguously confronted by experiment. The question of nucleon decay is now a purely experimental one, and theory awaits the guidance of present and future experiments.

The cosmic rays that produce the interfering muons also produce copious quantities of neutrinos (from the decays of pions, kaons, and muons). No amount of rock can block these neutrinos, and some of them interact in the water, mimicking the effects of proton decay. Estimates of this background as a function of energy are based on calculations of the flux of cosmic-ray-induced neutrinos from the known flux of primary cosmic rays. Although these calculations enjoy reasonable confidence, no accurate experimental data are available as a check. Full analyses of the neutrino backgrounds in the proton-decay experiments will provide the first such verification. Whether new effects in neutrino astronomy will be discovered from the spectrum of neutrinos incident on the earth remains to be seen. Thus nucleon-decay experiments may open a new field, that of neutrino astronomy.

The water Cerenkov experiments have detected several events that could possibly be interpreted as nucleon decays by modes other than $e^+\pi^0$ (Table 1). It is also possible that these events are induced by neutrinos. Although their configurations are not easily explained on that basis, their total number is consistent with the rate expected from the calculated neutrino flux.

A perusal of Table 1 shows that the IMB and Kamioka experiments yield different lifetime limits and do not see the same number of candidate events for the various decay modes. This is not surprising since the two also differ in aspects other than those already mentioned. The Kamioka experiment can more easily distinguish events with multiple tracks, such as $p \rightarrow \mu^+\eta$, which is immediately followed by decay of the $\eta$ meson

*Fig. 3. Photograph of the Kamioka nucleon-decay detector under construction at a depth of 2700 mwe in a lead-zinc mine about 300 kilometers west of Tokyo. Already installed are the bottom layer of photomultipliers and two ranks of photomultipliers on the sides of the cylindrical volume. The wire guards around the photomultipliers protect the workers from occasional implosions. The upper ranks and top layer of photomultipliers were installed from rafts as the water level was increased. The detector contains a total of 1000 20-inch photomultipliers. (Photo courtesy of the Kamioka collaboration.)*

## Table 1

**Some current results of the Kamioka and IMB experiments. Listed for each decay mode are the number of candidate events detected (in brackets) and the deduced lifetime limit.**

| Decay Mode | Number of Events and Lifetime Limit (years) | | | |
|---|---|---|---|---|
| | Kamioka | | IMB | |
| $p \rightarrow e^+\pi^0$ | [0] | $8 \times 10^{31}$ | [0] | $3 \times 10^{32}$ |
| $p \rightarrow \mu^+\pi^0$ | [0] | $2 \times 10^{31}$ | [0] | $1 \times 10^{32}$ |
| $p \rightarrow \mu^+K^0$ | [1] | $1 \times 10^{31}$ | [1] | $6 \times 10^{31}$ |
| $p \rightarrow \mu^+\eta$ | [1] | $8 \times 10^{30}$ | [0] | $9 \times 10^{31}$ |
| $p \rightarrow \nu K^+$ | [2] | $7 \times 10^{30}$ | [3] | $1 \times 10^{31}$ |
| $p \rightarrow \nu\pi^+$ | [5] | $3 \times 10^{30}$ | --- | |
| $n \rightarrow e^+\pi^-$ | [0] | $1 \times 10^{31}$ | [4] | $2 \times 10^{31}$ |
| $n \rightarrow \nu K^0$ | [0] | $3 \times 10^{30}$ | [3] | $8 \times 10^{30}$ |

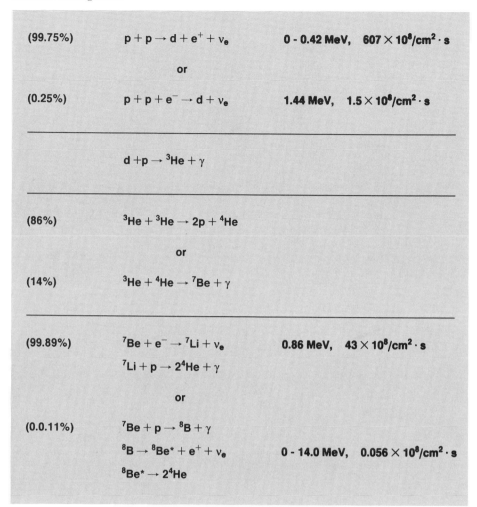

| | | | |
|---|---|---|---|
| (99.75%) | $p + p \rightarrow d + e^+ + \nu_e$ | 0 - 0.42 MeV, | $607 \times 10^8/cm^2 \cdot s$ |
| | or | | |
| (0.25%) | $p + p + e^- \rightarrow d + \nu_e$ | 1.44 MeV, | $1.5 \times 10^8/cm^2 \cdot s$ |
| | $d + p \rightarrow {}^3He + \gamma$ | | |
| (86%) | ${}^3He + {}^3He \rightarrow 2p + {}^4He$ | | |
| | or | | |
| (14%) | ${}^3He + {}^4He \rightarrow {}^7Be + \gamma$ | | |
| (99.89%) | ${}^7Be + e^- \rightarrow {}^7Li + \nu_e$ | 0.86 MeV, | $43 \times 10^8/cm^2 \cdot s$ |
| | ${}^7Li + p \rightarrow 2{}^4He + \gamma$ | | |
| | or | | |
| (0.0.11%) | ${}^7Be + p \rightarrow {}^8B + \gamma$ | | |
| | ${}^8B \rightarrow {}^8Be^* + e^+ + \nu_e$ | 0 - 14.0 MeV, | $0.056 \times 10^8/cm^2 \cdot s$ |
| | ${}^8Be^* \rightarrow 2{}^4He$ | | |

*Fig. 4. The proton-proton chain postulated by the standard solar model as the principal mechanism of energy production in the sun. The net result of this series of nuclear reactions is the conversion of four protons into a helium-4 nucleus, and the energy released is carried off by photons, positrons, and neutrinos. Predicted branching ratios for competing reactions are listed. Some of the reactions in this chain produce neutrinos; the energies of these particles and their predicted fluxes at the earth are listed at the right.*

by a number of modes. On the other hand, the IMB experiment has been in progress for a longer time and is thus more sensitive to decay modes with long lifetimes.

The IMB collaboration has recently installed light-gathering devices around each photomultiplier and will soon double the number of tubes with the goal of increasing the light-collection efficiency by a factor of about 6. At Kamioka accurate timing circuits are being installed on each photomultiplier to record the exact times of arrival of the light signals. As a result, more and better data can be expected from both experiments.

What else does the future hold? The European Fréjus collaboration (Aachen, Orsay, Palaiseau, Saclay, and Wuppertal) has completed construction of a 912-ton modular fine-grained tracking calorimeter. This detector is located at a depth of 4400 mwe in a 3300-cubic-meter laboratory excavated near the middle of the Fréjus Tunnel connecting Modane, France and Bardonnecchia, Italy. Its 114 modules consist of 6-meter by 6-meter planes of Geiger and flash chambers interleaved with thin iron-plate absorbers. The detector can pinpoint particle tracks with a resolution on the order of 2 millimeters, a 250-fold greater resolution than that of the water Cerenkov detectors. Data about energy losses of the particles along their tracks distinguish electrons and muons. To date the Fréjus collaboration has observed no candidate proton-decay events.

The Soudan II collaboration (Argonne National Laboratory, the University of Minnesota, Oxford University, Rutherford-Appleton Laboratory, and Tufts University) has excavated an 11,000-cubic-meter laboratory at 2200 mwe in the Soudan iron mine in northern Minnesota and is now constructing an 1100-ton dense fine-grained tracking calorimeter. The detector will contain 256 modules, each 1 meter by 1 meter by 2.5 meters, incorporating thin steel sheets and high-resolution drift tubes in hexagonal arrays. The spatial resolution of the detector will be about 3 millimeters. Information about the ionization deposited along the track lengths will provide excellent particle-identification capabilities. Completion of the detector is scheduled for 1988, but data collection will begin in 1987.

Because the Fréjus and Soudan II detectors view relatively small numbers of nucleons (fewer than $6 \times 10^{32}$), they can record reasonable event rates only for those decay modes (if any) with lifetimes considerably less than $10^{32}$ years. On the other hand, they have good resolution for high-energy cosmic-ray muons, and this feature will be put to good use in experiments of astrophysical interest.

Despite the hopes for these newer experiments, the IMB and Kamioka results to date imply that accurate investigation of most nucleon decay modes demands multikiloton detectors with very fine-grained resolution. These second-generation detectors will be multipurpose devices, sensitive to many other rare processes. Realistically, they can be operated to greatest advantage only in the environment of a dedicated facility capable of providing major technical support.

## The Solar Neutrino Mystery

The light from the sun so dominates our existence that all human cultures have marveled at its life-giving powers and have concocted stories explaining its origins. Scientists are no different in this regard. How do we explain the almost certain fact that the sun has been radiating energy at essentially the present rate of about $4 \times 10^{26}$ joules per second for some 4 to 5 billion years? Given a solar mass of $2 \times 10^{30}$ kilograms, chemical means are wholly inadequate, by many orders of magnitude, to support this rate of energy production. And the gravitational energy released in contracting the sun to its present radius of about $7 \times 10^5$ kilometers could provide but a tiny fraction of the radiated energy. The only adequate source is the conversion of mass to energy by nuclear reactions.

This answer has been known for a generation or two. Through the work of Hans Bethe and others in the 1930s and of many workers since, we have a satisfactory model for solar energy production based on the thermonuclear fusion of hydrogen, the most abundant element in the universe and in most stars. The product of this proton-proton chain (Fig. 4) is helium, but further nuclear reactions yield heavier and heavier elements. Detailed models of these processes are quite successful at explaining the observed abundances of the elements. Thus it is possible to say (with W. A. Fowler) that "you and your neighbor and I, each one of us and all of us, are truly and literally a little bit of stardust."

The successes of the standard solar model may, however, give us misplaced confidence in its reality. It is all very well to study nuclear reactions and energy transport in the laboratory and to construct elaborate computational models that agree with what we observe of the exteriors of stars. But what is the direct evidence in support of our story of what goes on deep within the cores of stars?

The difficulties presented by the demand for direct evidence are formidable, to say the least. Stars other than our sun are hopelessly distant, and even that star, although at least reasonably typical, cannot be said to lie conveniently at hand for the conduct of experiments. Moreover, the sun is optically so thick that photons require on the order of 10 million years to struggle from the deep interior to the surface, and the innumerable interactions they undergo on the way erase any memory of conditions in the solar core. Thus, all conventional astronomical observations of surface emissions provide no direct information about the stellar interior. The situation is not hopeless, however, for several of the nuclear reactions in the proton-proton chain give rise to neutrinos. These particles interact so little with matter that

*Fig. 5. A view of the solar neutrino experiment located at a depth of 4850 feet in the Homestake gold mine. The steel tank contains 380,000 liters of perchloroethylene, which serves as a source of chlorine atoms that interact with neutrinos from the sun. Nearby is a small laboratory where the argon atoms produced are counted. (Photo courtesy of R. Davis and Brookhaven National Laboratory.)*

they provide true testimony to conditions in the solar core.

The parameters incorporated in the standard solar model (such as nuclear cross sections, solar mass, radius, and luminosity, and elemental abundances, opacities (from the Los Alamos Astrophysical Opacity Library), and equations of state) are known with such confidence that a calculation of the solar neutrino spectrum is expected to be reasonably accurate. At the moment only one experiment in the world—that of Raymond Davis and his collaborators from Brookhaven National Laboratory—attempts to measure any portion of the solar neutrino flux for comparison with such a calculation. Located at a depth of 4400 mwe in the Homestake gold mine in Lead, South Dakota, this experiment (Fig. 5) detects solar neutrinos by counting the argon atoms from the reaction

$$\nu_e + {}^{37}\text{Cl} \rightarrow {}^{37}\text{Ar} + e^- \; ,$$

which is sensitive primarily to neutrinos from the beta decay of boron-8 (see Fig. 4). Since chlorine-37 occurs naturally at an abundance of about 25 percent, any compound containing a relatively large number of chlorine atoms per molecule and satisfying cost and safety criteria can serve as the target. The Davis experiment uses 380,000 liters of perchloroethylene ($C_2Cl_4$).

You might well ask why this reaction occurs at a detectable rate. All the solar neutrinos incident on the tank of percholoroethylene have made the journey from the solar core to the earth and then through 4850 feet of solid rock with essentially no interactions, and the neutrinos from the boron-8 decay constitute but a small fraction of the total neutrino flux. What is the special feature that makes this experiment possible? Apart from the large number of target chlorine atoms, it is the existence of an excited state in argon-37 that leads to an exceptionally high cross section for capture by chlorine-37 of neutrinos with energies greater than about 6 MeV. Figure 4 shows that the only branch of the proton-proton chain producing neutrinos with such energies is the beta decay of boron-8. The standard solar model predicts a rate for the reaction of about $7 \times 10^{-36}$ per target atom per

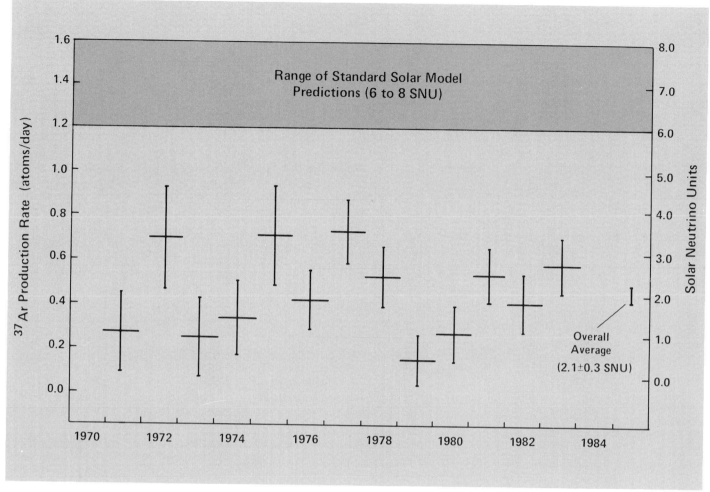

**Fig. 6.** *Yearly averages of the flux of boron-8 solar neutrinos, as measured by the Homestake experiment. The discrepancy between the experimental results and the predictions of the standard solar model has not yet been explained.*

*(From R. Davis, Jr., B. T. Cleveland, and J. K. Rowley, "Report on Solar Neutrino Experiments," in* Intersections Between Particle and Nuclear Physics *(Steamboat Springs, Colorado), New York: American Institute of Physics, 1984.)*

second (7 solar neutrino units, or SNUs), which corresponds in the Davis experiment to an expected argon-37 production rate of about forty atoms per month.

It may seem utterly miraculous that such a small number of argon-37 atoms can be detected in such a large volume of target material, but the technique is simple. About every two months helium is bubbled through the tank to sweep out any argon-37 that has been formed. The resulting sample is purified and concentrated by standard chemical techniques and is monitored for the 35-day decay of argon-37 by electron capture. Great care is taken to distinguish these events by pulse height, rise time, and half-life from various background-induced events. As part of the recovery technique argon-36 and -38 are inserted into the tank in gram quantities or less to monitor the recovery efficiency (about 95 percent). An artificially introduced sample of 500 argon-37 atoms has also been recovered successfully. Indeed, the validity of the technique has been verified by continual scrutiny over more than fifteen years.

The Homestake experiment has provided the scientific world with a long-standing mystery: its results are significantly and consistently lower than the predictions of the standard solar model (Fig. 6). So what's wrong?

The first possibility that immediately suggests itself, that the Davis experiment contains some subtle mistake, cannot be eliminated. But it must be dismissed as unlikely because of the careful controls incorporated in the experiment and because of the years of independent scrutiny that the experiment has survived. The possibility that the parameters employed in the calculation might be in error has been repeatedly examined by careful investigators seeking to explain the mystery (and thereby make reputations for themselves). However, no one has suggested corrections that are large enough to explain the discrepancy.

Another possibility is that the standard solar model is wrong. The reaction that gives rise to boron-8 is inhibited substantially by a Coulomb barrier and is thus extraordinarily sensitive to the calculated temperature at the center of the sun. A tiny change in this temperature or a small deviation from the standard-model value of the solar-core composition would be sufficient to change the rate of production of boron-8 and thus the neutrino flux to which the Davis experiment is primarily sensitive. Although many "nonstandard" solar models predict lower boron-8 neutrino fluxes, none of these are widely accepted. In general, the only experimentally testable distinction among the nonstandard models lies in their predictions of neutrino fluxes. A complete characterization of the solar neutrino spectrum is needed to provide quantitative constraints on the standard solar model of the future.

The explanation of the solar neutrino puzzle quite possibly lies in the realm of particle physics rather than solar physics, nuclear physics, or chemistry. The results of the Homestake experiment have generally been interpreted on the basis of conventional neutrino physics. It is, however, not known with certainty how many species of neutrinos exist, whether they are massless, or whether they are stable. New information about these issues could drastically influence the interpretation of solar neutrino experiments.

For example, Bahcall and collaborators have pointed out that it is possible for a more massive neutrino species to decay into a less massive neutrino species and a scalar particle (such as a Goldstone boson arising from spontaneous breaking of the symmetry associated with lepton number conservation). If a neutrino species less massive than the electron neutrino exists and if the lifetime of the electron neutrino is such that those with an energy of 10 MeV have a mean life of 500 seconds (the transit time to the earth), then lower-energy electron neutrinos would decay before reaching the earth. The resulting reduction in the solar neutrino flux could be sufficient to explain the Davis results. Note that this explanation for the solar neutrino puzzle, in direct contrast to explanations based on nonstandard solar models, involves a great reduction in the flux of essentially all *but* the boron-8 neutrinos.

Several other explanations of the solar neutrino puzzle are also based on speculated features of neutrino physics. One of these, "oscillations" among the various neutrino species, is discussed in the next section.

## Future Solar Neutrino Experiments

Among the nonstandard solar models alluded to above are some that allow long-term variations in the rate of energy production in the solar core. Such variations violate the constraint on steady-state solar models that hydrogen be burned in the core at a rate commensurate with the currently observed solar luminosity. To test the validity of these models, a Los Alamos group has devised an experiment for determining an average of the

solar neutrino flux over the past several million years.

The experiment, like Davis's, is based on an inverse beta decay induced by boron-8 solar neutrinos, namely,

$$\nu_e + {}^{97,98}\text{Mo} \rightarrow {}^{97,98}\text{Tc} + e^- \ .$$

The molybdenum target atoms must be located at depths such that the cosmic-ray-induced background of technetium isotopes is low compared to the solar neutrino signal. This condidtion is satisfied by a molybdenite ore body 1100 to 1500 meters below Red Mountain in Clear Creek County, Colorado. The ore is currently being mined by AMAX Inc. at a depth in excess of about 1150 meters. The long half-lives of technetium-97 and -98 (2.6 million and 4.2 million years, respectively) have permitted their accumulation to a level (calculated on the basis of the standard solar model) of about 10 million atoms each per 2000 metric tons of ore. Fortuitously, the initial large-scale concentration of the technetium (into a rhenium-selenium-technetium sludge) occurs during operations involved in producing molybdenum trioxide from the raw ore. The Los Alamos group has developed chemical and mass-spectrographic techniques for isolating and counting the technetium atoms in the sludge. The first results from the experiment should be available in late 1987.

Much more remains unknown about solar neutrinos. In particular, we completely lack information about the flux of neutrinos from other reactions in the proton-proton chain. According to the standard solar model, the preponderance of solar neutrinos arises from the first reaction in the chain, the thermonuclear fusion of two protons to form a deuteron. A thorough test of the solar model must include measurement of the neutrino flux from this reaction, the rate of which, although essentially independent of the details of the model (varying by at most a few percent), involves the basic assumption that hydrogen burning is the principal source of solar energy.

The preferred reaction for investigating

the initial fusion in the proton-proton chain is

$$\nu_e + {}^{71}\text{Ga} \rightarrow {}^{71}\text{Ge} + e^- \ ,$$

which has a threshold of 233 keV, well below the maximum energy of the *pp* neutrinos. Calculations based on the standard solar model and the relevant nuclear cross sections predict a capture rate of about 110 SNU, of which about two-thirds is due to the *pp* reaction, about one-third to the electron-capture reaction of beryllium-7, and a very small fraction to the other neutrino-producing reactions.

Several years ago members of the Homestake team, in collaboration with scientists from abroad, carried out a pilot experiment to assess a technique suggested for a solar neutrino experiment based on this reaction. Germanium-71 was introduced into a solution of over one ton of gallium (as $\text{GaCl}_3$) in hydrochloric acid. In such a solution germanium forms the volatile compound $\text{GeCl}_4$, which was swept from the tank with a gas purge. By fairly standard chemical techniques, a purified sample of $\text{GeH}_4$ was prepared for monitoring the 11-day decay of germanium-71 by electron capture. The pilot experiment clearly demonstrated the feasibility of the technique.

Why has the full-scale version of this important experiment not been done? The trouble, as usual, is money. The original estimates indicated that achieving an acceptable accuracy in the measured neutrino flux would require about one neutrino capture per day, which corresponded to 45 tons of gallium as a target. Gallium is neither common nor easy to extract, and the cost of 45 tons was about \$25,000,000, a sum that proved unavailable. Nor did the suggestion to "borrow" the required amount of gallium succeed (despite the fact that only one gallium atom per day was to be expended), and the collaboration disbanded.

The chances of mounting a gallium experiment seem brighter today, however, since recent Monte Carlo simulations have shown that an accuracy of 10 percent in the measured neutrino flux is possible from a four-year experiment incorporating improved counting efficiencies and reduced background rates and involving only 30 tons of gallium.

The European GALLEX collaboration (Heidelberg, Karlsruhe, Munich, Saclay, Paris, Nice, Milan, Rome, and Rehovot) has received approval to install a 30-ton gallium chloride experiment in the Gran Sasso Laboratory (this and other dedicated underground science facilities are described in the next section) and sufficient funding to acquire the gallium. The collaboration has achieved the low background levels required for monitoring the decay of germanium-71 and has the counting equipment in hand. Progress awaits acquisition of the gallium, which will take several years.

The Institute for Nuclear Research of the Soviet Academy of Sciences has 60 tons of gallium available for an experiment, and a chamber has been prepared in the Baksan Laboratory. As planned, this experiment uses metallic gallium as the target rather than $GaCl_3$. However, after a novel initial extraction of the germanium, the experiment is similar to the gallium chloride experiment. Pilot studies have demonstrated the chemical techniques necessary for separating the germanium from the gallium, and counters are being prepared. In November 1986 the Soviet group and scientists from Los Alamos and the University of Pennsylvania agreed to collaborate on the experiment, which will begin in late 1987.

The INR also plans to repeat the Davis experiment, increasing the target volume of perchloroethylene by a factor of 5. This will increase the signal proportionally.

As mentioned above, a gallium experiment detects neutrinos from both proton fusion and beryllium-7 decay. To determine the individual rates of the two reactions requires a separate measurement of the neutrinos from the latter. A reaction that satisfies the criterion of being sensitive primarily to the beryllium-7 neutrinos is

$$\nu_e + {}^{81}Br \rightarrow {}^{81}Kr + e^-  .$$

Results from this bromine experiment are important to an unambiguous test of the standard solar model.

The chemical techniques needed for the bromine experiment are substantially identical to those employed in the chlorine-37 experiment, and therefore the feasibility of this aspect of the experiment is assured. However, since krypton-81 has a half-life of 200,000 years, counting a small number of atoms by radioactive-decay techniques is out of the question. Fortunately, another technique has recently been developed by G. S. Hurst and his colleagues at Oak Ridge National Laboratory. In barest outline the technique involves selective ionization of atoms of the desired element by laser pulses of the appropriate frequency. The ionized atoms can then readily be removed from the sample and directed into a mass spectrometer, where the desired isotope is counted. Repetitive application of the technique to increase the selection efficiency has been demonstrated.

The standard solar model predicts that a few atoms of krypton-81 would be produced per day in a volume of bromine solution similar to that of the chlorine solution in the Davis experiment. This is a sufficient number for successful application of resonance ionization spectroscopy. However, two other problems must be addressed. Protons produced by muons, neutrons, and alpha particles may introduce a troublesome background via the $^{81}Br(p,n)^{81}Kr$ reaction, and naturally occurring isotopes of krypton may leak into the tank of bromine solution and complicate the mass spectrometry. Davis, Hurst, and their collaborators have undertaken a complete assessment of the feasibility of the bromine-81 experiment.

Other inverse beta decays have been suggested as bases for detecting solar neutrinos by radiochemical techniques. An experiment based on one such reaction,

$$\nu_e + {}^7Li \rightarrow {}^7Be + e^-  ,$$

is being actively developed in the Soviet Union by the INR. According to the standard solar model, the observed rate of the reaction will be about 46 SNU.

Particularly appealing is the inverse beta decay

$$\nu_e + {}^{115}In \rightarrow {}^{115}Sn^* + e^-  ,$$

which has an enormous predicted rate (700 SNU according to the standard solar model) and is dominated by $pp$, $ppe$, and beryllium-7 neutrinos. Moreover, the 3-microsecond half-life of the product, an excited state of tin-115, implies that the reaction could be the basis for real-time measurements of the solar neutrino flux. Unfortunately, indium-115 is not completely stable, decaying by beta emission with a half-life of about $5 \times 10^{14}$ years. Electrons from the beta decay of indium-115 give rise to signals that can mimic the signature of its interaction with a solar neutrino (a prompt electron followed 3 microseconds later by two coincident gamma rays). This background is difficult to overcome, and such an experiment has not yet been fully developed.

As mentioned above, the source of the solar neutrino puzzle may lie not in imperfections of solar models but in our limited knowledge of neutrino physics. Neutrino oscillations, for example, could provide an explanation for the Davis results. This phenomenon is a predicted consequence of nonzero neutrino rest masses, and no theory compels an assignment of zero mass to these particles.

If neutrinos are massive, the flavor eigenstates that participate in weak interactions need not be the same as the mass eigenstates that propagate in free space. The two types of eigenstates are related by a unitary matrix that mixes the various neutrino species. For the case of only two neutrino species, say electron and muon neutrinos, this relation is

$$\begin{pmatrix} \nu_e \\ \nu_\mu \end{pmatrix} = \begin{pmatrix} \cos\theta & \sin\theta \\ -\sin\theta & \cos\theta \end{pmatrix} \begin{pmatrix} \nu_1 \\ \nu_2 \end{pmatrix}  ,$$

where $\nu_e$ and $\nu_\mu$ and $\nu_1$ and $\nu_2$ are flavor and mass eigenstates, respectively. According to the Schrödinger equation, the wave functions of $\nu_1$ and $\nu_2$ acquire phase factors $e^{-iE_1t}$ and $e^{-iE_2t}$ as they propagate. Therefore a pure $\nu_e$ state (created by, say, the beta decay of boron-8) evolves with time ("oscillates") into a state with a nonzero $\nu_\mu$ component. The probability $P_{\nu_e}$ that $\nu_e$ remains at time $t$ is given by

$$P_{\nu_e} = 1 - \sin^2 2\varphi \sin^2[(E_2 - E_1)\, t/2] \ ,$$

where $E_1$ and $E_2$ are the energies of $\nu_1$ and $\nu_2$. Thus $P_{\nu_e}$ differs from unity if and only if $m_1 \neq m_2$, since, in units such that the speed of light and Planck's constant are unity, $E_i^2 = p^2 + m_i^2$. For $p \gg m_2 > m_1$,

$$E_2 - E_1 \approx \frac{m_2^2 - m_1^2}{2p} \equiv \frac{\Delta m^2}{2p} \approx \frac{\Delta m^2}{2E_\nu} \ ,$$

and the characteristic oscillation length (the distance over which $P_{\nu_e}$ undergoes one cycle of its variation) is proportional to $E_\nu/\Delta m^2$.

The failure of numerous experiments to detect neutrino oscillations in terrestrial neutrino sources places an upper limit on $\Delta m^2$ of about 0.02 (eV)$^2$. (The precise limits are joint limits on $\Delta m^2$ and the mixing angle $\theta$.) However, if $\Delta m^2 \ll 0.02$ (eV)$^2$ (as some theoretical considerations suggest), oscillations would be undetectable in most terrestrial experiments and would most profitably be sought in low-energy neutrinos at large distances from the source (distances comparable to the oscillation length). Unlike the terrestrial oscillation experiments to date, experiments designed to characterize the solar neutrino spectrum could effectively search for oscillations in solar neutrinos and be capable of lowering the upper limit on $\Delta m^2$ to perhaps $10^{-11}$ (eV)$^2$.

Vacuum oscillations consistent with the standard solar model and the Davis experiment would require a rather large value of the mixing angle $\theta$. However, Wolfenstein, Mikhaev, and Smirnov have recently pointed out a feature of neutrino oscillations, namely, their amplification by matter, that could accommodate the Davis results even if $\theta$ is small, since it would greatly increase the probability that an electron neutrino produced in the high-density core of the sun emerge as a muon neutrino. The amplification is due to scattering by electrons and is therefore dependent upon electron density. (Scattering changes the phase of the propagating neutrino; its effect can be viewed as a change in either the index of refraction of the matter for neutrinos or in the potential energy (that is, effective mass) of the neutrino.) Observation of matter-enhanced oscillations should be possible for values of $\Delta m^2$ between $10^{-4}$ and $10^{-8}$ (eV)$^2$, a range inaccessible to experiments on terrestrial neutrino sources.

The importance of the solar neutrino puzzle and the exciting possibility that its solution may involve fundamental properties of neutrinos have led to a number of recent proposals for real-time flux measurements. The Japanese proton-decay group, together with researchers from Caltech and the University of Pennsylvania, is improving the Kamioka detector to observe the most energetic of the boron-8 solar neutrinos. The signal detected will be the Cerenkov radiation emitted by electrons in the water that recoil from neutrino scattering, receiving on average about half the neutrino energy. If the goal of a 7-MeV threshold for the detector is achieved, about 1 scattering event should be observed every 2 days (as predicted on the basis of the Davis flux measurements). The directionality of the signal relative to the sun will help distinguish scattering events from the isotropic background. Similar real-time flux measurements will also be possible with several of the second-generation detectors being built or planned for the Gran Sasso Laboratory.

The Sudbury Neutrino Observatory collaboration (Queen's, Irvine, Oxford, NRCC, Chalk River, Guelph, Laurentian, Princeton, Carleton) has proposed installing a 1000-ton heavy-water Cerenkov detector in the Sudbury Facility for real-time flux measurements of a different type. Here the source of the Cerenkov radiation will be the electrons produced in the inverse beta decay $\nu_e + d \rightarrow p + p + e^-$. Since the energy imparted to the electron is $E_\nu - 1.44$ MeV and the hoped-for threshold of the detector is about 7 MeV, the experiment will provide data on the higher energy portion of the boron-8 spectrum. About 8 events per day are expected to be recorded. The detector will be sensitive also to proton decay and to events induced by neutrinos from astrophysical sources and by muon neutrinos.

## Dedicated Underground Science Facilities

For at least two decades scientists with experiments demanding the enormous shielding from cosmic rays afforded by deep underground sites have been setting up their apparatus in working mines. We owe a great debt to the enlightened mine owners who have allowed this pursuit of knowledge to take place alongside their search for valuable minerals. However, as the experiments increase in complexity, the need for more supportive, dedicated facilities becomes more obvious.

One argument in favor of a dedicated facility is simple but compelling: the need to have access to the experimental area controlled not by the operation of a mine or a tunnel but by the schedules of the experiments themselves. Another is the need for technical support facilities adequate to experiments that will rival in complexity those mounted at major accelerators. And not to be ignored is the need for accommodations for the scientists and graduate students from many institutions who will participate in the experiments.

What should such a facility be like? The entryway should be large, and the experimental area should include at least several rooms in which different experiments can be in progress simultaneously. Provisions for easy expansion, ideally not only at the principal depth but also at greater and lesser depths, should be available. Another aspect that must be carefully planned for is safety. The underground environment is intrinsically hostile, and in addition some experiments may, like the Homestake experiment, involve large quantities of

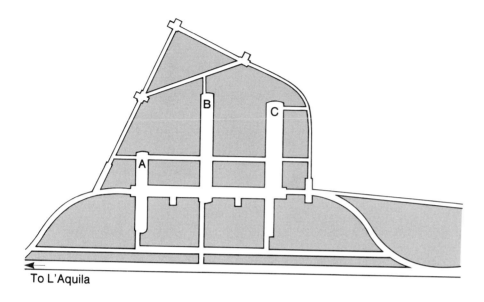

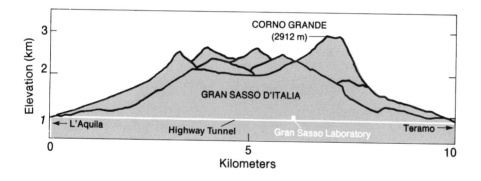

*Fig. 7. The three large (~35,000-cubic-meter) experimental halls planned for the Gran Sasso Laboratory are shown in a floor plan of the facility (top). Two of the three halls are now excavated. Also shown are the locations of the laboratory off the highway tunnel under the Gran Sasso d'Italia and of the tunnel in central Italy.*

materials that pose hazards in enclosed spaces. Materials being considered for the bromine experiment, for example, include dibromoethane, and other experiments being planned involve cryogenic materials under high pressure and toxic or inflammable materials. Excellent ventilation and gas-tight entries to some areas are obvious requirements.

Such dreams of dedicated facilities for underground science are now being realized. Italy, for example, recognized the opportunity offered by the construction several years ago of a new highway tunnel in the Apennines and incorporated a major underground laboratory (Fig. 7) under the Gran Sasso d'Italia near L'Aquila, which is about 80 kilometers east of Rome. This location offers an overburden of about 5000 mwe in rock of high strength and low background radioactivity. Two of three large rooms (each about 120 meters by 20 meters by 15 meters) have been completed. Support laboratories and offices are located above ground at the west end of the tunnel.

Because of its size, depth, support facilities, and ready access by superhighway, the Gran Sasso Laboratory is unrivaled as a site for underground science. In the spring of 1985, about a dozen new experiments were approved for installation. Among these are experiments on geophysics, gravity waves, and double beta decay; the GALLEX solar neutrino experiment; the large-area (1400-square-meter) MACRO detector, which can be used in studies of rare cosmic-ray phenomena, high-energy neutrino and gamma-ray astronomy, and searches for magnetic monopoles; and the 6500-ton liquid-argon ICARUS detector, which will have unprecedented sensitivity to neutrinos of solar and galactic origin, proton decay, high-energy muons, and many other rare phenomena. As an example of the capabilities of ICARUS, in one year of operation, it will detect, with an accuracy of 10 percent, a flux of boron-8 neutrinos more than twenty times smaller than the Davis limit, far below that allowed by any nonstandard solar model.

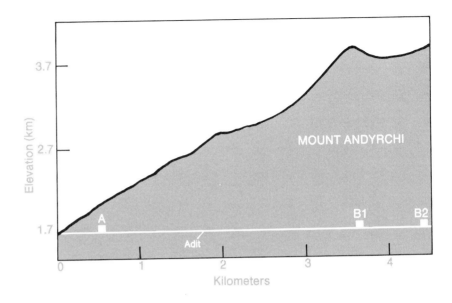

For more than ten years the Soviet Union has maintained an underground laboratory for cosmic-ray experiments in the Baksan River valley near Mt. Elbrus, the highest peak of the Caucasus Mountains. A 460-ton cosmic-ray telescope and a double beta decay experiment are in place at about 800 mwe. This laboratory is being greatly expanded (Fig. 8). The horizontal entry has been extended 3.6 kilometers under Mt. Andyrchi. There a 60-meter by 5-meter laboratory has been constructed to accommodate the Soviet gallium solar neutrino experiment and other smaller experiments. Further excavations are in progress to extend the adit an additional 700 meters and provide a large room for the 3000-ton chlorine solar neutrino experiment.

On a more modest scale Canada has proposed creation of an underground laboratory within the extensive and very deep excavations of the INCO Creighton No. 9 nickel mine near Sudbury, Ontario. The company has suggested available sites at about 2100 meters where rooms as large as 20 meters in diameter can be constructed.

Within the United States all underground experiments are in working or abandoned mines. None of these sites offers any prospect for expansion into a full-scale underground laboratory to rival Gran Sasso, Baksan, or even Sudbury. In 1981 and 1982 Los Alamos conducted a site survey and developed a detailed proposal to create a dedicated National Underground Science Facility at the Department of Energy's Nevada Test Site. The proposal called for vertical entry by a 14-foot shaft extending initially to 3600 feet (approximately 2900 mwe) and optionally to 6000 feet, excavation of two large experimental chambers, and provision of surface laboratories and offices. The proposal was not funded, and there is no other plan to provide a dedicated site in the United States for the next generation of underground searches for rare events.

## Conclusion

We have touched in detail upon only two of the fascinating experiments that drive

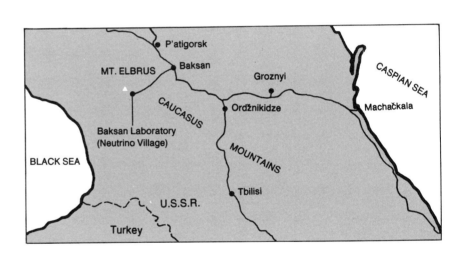

*Fig. 8. The main experimental areas of the Baksan Laboratory are shown in a profile of Mt. Andyrchi through the adit (top). Area A houses a large cosmic-ray telescope, area B1 has been excavated for the gallium solar neutrino experiment, and area B2, when excavated, will house the 3000-ton chlorine solar neutrino experiment. Also shown is the location of the facility near Mt. Elbrus in the Kabardino-Balkarian Autonomous Soviet Socialist Republic.*

scientists deep underground. Such experiments are not new on the scene, but the large and sophisticated second-generation detectors being built open up a new era. These devices should not be regarded as apparatus for a single experiment but as facilities useful for a variety of observations. They may be able to monitor continuously the galaxy for rare neutrino-producing events or the sun for variations in neutrino flux and hence in energy production. The day may be approaching, as Alfred Mann is fond of saying, where we will be able, from underground laboratories, to take the sun's temperature each morning to see how our nearest star is feeling. ■

**L. M. Simmons, Jr.,** is Associate Division Leader for Research in the Laboratory's Theoretical Division and was until 1985 Program Manager for the proposed National Underground Science Facility. He received a B.A. in physics from Rice University in 1959, an M.S. from Louisiana State University in 1961, and, in 1965, a Ph.D. in theoretical physics from Cornell University, where he studied under Peter Carruthers. He did postdoctoral work in elemetary particle theory at the University of Minnesota and the University of Wisconsin before joining the University of Texas as Assistant Professor. In 1973 he left the University of New Hampshire, where he was Visiting Assistant Professor, to join the staff of the Laboratory's Theoretical Division Office. There he worked closely with Carruthers, as Assistant and as Associate Division Leader, to develop the division as an outstanding basic research organization while continuing his own research in particle theory and the quantum theory of coherent states. He has been, since its inception, co-editor of the University of California's "Los Alamos Series in Basic and Applied Sciences." In 1979 he originated the idea for the Center for Nonlinear Studies and was instrumental in its establishment. In 1980 he took leave, as Visiting Professor of Physics at Washington University, to work on strong-coupling field theories and their large-order behavior, returning in 1981 as Deputy Associate Director for Physics and Mathematics. While in that position, he developed an interest in underground science and began work as leader of the NUSF project. He is President of the Aspen Center for Physics and has also served that organization as Trustee and Treasurer.

# Further Reading

Michael Martin Nieto, W. C. Haxton, C. M. Hoffman, E. W. Kolb, V. D. Sandberg, and J. W. Toevs, editors. *Science Underground (Los Alamos, 1982)*. New York: American Institute of Physics, 1983.

F. Reines. "Baryon Conservation: Early Interest to Current Concern." In *Proceedings of the 8th International Workshop on Weak Interactions and Neutrinos,* A Morales, editor. Singapore: World Scientific, 1983.

D. H. Perkins. "Proton Decay Experiments." *Annual Review of Nuclear and Particle Science* 34(1984):1.

R. Bionta et al. "The Search for Proton Decay." In *Intersections Between Particle and Nuclear Physics (Steamboat Springs, 1984)*. New York: American Institute of Physics, 1984.

R. Davis, Jr., B. T. Cleveland, and J. K. Rowley, "Report on Solar Neutrino Experiments." In *Intersections Between Particle and Nuclear Physics (Steamboat Springs, 1984)*. New York: American Institute of Physics, 1984.

J. N. Bahcall, W. F. Huebner, S. H. Lubow, P. D. Parker, and R. K. Ulrich. "Standard Solar Models and the Uncertainties in Predicted Capture Rates of Solar Neutrinos." *Reviews of Modern Physics* 54(1982):767.

R. R. Sharp, Jr., R. G. Warren, P. L. Aamodt, and A. K. Mann. "Preliminary Site Selection and Evaluation for a National Underground Physics Laboratory." Los Alamos National Laboratory unclassified release LAUR–82–556.

S. P. Rosen, L. M. Simmons, Jr., R. R. Sharp, Jr., and M. M. Nieto. "Los Alamos Proposal for a National Underground Science Facility. In *ICOBAN 84: Proceedings of the International Conference on Baryon Nonconservation (Park City, January 1984),* D. Cline, editor. Madison, Wisconsin: University of Wisconsin, 1984.

R. E. Mischke, editor. *Intersections Between Particle and Nuclear Physics (Steamboat Springs, 1984)*. New York: American Institute of Physics, 1984.

C. Castagnoli, editor. "First Symposium on Underground Physics (St.-Vincent, 1985)." *Il Nuovo Cimento* 9C(1986): 111–674.

J. C. Van der Velde. "Experimental Status of Proton Decay." In *First Aspen Winter Physics Conference,* M. Block, editor. *Annals of the New York Academy of Sciences* 461(1986).

M. L. Cherry, K. Lande, and W. A. Fowler, editors. *Solar Neutrinos and Neutrino Astronomy (Homestake, 1984)* AIP Conference Proceedings No. 126. New York: American Institute of Physics, 1984.

G. A. Cowan and W. C. Haxton. "Solar Neutrino Production of Technetium-97 and Technetium-98." *Science* 216(1982):51.

# *Quarks & Quirks among friends*

**"W**hat could be worse than a bunch of physicists gathering in a corner at a cocktail party to discuss physics?" asks Pete Carruthers. We at *Los Alamos Science* frankly didn't know what could be worse. . .or better, for that matter. However we did find the idea of "a bunch of physicists gathering in a corner to discuss physics" quite intriguing. We felt we might gain some insight and, at the same time, provide them with an opportunity to say things that are never printed in technical journals. So we gathered together a small bunch of four, Pete Carruthers, Stuart Raby, Richard Slansky, and Geoffrey West, found them a corner in the home of physicist and neurobiologist George Zweig and turned them loose. We knew it would be informative; we didn't know it would be this entertaining.

*WEST: I have here a sort of "fractalized" table of discussion, the first topic being, "What is particle physics, and what are its origins?" Perhaps the older gentlemen among us might want to answer that.*

**CARRUTHERS:** Everyone knows that older gentlemen don't know what particle physics is.

**ZWEIG:** Particle physics deals with the structure of matter. From the time people began wondering what everything was made of, whether it was particulate or continuous, from that time on we had particle physics.

**WEST:** In that sense of wondering about the nature of matter, particle physics started with the Greeks, if not observationally, at least philosophically.

**ZWEIG:** I think one of the first experimental contributions to particle physics came around 1830 with Faraday's electroplating experiments, where he showed that it would take certain quantities of electricity that were integral multiples of each other to plate a mole of one element or another onto his electrodes.

An even earlier contribution was Brown's observation of the motion of minute particles suspended in liquid. We now know the chaotic motion he observed was caused by the random collision of these particles with liquid molecules.

**RABY:** So Einstein's study of Brownian motion is an instance of somebody doing particle physics?

**ZWEIG:** Absolutely. There's a remarkable description of Brown's work by Darwin, who was a friend of his. It's interesting that Darwin, incredible observer of nature though he was, didn't recognize the chaotic nature of the movement under Brown's microscope; instead, he assumed he was see-

ing "the marvelous currents of protoplasm in some vegetable cell." When he asked Brown what he was looking at, Brown said, "That is my little secret."

**SLANSKY:** Quite a bit before Brown, Newton explained the sharp shadows created by light as being due to its particulate nature. That's really not the explanation from our present viewpoint, but it was based on what he saw.

**CARRUTHERS:** Newton was only half wrong. Light, like everything else, does have its particulate aspect. Newton just didn't have a way of explaining its wave-like behavior. That brings us to the critical concept of field, which Faraday put forward so clearly. You can speak of particulate structure, but when you bring in the field concept, you have a much richer, more subtle structure: fields are things that propagate like waves but materialize themselves in terms of quanta. And that is the current wisdom of what particle physics is, namely, quantized fields.

Quantum field theory is the only conceptual framework that pieces together the concepts of special relativity and quantum theory, as well as the observed group structure of the elementary particle spectrum. All these things live in this framework, and there's nothing to disprove its structure. Nature looks like a transformation process in the framework of quantum field theory. Matter is not just pointy little particles; it involves the more ethereal substance that people sometimes call waves, which in this theory are subsumed into one unruly construct, the quantized field.

**ZWEIG:** Particle physics wasn't always quantized field theory. When I was a graduate student, a different philosophy governed: S-matrix theory and the bootstrap hypothesis.

**CARRUTHERS:** That was a temporary aberration.

**ZWEIG:** But a big aberration in our lives! S-matrix theory was not wrong, just largely irrelevant.

**RABY:** If particle physics is the attempt to understand the basic building blocks of nature, then it's not a static thing. Atomic physics at one point was particle physics, but once you understood the atom, then you moved down a level to the nucleus, and so forth.

**WEST:** Let's bring it up to date, then. When would you say particle physics turned into high-energy physics?

**ZWEIG:** With accelerators.

**SLANSKY:** Well, it really began around 1910 with the use of the cloud chambers to detect cosmic rays; that is how Anderson detected the positron in 1932. His discovery straightened out a basic concept in quantized field theory, namely, what the antiparticle is.

**CARRUTHERS:** Yes, in 1926 Dirac had quantized the electromagnetic field and had given wave/particle duality a respectable mathematical framework. That framework predicted the positron because the electron had to have a positively charged partner. Actually, it was Oppenheimer who predicted the positron. Dirac wanted to interpret the positive solution of his equation as a proton, since there were spare protons sitting around in the world. To make this interpretation plausible, he had to invoke all that hanky-panky about the negative energy sea being filled—you could imagine that something was screwy.

**SLANSKY:** Say what you will, Dirac's idea was a wonderful unification of all nature, much more wonderful than we can envisage today.

**ZWEIG:** Ignorance is bliss.

**SLANSKY:** There were two particles, the proton and the electron, and they were the basic structure of all matter, and they were, in fact, manifestations of the same thing in field theory. We have nothing on the horizon that promises such a magnificent unification as that.

**RABY:** Weren't the proton and the electron supposed to have the same mass according to the equation?

**WEST:** No, the negative energy sea was supposed to take care of that.

**CARRUTHERS:** It was not unlike the present trick of explaining particle masses through spontaneous symmetry breaking.

Dirac's idea of viewing the proton and the electron as two different charge states of the same object was a nice idea that satisfied all the desires for symmetries that lurk in the hearts of theorists, but it was wrong. And the reason it was wrong, of course, is that the proton is the wrong object to compare with the electron. It's the quark and the electron that may turn out to be different states of a single field, a hypothesis we call grand unification.

**WEST:** Well, it is certainly true that high-energy particle physics now is cloaked in the language of quantized field theory, so much so that we call these theories the standard model.

**CARRUTHERS:** But I think we're overlooking the critical role of Rutherford in inventing particle physics.

**WEST:** The experiments of alpha scattering on gold foils to discern the structure of the atom.

**ZWEIG:** Rutherford established the paradigm we still use for probing the structure of matter: you just bounce one particle off another and see what happens.

**CARRUTHERS:** In fact, particle physics is a continuing dialogue (not always friendly) between experimentalists and theorists. Sometimes theorists come up with something that is interesting but that experimentalists suspect is wrong, even though they will win a Nobel prize if they can find the thing. And what the experimentalists do discover is frequently rather different from what the theorists thought, which makes the theorists go back and work some more. This is the way the field grows. We make lots of mistakes, we build the wrong machines, committees decide to do the wrong experiments, and journals refuse to publish the right theories. The process only works because there are so many objective entrepreneurs in the world who are trying to find out how matter behaves under these rather extreme conditions. It is marvelous to have great synthetic minds like those of Newton and Galileo, but they build not only on the work of unnamed thousands of theorists but also on these countless experiments.

> *"To understand the universe that we feel and touch, even down to its minutiae, you don't have to know a damn thing about quarks."*

*WEST: Perhaps we should tell how we personally got involved in physics, what drives us, why we stay with it. Because it is an awfully difficult field and a very frustrating field. How do we find the reality of it compared to our early romantic images? Let's start with Pete, who's been interviewed many times and should be in practice.*

CARRUTHERS: I was enormously interested in biology as a child, but I decided that it was too hard, too formless. So I thought I'd do something easy like physics. Our town library didn't even have modern quantum mechanics books. But I read the old quantum mechanics, and I read Jeans and Eddington and other inspirational books filled with flowery prose. I was very excited about the mysteries of the atom. It was ten years before I realized that I had been tricked. I had imagined I would go out and learn about the absolute truth, but after a little bit of experience I saw that the "absolute truth" of this year is replaced next year by something that may not even resemble it, leaving you with only some small residue of value. Eventually I came to feel that science, despite its experimental foundation and reference frame, shares much with other intellectual disciplines like music, art, and literature.

WEST: Dick, what about you?

SLANSKY: In college I listed myself as a physics major, but I gave my heart to philosophy and writing fiction. I had quite a hard time with them, too, but physics and mathematics remained easy. However, since I didn't see physics as very deep, I decided after I graduated to look at other fields. I spent a year in the Harvard Divinity School, where I found myself inadvertently a spokesman for science. I took Ed Purcell's quantum mechanics course in order to be able to answer people's questions, and it was there that I found myself, for the first time, absolutely fascinated by physics.

During that year I had been accepted at Berkeley as a graduate student in philosophy, but in May I asked them whether I could switch to physics. They wrote back saying it

*GEOFFREY B. WEST: "One of the great things that has happened in particle physics is that some of . . . the wonderful, deep questions . . . are being asked again . . . Somehow we have to understand why there is a weak scale, why there is an electromagnetic scale, why a strong scale, and ultimately why a grand scale."*

would be fine. I don't know that one can be such a dilettante these days.

SCIENCE: Why were people in Divinity School asking about quantum mechanics?

SLANSKY: People hoped to gain some insight into the roles of theology and philosophy from the intellectual framework of science. In the past certain philosophical systems have been based on physical theories. People were wondering what had really happened with quantum mechanics, since no philosophical system had been built upon it. Efforts have been made, but none so successful as Kant's with Newtonian physics, for example.

CARRUTHERS: Particle physics doesn't

stand still for philosophy. The subject is such that as soon as you understand something, you move on. I think restlessness characterizes this particular branch of science, in fact.

SLANSKY: I never looked at science as something I wanted to learn that would be absolutely permanent for all the rest of the history of mankind. I simply enjoy the doing of the physics, and I enjoy cheering on other people who are doing it. It is the intellectual excitement of particle physics that draws me to it.

ZWEIG: Dick, was there some connection, in your own mind, between religion and physics?

183

> *"The real problem was that you had a zoo of particles, with none seemingly more fundamental than any other."*

---

SLANSKY: Some. One of the issues that concerned me was the referential mechanisms of theological language. How we refer to things. In science we also have that concern, very much so.

ZWEIG: What do you mean by "How we refer to things"?

SLANSKY: When we use a word to refer to *God* or to refer to great generalizations in our experience, how does the word work to refer beyond the language? Language is just a sound. How does the word refer beyond just the mere word to the total experience? I've never really solved that problem in my own mind.

CARRUTHERS: When you mention the word *God,* isn't there a pattern of signals in your mind that corresponds to the pattern of sound? Doesn't *God* have a peculiar pattern?

SLANSKY: The referential mechanisms of theological language became a major concern around 1966, after I'd left Harvard Divinity. Before that the school was under the influence of the two great theologians Paul Tillich and Reinhold Niebuhr. Their concern was with the eighteenth and nineteenth century efforts to put into some sort of theoretical or logical framework all of man and his nature. I found myself swept up much more into theological and philosophical issues than into the study of ethics.

ZWEIG: Do you think these issues lie in the domain of science now? Questions about what man is, what his role in nature is, and what nature itself is, are being framed and answered by biologists and physicists.

SLANSKY: I don't view what I am trying to do in particle physics as finding man's place in nature. I think of it as a puzzle made of a lot of experimental data, and we are trying to assemble the pieces.

CARRUTHERS: But the attitudes are very theological, and often they tend to be dogmatic.

SLANSKY: I would like to make a personal statement here. That is, when I go out for a walk in the mountains, enjoying the beauties of nature with a capital *N,* I don't feel that that has any very direct relationship to formulating a theory of nature. While my per-

sonal experience may set my mind in motion, may provide some inspiration, I don't feel that seeing the Truchas peaks or seeing wild flowers in the springtime is very closely related to my efforts to build a theory.

WEST: Along that line I have an apocryphal story about Hans and Rose Bethe. One summer's evening when the stars were shining and the sky was spectacular, Rose was exclaiming over their beauty. Allegedly Hans replied, "Yes, but you know, I think I am the only man alive that knows why they shine." There you have the difference between the romantic and the scientific views.

RABY: Particle physics to me is a unique marriage of philosophy and reality. In high school I read the philosopher George Berkeley, who discusses space and time and tries to imagine what space would be like were there nothing in it. Could there be a force on a particle were there nothing else in space? Obviously a particle couldn't move because it would have nothing to move with respect to. Particle physics has the beauty of

philosophy constrained by the fact you are working with observable reality. For a science fair in high school I built a cloud chamber and tried to observe some alpha particles and beta particles. That's the reality part: you can actually build an experiment and actually see some of these fundamental objects. And there are people who are brilliant enough, like Einstein, to relate ideas and thought to reality and then make predictions about how the world must be. Special relativity and all the Gedanken experiments, which are basically philosophical, say how the world is. To me what particle physics means is that you can have an idea, based on some physical fact, that leads to some experimental prediction. That is beautiful, and I don't know how you define beauty except to say that it's in the eye of the beholder.

ZWEIG: How was science viewed in your family?

RABY: No one understood science in my family.

ZWEIG: Well, did they respect it even if they

STUART A. RABY: *"I think what particle physics means to me is this unique intermarriage of philosophy and reality . . . . Particle physics has the beauty of philosophy constrained by the fact that you are working with observable reality . . . . If you have a beautiful idea and it leads to a prediction that, in fact, comes true, that would be the most amazing thing. That you can understand something on such a fundamental level!"*

> *". . . . one thing that distinguishes physics from philosophy is predictive power. The quark model had a lot of predictive power."*

didn't understand it?

**RABY:** I guess they accepted the fact that I would pursue what interested me. I'm the first one in my family to finish college, and that in itself is something big to them. My grandfather, who does understand a little, has read about Einstein. My grandfather's interest in science doesn't come from any particular training, but from the fact that he is very inventive and intuitive and puts radios together and learns everything by himself.

**ZWEIG:** Was he respected for it?

**RABY:** By whom? My grandfather owned a chicken market, so he did these things in his spare time.

**WEST:** That's interesting. I have to admit I am another person who got into physics in spite of himself. I was facile in mathematics but more keen on literature. I turned to natural sciences when I went to Cambridge only because I had begun reading Jeans and Eddington and all those early twentieth century visionaries. They were describing that wonderful time of the birth of quantum mechanics, the birth of relativity, the beginning of thinking about cosmology and the origin of the universe. Wonderful questions! Really important questions that dovetailed into the big questions raised by literature. What is it all really about, this mysterious universe?

The other crucial reason that I went into science was that I could not stand the world of business, the world of the wheelerdealer, that whole materialistic world. Somehow I had an image of the scientist as removed from that, judged only by his work, his only criteria being proof, knowledge, and wisdom. I still hold that romantic image. And that has been my biggest disappointment, because, of course, science, like everything else that involves millions of dollars, has its own wheeler-dealers and salesmen and all the rest of it.

My undergraduate experience at Cambridge was something of a disaster in terms of physics education, and I was determined to leave the field. I had become very interested in West Coast jazz and managed to obtain a fellowship to Stanford where, for a

*RICHARD A. SLANSKY: "It is the intellectual excitement of particle physics that draws me to it, really . . . . I find particle physics an intriguing effort to try to explain and understand, in a very special way, what goes on in nature . . . . I enjoy the effort . . . . I enjoy cheering on other people who are trying . . . . I think of it as a puzzle made of a lot of experimental data, and we are trying to assemble all the pieces."*

year, I could be near San Francisco, North Beach, and that whole scene. Although at first I hated Palo Alto, my physics courses were on so much more a professional level, so much more an exciting level, that my attitude eventually changed. Somehow the whole world opened up. But even in graduate school I would go back to reading Eddington, whether he were right or not, because his language and way of thinking were inspirational, as of course, were Einstein's.

**CARRUTHERS:** Do you think our visions have become muddied in these modern times?

**WEST:** I don't think so at all. One of the great things that has happened in particle physics is that some of the deep questions are being asked again. Not that I like the proposed answers, particularly, but the questions are being asked. George, what do you say to all this? You often have a different slant.

**ZWEIG:** My parents came from eastern Europe—they fled just before the second World War. I was born in Moscow and came to this country when I was less than two years old. Most of my family perished in the war, probably in concentration camps. I learned

> *"It is an old Jewish belief that ideas are what really matter. If you want to create things that will endure, you create them in the mind of man."*

at a very early age from the example of my father, who was wise enough to see the situation in Germany for what it really was, that it is very important to understand reality. Reality is the bottom line. Science deals with reality, and psychology with our ability to accept it.

I grew up in a rough, integrated neighborhood in Detroit. Much of it subsequently burned down in the 1967 riot. I hated school and at first did very poorly. I was placed in a "slower" non-college preparatory class and took a lot of shop courses. Although I did not like being viewed as a second class citizen, I thought that operating machines was a hell of a lot more interesting than discussing social relations with my classmates and teachers.

Eventually I was able to do everything that was asked of me very quickly, but the teachers were not knowledgeable, and classes were boring. In order to get along I kept my mouth shut. Occasionally I acted as an expediter, asking questions to help my classmates.

At that time science and magic were really one and the same in my mind, and what child isn't fascinated by magic? At home I did all sorts of tinkering. I built rockets that flew and developed my own rocket fuels. The ultimate in magic was my tesla coil with a six foot corona emanating from a door knob.

College was a revelation to me. I went to the University of Michigan and majored in mathematics. For the first time I met teachers who were smart. And then I went to Caltech, a place I had never even heard of six months before I arrived. At Caltech I was very fortunate to work with Alvin Tollestrup, an experimentalist who later designed the superconducting magnets that are used at Fermilab. And I was exposed to Feynman and Gell-Mann, who were unbelievable individuals in their own distinctive ways. That was an exciting time. Shelly Glashow was a postdoc. Ken Wilson, Hung Cheng, Roger Dashen, and Sidney Coleman were graduate students. Rudy Mössbauer was down the hall. He was still a research fellow one month before he got the Nobel prize. The board of trustees called a

GEORGE ZWEIG: *"I learned at a very early age. . .that reality is the bottom line. Science deals with reality, and psychology with our ability to accept it."*

crash meeting and promoted him to full professor just before the announcement. I remember pleading with Dan Kevles in the history department to come over to the physics department and record the progress, because science history was in the making, but he wouldn't budge. "You can never tell what is important until many years later," he said.

**CARRUTHERS:** I've forgotten whether I first met all of you at Cal Tech or at Aspen.

**ZWEIG:** Wherever Pete met us, I know we're all here because of him. He was always very gently asking me, "How about coming to Los Alamos?" Eventually I took him up on his offer.

**WEST:** Before we leave this more personal side of the interview, I want to ask a question or two about families. Is it true that physicists generally come from middle class and lower backgrounds? Dick, what about your family?

**SLANSKY:** My father came from a farming family. Since he weighed only ninety-seven pounds when he graduated from high school, farm work was a little heavy for him. He entered a local college and eventually earned a graduate degree from Berkeley as a physical chemist. My mother wanted to attend medical school and was admitted, but back in those days it was more important to have children. So I am the result rather than her becoming a doctor.

**CARRUTHERS:** My father grew up on a farm in Indiana, was identified as a bright kid, and was sent off to Purdue, where he became an engineer. So I at least had somebody who believed in a technical world. However, when I finally became a professor at Cornell, my parents were a bit disappointed because in their experience only those who couldn't make it in the business world became faculty members.

**WEST:** What about your parents, George?

**ZWEIG:** Both my parents are intellectuals, people very much concerned with ideas. To me one of the virtues of doing science is that you contribute to the construction of ideas, which last in ways that material monuments don't. It is an old Jewish belief that ideas are what really matter. If you want to create things that will endure, you create them in the mind of man.

**WEST:** What did your parents do?

**ZWEIG:** My mother was a nursery school teacher. She studied in Vienna in the '20s, an exciting time. Montessori was there; Freud was there. My father was a structural engineer. He chose his profession for political reasons, because engineering was a useful thing to do.

**WEST:** Then all three of you have scientific or engineering backgrounds. My mother is a dressmaker, and my father was a professional gambler. But he was an intellectual in many ways, even though he left school at fifteen. He read profusely, knew everything superficially very well, and was brilliant in languages. He wasted his life gambling, but it was an interesting life. I think I became facile in mathematics at a young age just because he was so quick at working out odds, odds on dogs and horses, how to do triples and doubles, and so on.

**CARRUTHERS:** Are we all firstborn sons? I think we are, and that's an often quoted statistic about scientists.

**WEST:** Have we all retreated into science for solace?

**RABY:** It's more than that. At one time I felt divided between going into social work in

**PETER A. CARRUTHERS:** *"There's no point in a full-blown essay on quantum field theory because it's probably wrong anyway. That's what fundamental science is all about—whatever you're doing is probably wrong. That's how you know when you're doing it. Once in a while you're right, and then you're a great man, or woman nowadays. I've tried to explain this before to people, but they're very slow to understand. What you have to do is look back and find what has been filtered out as correct by experiments and a lot of subsequent restructuring. Right? But when you're actually doing it, almost every time you're wrong. Everybody thinks you sit on a mountaintop communing with Jung's collective unconscious, right? Well you try, but the collective unconscious isn't any smarter than you are."*

*"Why do the forces in nature have different strengths . . . That's one of those wonderful deep questions that has come back to haunt us."*

---

order to be involved with people or going into science and being involved with ideas. It was continually on my mind, and when I graduated from college, I took a year off to do social work. I worked in a youth house in the South Bronx as a counselor for kids between the ages of seven and seventeen. They were all there waiting to be sentenced, and they were very self-destructive kids. The best thing you could do was to show them that they should have goals and that they shouldn't destroy themselves when the goals seemed out of reach. For example, a typical goal was to get out of the place, and a typical reaction was to end up a suicide. I kept trying to tell these kids, "Do what you enjoy doing and set a goal for yourself and try to fulfill that goal in positive ways." In the end I was convinced by my own logic that I should return to physics.

WEST: Let's discuss the way physics affects our personal lives now that we are grown men. Suppose you are at a cocktail party, and someone asks, "What do you do?" "I am a physicist," you say, "High-energy physics," or "Particle physics." Then there is a silence and it is very awkward. That is one response, and here is the other. "Oh, you do particle physics? My God, that's exciting stuff! I read about quarks and couldn't understand a word of it. But then I read this great book, *The Tao of Physics.* Can you tell me what you do?" I groan inwardly and sadly reflect on how great the communication gap is between scientists such as ourselves and the general public that supports us. We seem to have shirked our responsibility in communicating the fantastic ideas and concepts involved in our enterprise to the masses. It is a sobering thought that Capra's book, which most of us don't particularly like because it represents neither particle physics nor Zen accurately, is probably unique in turning on the layman to some aspects of particle physics. Whatever your views of that book may be, you've certainly got to appreciate what he's done for the publicity of the field. As for me, I find it difficult to talk about this life that I love in two-line sentences.

Now, the cocktail party is just a superficial

aspect of my social life, but the problem enters in a more crucial way in my relationship with my family, the people dear to me. Here is this work which I love, which I spend a majority of my time in doing, and from which a large number of the frustrations and disappointments and joys in my life come, and I cannot communicate it to my family except in an incredibly superficial way.

SLANSKY: The cocktail party experiences that Geoffrey describes are absolutely perfect, and I know what he means about the family. Now that my children are older, they are into science, and sometimes they ask me questions at the dinner table. I try to give clear explanations, but I'm never sure I've succeeded even superficially. And my wife, who is very bright but has no science background, doesn't hesitate to say that science in more than twenty-five words is boring. Sometimes, in fact, I feel that my doing physics is viewed by them as a hobby.

CARRUTHERS: Socially, what could be worse than a bunch of physicists gathering in a corner at a cocktail party to discuss physics?

RABY: I find there are two types of people. There are people who ask you a question just to be polite and who don't really want an answer. Those people you ignore. Then there are people who are genuinely interested, and

you talk to them. If they don't understand what a quark is, you ask them if they understand what a proton or an electron is. If they don't understand those, then you ask them if they know what an atom is. You describe an atom as electrons and a nucleus of protons and neutrons. You go down from there, and you eventually get to what you are studying—particle physics.

WEST: Does particle physics affect your relationship with your wife?

RABY: My wife is occasionally interested in all this. My son, however, is genuinely interested in all forms of physical phenomena and is constantly asking questions. He likes to hear about gravity, that the gravity that pulls objects to the earth also pulls the moon around the earth. I have to admit that I find his interest very rewarding.

*WEST: Maybe, since we've been given the opportunity today, we should start talking about physics. Particle physics has gone through a minirevolution since the discovery of the psi/J particle at SLAC and at Brookhaven ten years ago. Although not important in itself, that discovery confirmed a whole way of thinking in terms of quarks, symmetry principles, gauge theories, and unification. It was a bolt out of the blue at a time when the direction of particle physics was uncertain. From then on, it became clear that non-Abelian gauge theories and unification were going to form the fundamental principles for research. Sociologically, there developed a unanimity in the field, a unanimity that has remained. This has led us to the standard model, which incorporates the strong, weak, and electromagnetic interactions.*

SLANSKY: Yes, the standard model is a marvelous synthesis of ideas that have been around for a long time. It derives all interactions from one elegant principle, the principle of local symmetry, which has its origin in the structure of electromagnetism. In the 1950s Yang and Mills generalized this structure to the so-called non-Abelian gauge theories and then through the '60s and '70s we learned enough about these field theories to feel confident describing all the forces of nature in terms of them.

**RABY:** I think we feel confident with Yang-Mills theories because they are just a sophisticated version of our old concept of force. The idea is that all of matter is made up of quarks and leptons (electrons, muons, etc.) and that the forces or interactions between them arise from the exchange of special kinds of particles called gauge particles: the photon in electromagnetic interactions, the $W^\pm$ and $Z^0$ in the weak interactions responsible for radioactive decay, and the gluons in strong interactions that bind the nucleus. (It is believed that the graviton plays a similar role in gravity.) [The local gauge theory of the strong forces is called quantum chromodynamics (QCD). The local gauge theory that unifies the weak force and the electromagnetic force is the electroweak theory that predicted the existence of the $W^\pm$ and $Z^0$.]

**ZWEIG:** You are talking about a very limited aspect of what high-energy physics has been. Our present understanding did not develop in an orderly manner. In fact, what took place in the early '60s was the first revolution we have had in physics since quantum mechanics. At that time, if you were at Berkeley studying physics, you studied S-matrix, not field theory, and when I went to Caltech, I was also taught that field theory was not important.

**SLANSKY:** Yes, the few people that were focusing on Yang-Mills theories in the '50s and early '60s were more or less ignored. Perhaps the most impressive of those early papers was one by Julian Schwinger in which he tried to use the isotopic spin group as a local symmetry group for the weak, not the strong, interactions. (Schwinger's approach turned out to be correct. The Nobel prize-winning $SU(2) \times U(1)$ electroweak theory that predicted the $W^\pm$ and $Z^0$ vector mesons to mediate the weak interactions is an expanded version of Schwinger's $SU(2)$ model.)

**WEST:** In retrospect Schwinger is a real hero in the sense that he kept the faith and made

some remarkable discoveries in field theory during a period when everybody was basically giving him the finger. He was completely ignored and, in fact, felt left out of the field because no one would pay any attention.

**SCIENCE:** Why was field theory dropped?

**CARRUTHERS:** Now we reach a curious sociological phenomenon.

**RABY:** Sociological? I thought the theory was just too hard to understand. There were all those infinities that cropped up in the calculations and had to be renormalized away.

**CARRUTHERS:** I am afraid there is a phase transition that occurs in groups of people of whatever IQ who feverishly follow each new promising trend in science. They go to a conference, where a guru raises his hands up and waves his baton; everyone sits there, their heads going in unison, and the few heretics sitting out there are mostly intimidated into keeping their heresies to themselves. After a while some new religion comes along, and a new faith replaces the old. This is a curious thing, which you often see at football games and the like.

**WEST:** The myth perpetrated about field theory was, as Stuart said, that the problems were too hard. But if you look at Yang-Mills and Julian Schwinger's paper, for example, there was still serious work that could have been done. Instead, when I was at Stanford, Sidney Drell taught advanced quantum mechanics and gave a whole lecture on why you didn't need field theory. All you needed were Feynman graphs. That was the theory.

**RABY:** The real problem was that you had a zoo of particles, with none seemingly more fundamental than any other. Before people knew about quarks, you didn't feel that you were writing down the fundamental fields.

**WEST:** In 1954 we had all the machinery necessary to write down the standard model. We had the renormalization group. We had local gauge theories.

**SLANSKY:** But nobody knew what to apply them to.

**SCIENCE:** George, in 1963, when you came up with the idea that quarks were the constit-

uents of the strongly interacting particles, did you think at all about field theory?

**ZWEIG:** No. The history I remember is quite different. The physics community responded to this proliferation of particles by embracing the bootstrap hypothesis. No particle was viewed as fundamental; instead, there was a nuclear democracy in which all particles were made out of one another. The idea had its origins in Heisenberg's S-matrix theory. Heisenberg published a paper in 1943 reiterating the philosophy that underlies quantum mechanics, namely, that you should only deal with observables. In the case of quantum mechanics, you deal with spectral lines, the frequencies of light emitted from atoms. In the case of particle physics, you go back to the ideas of Rutherford. Operationally, you study the structure of matter by scattering one particle off another and observing what happens. The experimental results can be organized in a kind of a matrix that gives the amplitudes for the incoming particles to scatter into the outgoing ones. Measuring the elements of this scattering, or S-matrix, was the goal of experimentalists. The work of theorists was to write down relationships that these S-matrix elements had to obey. The idea that there was another hidden layer of reality, that there were objects inside protons and neutrons that hadn't been observed but were responsible for the properties of these particles, was an idea that was just totally foreign to the S-matrix philosophy; so the proposal that the hadrons were composed of more fundamental constituents was vigorously resisted. Not until ten years later, with the discovery of the psi/$J$ particle, did the quark hypothesis become generally accepted. By then the evidence was so dramatic that you didn't have to be an expert to see the underlying structure.

**RABY:** The philosophy of the bootstrap, from what I have read of it, is a very beautiful philosophy. There is no fundamental particle, but there are fundamental rules of how particles interact to produce the whole spectrum. But one thing that distinguishes physics from philosophy is predictive power. The quark model had a lot of predictive power. It predicted the whole spectrum of hadrons observed in high energy experiments. It is not because of sociology that the bootstrap went out; it was the experimental evidence of *J*/psi that made people believe there really are objects called quarks that are the building blocks of all the hadrons that we see. It is this reality that turned people in the direction they follow today.

**CARRUTHERS:** And because of the very intense proliferation of unknowns, it is unlikely that the search for fundamental constituents will stop here. In the standard model you have dozens of parameters that are beyond any experimental reach.

**SCIENCE:** But you have fewer coordinates now than you had originally, right?

**CARRUTHERS:** If you are saying the coordinates have all been coordinated by group symmetry, then of course there are many fewer.

**WEST:** I think the deep inelastic scattering experiments at SLAC played an absolutely crucial role in convincing people that quarks are real. It was quite clear from the scaling behavior of the scattering amplitudes that you were doing a classic Rutherford type scattering experiment and that you were literally seeing the constituents of the nucleon. I think that was something that was extremely convincing. Not only was it qualitatively correct, but quantitatively numbers were coming out that could only come about if you believed the scattering was taking place from quarks, even though they weren't actually being isolated. But let me say one other thing about the S-matrix approach. That approach is really quantum mechanics in action. Everything is connected with everything else by this principle of unitarity or conservation of probability. It is a very curious state of affairs that the quark model, which requires less quantum mechanics to predict, say, the spectrum of particles, has proven to be much more useful.

**SLANSKY:** Remember, though, there were some important things missing in the bootstrap approach. There was no natural way to incorporate the weak and the elec-

tromagnetic forces.

**WEST:** That picks up another important point; the S-matrix theory could not cope with the problem of scale. And that brings us back to the standard model and then into grand unification. The deep inelastic scattering experiments focused attention on the idea that physical theories exhibit a scale invariance similar to ordinary dimensional analysis.

One of the wonderful things that happened as a result was that all of us began to accept renormalization (the infinite rescaling of field theories to make the answers come out finite) as more than just hocus-pocus. Any graduate student first learning the renormalization procedure must have thought that a trick was being pulled and that the procedure for getting finite answers by subtracting one infinity from another really couldn't be right. An element of hocus-pocus may still remain, but the understanding that renormalization was just an exploitation of scale invariance in the very complicated context of field theory has raised the procedure to the level of a principle.

The focus on scale also led to the feeling that somehow we have to understand why

the forces in nature have different strengths and become strong at different energies, why there are different energy scales for the weak, for the electromagnetic, and for the strong interactions, and ultimately whether there may be a grand scale, that is, an energy at which all the forces look alike. That's one of those wonderful deep questions that has come back to haunt us.

**RABY:** I guess we think of quantum electrodynamics (QED) as being such a successful theory because calculations have been done to an incredibly high degree of accuracy. But it is hard to imagine that we will ever do that well for the quark interactions. The whole method of doing computations in QED is perturbative. You can treat the electromagnetic interaction as a small perturbation on the free theory. But, in order to understand what is going on in the strong interactions of quantum chromodynamics (QCD), you have to use nonperturbative methods, and then you get a whole new feeling about the content of field theory. Field theory is much richer than a perturbative analysis might lead one to believe. The study of scaling by M. Fisher, L. Kadanoff, and K. Wilson emphasized the interrelation of statistical mechanics and field theory. For example, it is now understood that a given field theoretic model may, as in statistical mechanics systems, exist in several qualitatively different phases. Statistical mechanical methods have also been applied to field theoretic systems. For example, gauge theories are now being studied on discrete space-time lattices, using Monte Carlo computer simulations or analog high temperature expansions to investigate the complicated phase structure. There has now emerged a fruitful interdisciplinary focus on the non-linear dynamics inherent in the subjects of field theory, statistical mechanics, and classical turbulence.

ZWEIG: Isn't it true to say that the number of things you can actually compute with QCD is far less than you could compute with S-matrix theory many years ago?

WEST: I wouldn't say that.

ZWEIG: What numbers can be experimentally measured that have been computed cleanly from QCD?

RABY: What is your definition of clean?

ZWEIG: A clean calculation is one whose assumptions are only those of the theory. Let me give you an example. I certainly will accept the numerical results obtained from lattice gauge calculations of QCD as definitive if you can demonstrate that they follow directly from QCD. When you approximate space-time as a discrete set of points lying in a box instead of an infinite continuum, as you do in lattice calculations, you have to show that these approximations are legitimate. For example, you have to show that the effects of the finite lattice size have been properly taken into account.

RABY: To return to the question, this is the first time you can imagine calculating the spectrum of strongly interacting particles from first principles.

ZWEIG: The spectrum of strongly interacting particles has not yet been calculated in QCD. In principle it should be possible, and much progress has been made, but operationally the situation is not much better than it was in the early '60s when the bootstrap was gospel.

SLANSKY: Yes, but that was a very dirty calculation. The agreement got worse as the calculations became more cleverly done.

WEST: The numbers from lattice gauge theory calculations of QCD are not necessarily meaningful at present. There is a serious question whether the lattice gauge theory, as formulated, is a real theory. When you take the lattice spacing to zero and go to the continuum limit, does that give you the theory you thought you had?

RABY: That's the devil's advocate point of view, the view coming from the mathematical physicists. On the other hand, people have made approximations, and what you can say is that any approximation scheme that you use has given the same results. First, there are hadrons that are bound states of quarks, and these bound states have finite size. Second, there is no scale in the theory, but everything, all the masses, for example, can be defined in terms of one fundamental scale. You can get rough estimates of the whole particle spectrum.

WEST: You can predict that from the old quark model, without knowing anything about the local color symmetry and the eight colored gluons that are the gauge particles of the theory. There is only one clean calculation that can be done in QCD. That is the calculation of scattering amplitudes at very high energies. Renormalization group analysis tells us the theory is asymptotically free at high energies, that is, at very high energies quarks behave as free point-like particles so the scattering amplitudes should scale with energy. The calculations predict logarithmic corrections to perfect scaling. These have been observed and they seem to be unique to QCD. Another feature unique to quantum chromodynamics is the coupling of the gluon to itself which should predict the existence of glueballs. These exotic objects would provide another clean test of QCD.

ZWEIG: I agree. The most dramatic and interesting tests of quantum chromodynamics follow from those aspects of the theory that have nothing to do with quarks directly. The theory presumably does predict the existence of bound states of gluons, and

furthermore, some of those bound states should have quantum numbers that are not the same as those of particles made out of quark-antiquark pairs. The bound states that I would like to see studied are these "oddballs," particles that don't appear in the simple quark model. The theory should predict quantum numbers and masses for these objects. These would be among the most exciting predictions of QCD.

RABY: People who are calculating the hadronic spectrum are doing those sorts of calculations too.

ZWEIG: It's important to pick one fundamental question, push on it, and get the right answer. You may differ as to whether you want to use the existence of oddballs as a crucial test or something else, but you should accept responsibility for performing calculations that are clean enough to provide meaningful comparison between theory and experiment. The spirit of empiricism does not seem to be as prevalent now as it was when people were trying different approaches in particle physics, that is, S-matrix theory, field theory, and the quark model. The development of the field was much more Darwinian then. People explored many different ideas, and natural selection picked the winner. Now evolution has changed; it is Lamarckian. People think they know what the right answer is, and they focus and build on one another's views. The value of actually testing what they believe has been substantially diminished.

SLANSKY: I don't think that is true. The technical problems of solving QCD have proved to be harder than any other technical problems faced in physics before. People have had to back off and try to sharpen their technical tools. I think, in fact, that most do have open minds as to whether it is going to be right or wrong.

WEST: What do you think about the rest of the standard model? Do we think the electroweak unification is a closed book, especially now that $W^{\pm}$ and $Z^0$ vector bosons have been discovered?

SLANSKY: It is to a certain level of accuracy, but the theory itself is just a

> *"It may be that all this matter is looped together in some complex topological web and that if you tear apart the Gordian knot with your sword of Damocles, something really strange will happen."*

phenomenology with some twenty or so free parameters floating around. So it is clearly not the final answer.

**SCIENCE:** What are these numbers?

**RABY:** All the masses of the quarks and leptons are put into the theory by hand. Also, the mixing angle, the so-called Cabibbo angle, which describes how the charmed quark decays into a strange quark and a little bit of the down quark, is not understood at all.

**ZWEIG:** Operationally, the electroweak theory is solid. It predicted that the $W^\pm$ and $Z^0$ vector bosons would exist at certain masses, and they actually do exist at those masses.

**SLANSKY:** The theory also predicted the coupling of the $Z^0$ to the weak neutral current. People didn't want to have to live with neutral currents because, to a very high degree of experimental accuracy, there was no evidence for strangeness-changing weak neutral currents. The analysis through local symmetry seemed to force on you the existence of weak neutral currents, and when they were observed in '73 or whenever, it was a tremendous victory for the model. The electron has a weak neutral current, too, and this current has a very special form in the standard model. (It is an almost purely axial current.) This form of the current was established in polarized electron experiments at SLAC. Very shortly after those experimental results, Glashow, Weinberg, and Salam received the Nobel prize for their work on the standard model of electroweak interactions. I think that was the appropriate time to give the Nobel prize, although a lot of my colleagues felt it was a little bit premature.

**RABY:** However, the Higgs boson required for the consistency of the theory hasn't been seen yet.

**SLANSKY:** A little over a year ago there were four particles that needed to be seen—now there is only one. The standard model theory has had some rather impressive successes.

**WEST:** Can we use this as a point of departure to talk about grand unification? Unification of the weak and electromagnetic interactions, which had appeared to be quite separate forces, has become the prototype for attempts to unify those two with the strong interactions.

**RABY:** In the standard model of the weak interactions, the quarks and the leptons are totally separate even though phenomenologically they seem to come in families. For example, the up and the down quarks seem to form a family with the electron and its neutrino. Grand unification is an attempt to unify quarks and leptons, that is, to describe them as different aspects of the same object. In other words, there is a large symmetry group within which quarks and leptons can transform into each other. The larger group includes the local symmetry groups of the strong and electroweak interaction and thereby unifies all the forces. These grand unified theories also predict new interactions that take quarks into leptons and vice versa. One prediction of these grand unified theories is proton decay.

*WEST: The two most crucial predictions of grand unified theories are, first, that protons are not perfectly stable and can decay and, second, that magnetic monopoles exist. Neither of these has been seen so far. Suppose they are never seen. Does that mean the question of grand unification becomes merely philosophical? Also, how does that bear on the idea of building a very high-energy accelerator like the SSC (superconducting super collider) that will cost the taxpayer $3 billion?*

**CARRUTHERS:** Why should we build this giant accelerator? Because in our theoretical work we don't have a secure world view; we need answers to many critical questions raised by the evidence from the lower energies. Even though I know that as soon as you do these new experiments, the number of questions is likely to multiply. This is part of my negative curvature view of the progress of science. But there are some rather primitive questions which can be answered and which don't require any kind of sophistication. For instance, are there any new particles of well-defined mass of the old-fashioned type or new particles with different properties, perhaps? Will we see the Higgs particle that people stick into theories just to make the clock work? If you talk to people who make models, they will give you a panorama of predictions, and those predictions will become quite vulnerable to proof if we increase the amount of accelerator energy by a factor of 10 to 20. Those people are either going to be right, or they're going to have to retract their predictions and admit, "Gee, it didn't work out, did it?"

There is a second issue to be addressed, and that is the question of what the fundamental constituents of matter are. We messed up thirty years ago when we thought protons and neutrons were fundamental. We know now that they're structured objects, like atoms: they're messy and squishy and all kinds of things are buzzing around inside. Then we discovered that there are quarks and that the quarks must be held together by glue. But some wise guy comes along and says, "How do you know those quarks and gluons and leptons are not just as messy as those old protons were?" We need to test whether or not the quark itself has some composite structure by delivering to the quarks within the nucleons enough energy and momentum transfer. The accelerator acts like a microscope to resolve some fuzziness in the localization of that quark, and a whole new level of substructure may be discovered. It may be that all this matter is looped together in some complex topological web and that if you tear apart the Gordian knot with your sword of Damocles, something really strange will happen. A genie may pop out of the bottle and say, "Master, you have three wishes."

A third issue to explore at the SSC is the dynamics of how fundamental constituents interact with one another. This takes you into the much more technical area of analyzing numbers to learn whether the world view you've constructed from evidence and theory makes any sense. At the moment we have no idea why the masses of anything are what they are. You have a theory which is attractive, suggestive, and can explain many, many things. In the end, it has twenty or thirty

> *"If we can get people to agree on why we should be doing high-energy physics, then I think we can solve the problem of price."*

parameters. You can't be very content that you've understood the structure of matter.

**SLANSKY:** To make any real progress both in unification of the known forces and in understanding anything about how to go beyond the interactions known today, a machine of something like 20 to 40 TeV center of mass energy from proton-proton collisions absolutely must be built.

**SCIENCE:** Will these new machines test QCD at the same time they test questions of unification?

**SLANSKY:** The pertinent energy scale in QCD is on the order of GeV, not TeV, so it is not clear exactly what you could test at very high energies in terms of the very nonlinear structure of QCD. Pete feels differently.

**CARRUTHERS:** All I say is that you may be looking at things you don't think you are looking at.

**WEST:** Obviously all this is highly speculative. A question you are obligated to ask is at what stage do you stop the financing. I think we have to put the answer in terms of a realistic scientific budget for the United States, or for the world for that matter.

**CARRUTHERS:** Is there a good reason why the world can't unify its efforts to go to higher energies?

**WEST:** Countries are mostly at war with one another. They couldn't stop to have the Olympic games together, so certainly not for a bloody machine.

**SLANSKY:** The Europeans themselves have gotten together in probably one of the most remarkable examples of international collaboration that has ever happened.

**WEST:** Yes, I think the existence of CERN is one of the greatest contributions of particle physics to the world.

**RABY:** But the next step is going to have to be some collaborative effort of CERN, the U.S., and Japan.

**WEST:** Our SSC is going to be the next step. But you are still not answering the question. Should we expect the government to support this sort of project at the $3 billion level?

**SLANSKY:** That's $3 billion over ten years.

**RABY:** You can ask that same question of any fundamental research that has no direct application to technology or national security, and you will get two different answers. The "practical" person will say that you do only what you conceive to have some benefits five or ten years down the line, whereas the person who has learned from history will say that all fundamental research leads eventually either to new intellectual understanding or to new technology. Whether technology has always benefited mankind is debatable, but it has certainly revolutionized the way people live. I think we should be funded purely on those grounds.

**WEST:** Where do you stop? If you decide that $3 billion is okay or $10 billion, then do you ask for $100 billion?

**ZWEIG:** This is a difficult question, but if we can get people to agree on why we should be doing high-energy physics, then I think we can solve the problem of price. Although what we have been talking about may sound very obscure and possibly very ugly to an outside observer (quantum chromodynamics, grand unification, and twenty or thirty arbitrary parameters), the bottom line is that all of this really deals with a fundamental question, "What is everything made of?"

It has been our historical experience that answers to fundamental questions always lead to applications. But the time scale for those applications to come forward is very, very long. For example, we talked about Faraday's experiments which pointed to the quantal nature of electricity in the early 1800s; well, it was another half century before the quantum of electricity, the electron, was named and it was another ten years before electrons were observed directly as cathode rays; and another quarter century passed before the quantum of electric charge was accurately measured. Only recently has the quantum mechanics of the electron found application in transistors and other solid state devices.

Fundamental laws have always had application, and there's no reason to believe this will not hold in the future. We need to insist that our field be supported on that basis. We need ongoing commitment to this potential for new technology, even though technology's future returns to society are difficult to assess.

**CARRUTHERS:** Whenever support has to be ongoing, that's just when there seems to be a tendency to put it off.

**WEST:** What's another few years, right? Now I would like to play devil's advocate. One of the unique things about being at Los Alamos is that you are constantly being asked to justify yourself. In the past, science has dealt with macroscopic phenomena and natural phenomena. (I am a little bit on dangerous ground here.) Even when it dealt with the quantum effects, the effects were macroscopic: spectroscopic lines, for example, and the electroplating phenomena. The crucial difference in high-energy physics is that what we do is artificial. We create rare states of matter: they don't exist except possibly in some rare cosmic event, and they have little impact on our lives. To understand the universe that we feel and touch, even down to its minutiae, you don't have to know a damn thing about quarks.

**ZWEIG:** Maybe our experience is limited. Let me give you an example. Suppose we had stable heavy negatively-charged leptons, that is, heavy electrons. Then this new form of matter would revolutionize our technology because it would provide a sure means of catalyzing fusion at room temperature. So it is not true that the consequences of our work are necessarily abstract, beyond our experience, something we can't touch.

**WEST:** This discussion reminds me of something I believe Robert Wilson said during his first years as director of Fermilab. He was before a committee in Congress and was asked by some aggressive Congressman, "What good does the work do that goes on at your lab? What good is it for the military defense of this country?" Wilson replied something to the effect that he wasn't sure it helped directly in the defense of the country, but it made the country worth defending. Certainly, finding applications isn't

predominantly what drives people in this field. People don't sit there trying to do grand unification, saying to themselves that in a hundred years' time there are going to be transmission lines of Higgs particles. When I was a kid, electricity was going to be so cheap it wouldn't be metered. And that was the kind of attitude the AEC took toward science. I, at least, can't work that way.

**SCIENCE:** George, do you work that way?

**ZWEIG:** I was brought up, like Pete, at a time when the funding for high-energy physics was growing exponentially. Every few years the budget doubled. It was absolutely fabulous. As a graduate student I just watched this in amazement. Then I saw it turn off, overnight. In 1965, two years after I got my degree from Caltech, I was in Washington and met Peter Franken. Peter said, "It's all over. High-energy physics is dead." I looked at him like he was crazy. A year later I knew that, in a very real sense, he was absolutely right.

It became apparent to me that if I were going to get support for the kind of research I was interested in doing, I would have to convince the people that would pay for it that it really was worthwhile. The only common ground we had was the conviction that basic research eventually will have profound applications.

The same argument I make in high-energy physics, I also make in neurobiology. If you understand how people think, then you will be able to make machines that think. That, in turn, will transform society. It is very important to insist on funding basic research on this basis. It is an argument you can win.

There are complications, as Pete says; if applications are fifty years off, why don't we think about funding twenty-five years from now? In fact, that is what we have just heard: they have told us that we can have another accelerator, maybe, but it is ten or fifteen years down the road.

**SLANSKY:** We really can't build the SSC any faster than that.

**ZWEIG:** They could have built the machine at Brookhaven.

**WEST:** Let's talk about that. How can you

explain why a community who agreed that building the Isabelle machine was such a great and wonderful thing decided, five years later, that it was not worth doing.

**SLANSKY:** It is easy to answer that in very few words. The Europeans scooped the U.S. when they got spectacular experimental data confirming the electroweak unification. That had been one of our main purposes for building Isabelle.

**CARRUTHERS:** If you want to stay on the frontier, you have to go to the energies where the frontier is going to be.

**ZWEIG:** Some interesting experiments were made at energies that were not quite what you would call frontier at the time. CP violation was discovered at an embarrassingly low energy.

**SLANSKY:** The Europeans already have the possibility of building a hadron collider in a tunnel already being dug, the large electron-positron collider at CERN. It is clear that the U.S., to get back into the effort, has to make a big jump. Last spring the High Energy Physics Advisory Panel recommended cutting off Isabelle so the U.S. could go ahead in a timely fashion with the building of the SSC.

**WEST:** If you were a bright young scientist, would you go into high-energy physics now? I think you could still say there is a glamour in doing theory and that great cosmic questions are being addressed. But what is the attraction for an experimentalist, whose talents are possibly more highly rewarded in Silicon Valley?

**RABY:** It will become more and more difficult to get people to go into high-energy physics as the time scale for doing experiments grows an order of magnitude equal to a person's lifetime.

**ZWEIG:** Going to the moon was a successful enterprise even though it took a long time and required a different state of mind for the participating scientists.

**WEST:** Many of the great creative efforts of medieval life went into projects that lasted more than one generation. Building a great cathedral lasted a hundred, sometimes two hundred years. Some of the great craftsmen, the great architects, didn't live to see their

> *"I consider doing physics something that causes me an enormous amount of emotional energy. I get upset. I get depressed. I get joyful."*

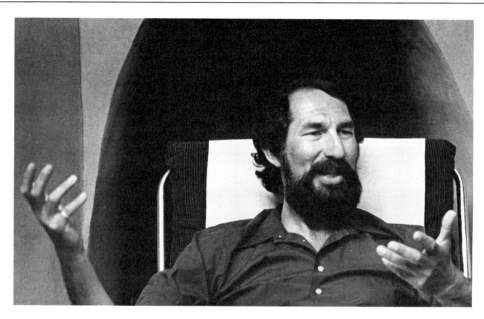

work completed.

As for going to high energies, I see us following Fermi's fantasy: we will find the hydrogen atom of hadronic physics and things will become simpler. It is a sort of Neanderthal approach. You hit as hard as you can and hope that things break down into something incredibly small. Somewhere in those fragments will be the "hydrogen" atom. That's the standard model. Some people may decide to back off from that paradigm. Lower energies are actually amenable.

**CARRUTHERS:** I think that people have already backed off. Wasn't Glashow going around the country saying we should do low-energy experiments?

**WEST:** Just to bring it home, the raison d'etre for LAMPF II is to have a low-energy, high-intensity machine to look for interesting phenomena. It is again this curious thing. We are looking at quantum effects by using a classical mode—hitting harder. The idea of high accuracy still uses quantum mechanics. I suppose it is conceivable that one would reorient the paradigm toward using the quantum mechanical nature of things to learn about the structure of matter.

**SLANSKY:** Both directions are very important.

**SCIENCE:** Is high-energy physics still attracting the brightest and the best?

**SLANSKY:** Some of the young guys coming out are certainly smart.

**CARRUTHERS:** I think there is an increasing array of very exciting intellectual challenges and new scientific areas that can be equally interesting. Given a limited pool of intellectual talent, it is inevitable that many will be attracted to the newer disciplines as they emerge.

**ZWEIG:** Computation, for example. Stephen Wolfram is a great example of someone who was trained in high-energy physics but then turned his interest elsewhere, and profitably so.

**CARRUTHERS:** Everything to do with conceptualization—computers or theory of the mind, nonlinear dynamics advances. All of these things are defining new fields that are very exciting—and that may in turn help us solve some of the problems in particle physics.

**ZWEIG:** That's optimistic. What would physics have been like without your two or three favorite physicists? I think we would all agree that the field would have been much the poorer. The losses of the kind we are talking about can have a profound effect on a field. Theoretical physics isn't just the

cumulative efforts of many trolls pushing blocks to build the pyramids.

**WEST:** But my impression is that the work is much less individualized than it ever was. The fact that the electroweak unification was shared by three people, and there were others who could have been added to that list, is an indication. If you look at QCD and the standard model, it is impossible to write a name, and it is probably impossible to write ten names, without ignoring large numbers of people who have contributed. The grand unified theory, if there ever is one, will be more the result of many people interacting than of one Einstein, the traditional one brilliant man sitting in an armchair.

**SCIENCE:** Was that idea ever really correct?

**WEST:** It was correct for Einstein. It was correct for Dirac.

**SCIENCE:** Was their thinking really a total departure?

**ZWEIG:** The theory of general relativity is a great example, and almost a singular example, of someone developing a correct theoretical idea in the absence of experimental information, merely on the basis of intuition. I think that is what people are trying to do now. This is very dangerous.

**RABY:** Another point is that Einstein in his later years was trying to develop the grand unified theory of all known interactions, and he was way off base. All the interactions weren't even known then.

**WEST:** Theorizing in the absence of supportive data is still dangerous.

**CARRUTHERS:** Particle physics, despite all of its problems, remains one of the principal frontiers of modern science. As such it combines a ferment of ideas and speculative thoughts that constantly works to reassess the principles with which we try to understand some of the most basic problems in nature. If you take away this frothy area in which there's an enormous interface between the academic community and all kinds of visitors interacting with the laboratory, giving lectures on what is the latest excitement in physics, then you won't have much left in the way of an exciting place to work, and people here won't be so good after awhile. ■

# Index

accelerators 150-157
  AGS 52, 150, 152, 153, 158, 188, 194
  Bevatron 150, 152
  CERN 130, 150, 152, 156, 157, 159, 193, 194
  CESR 153
  Cosmotron 152
  cyclotron 152, 153
  DORIS 153
  Fermilab 152, 155
  HERA 156, 157
  LAMPF 136, 137, 141-145, 147, 148, 154, 158-163, 195
  LEP 154, 156, 157
  linac 152, 154
  PEP 153
  PETRA 153
  SIN 137, 141, 154
  SLAC 18, 140, 152, 159, 188, 190, 192
  SLC 154, 156, 157
  SPEAR 153
  SSC 152, 157, 164-165
  synchrocyclotron 152
  synchrotron 152
  Tevatron 152, 155, 156, 157
  Tristan 156
  TRIUMF 137, 146, 154
antiparticles 108, 131, 182
  antilepton 160
  antimuon 160
  antineutrino 100, 112, 132
  antiproton 28, 109, 150, 158-160, 162
  antiquark 108, 160, 191
  antisquark 108, 112, 113
asymmetry, matter-antimatter 168
asymptotic freedom 16-18, 19, 40, 41
axial vector 145, 146

bare parameters 16, 17
bare theory 19

baryon 161
  Lambda 160, 161
  number 37, 167
    conservation of 34, 154, 167, 168
  Sigma 161
beta decay 43, 128, 130, 146, 167
  intrinsic linewidth 132
beta-electron spectrometer 133, 134
  toroidal 132
  resolution function 132-134
beta-function 17
big bang theory 168
birds' eggs 5-7
bone structure 4-5
bosonic field 102, 103, 105
boson 25, 28, 40-43, 69, 76, 79, 98-112, 131, 141, 142, 146, 153, 158
  charged-current 141, 142, 145
  charged-vector 141, 142
  heavy vector 130
  Goldstone 59, 63, 121
  Higgs 47, 50, 66, 101, 105-107, 131, 156, 157, 192, 194
  intermediate vector 136
  massless gauge 131
  neutral vector 131, 141, 142
  vector 25, 40, 43, 69, 76, 130, 131, 145, 158, 191, 192
  $W^{\pm}$ 44, 45, 50, 67, 77-78, 101, 105-107, 112, 130, 141, 142, 146, 148, 150, 154, 156, 157, 189, 191, 192
  $Z^0$ 46-48, 50, 67, 77-78, 112, 130, 141, 142, 148, 150, 154, 156, 157, 189, 191, 192
brain size 10

Cabibbo
  angle 51, 71
  matrix 120
charged currents 130, 131, 141
  interference effects 130, 141, 142
chirality 44
color 30, 39, 40, 131, 191
  charge 69, 131
  force 131
  gluons 130
conservation 34, 100, 136, 137
  of energy 100, 132
  of leptons 128, 136, 137, 145
continuity equation 57
correlation functions 14, 18
cosmic ray 144, 145
cosmic-ray astronomy 167

cosmological constant problem 84, 91
CP violation, 33, 121, 123, 131, 154-157, 194
  in $B^0$-$\bar{B}^0$ system 123
  in $K^0$-$\bar{K}^0$ system 121, 156, 157
  relation to family symmetry breaking 121, 123
Crystal Ball 140
Crystal Box 137-141

deBroglie relation 27
decuplet 37
detector properties 138, 139
dimension 3
  anomalous 17
dimensional analysis 4, 6-14, 17, 190
dimensionless variables 7-19
doublets 131
drag, viscous 7-10

Eightfold way 37, 38
electrodynamics 26
electromagnetic
  coupling constant 131
  current 142
  field 14, 34
  force 19, 23, 24, 135
  interaction 25, 128
  shower 139
electromagnetism 23, 28, 188
electron 27, 29, 100, 128, 131
  number 34, 136
  scattering
    deep inelastic 17, 18, 42, 190
    electron-electron 142
    electron-neutrino elastic 142
    polarized 141
electron-positron colliders 141
electron-positron pair 15
electroweak theory 19, 45-50, 65-68, 76-79, 128, 130, 131, 141, 142, 168, 189, 192, 194, 195. *See also* forces and interactions, basic
elementary particles, representations of
  in quantum chromodynamics 115
  in electroweak theory 115, 116
EMC effect 159, 160
end-point energy 132, 133
endocranial volume 10

families, quark-lepton 51, 71, 115, 117, 136, 157, 158
family-changing interactions 116-122
family problem 31, 128, 130, 136
family-symmetry breaking 116-122
Fermi 130
    constant 43
    theory 43
fermionic field 102, 103, 105
fermion 28, 98-112, 131
    generations 136
    Goldstone 101
Feynman diagram 14-16, 29, 189
fields in higher dimensions 86-87
fine-structure constant 130
fixed points 18
flavor 130
    symmetry 39
forces and interactions
    basic 24, 74, 78-79
    electromagnetic 19, 23-26, 28-30, 31, 36, 42, 98, 110, 128, 130, 135, 141, 142, 182, 188-190, 192
    electroweak 65-68, 70-71, 128, 130, 131, 189, 192
    gravitational 23, 24, 26, 59, 98, 110, 130, 188, 189
    neutral current 46
    strengths of 24, 28, 30, 40, 43-45, 66, 70
    strong 19, 23-25, 28, 30, 36, 38-40, 69-70, 98, 110, 128, 130, 136, 188-192, 192
    unification of 23, 25, 28, 30, 44-46, 53, 72-95, 152, 153, 168
    weak 19, 23, 24, 28, 42-45, 98, 107, 110, 112, 128, 130, 136, 141-143, 145, 146, 188-190, 192
        beta decay 42-43
        Fermi theory of 43-44, 76, 77, 79, 155
        charged-current 43, 46-49, 71
        neutral-current 46-47, 48, 49, 68, 71, 152, 153
        right-handed 157
form invariance 11, 14
fundamental
    constants, 12, 13
    scales 11
    symmetries 136

gauge
    fields 142
    invariance 36
    prticles 189, 191
    theory 25, 39, 45, 130, 155, 188, 191
        non-Abelian (Yang-Mills) 40, 45, 188, 189
gauginos 107
geometry box 140
general relativity. *See* gravity, Einstein's theory of
global invariance 34
glueballs 41, 161, 191
gluino 108, 109, 112, 131
gluons 19, 25, 40, 42, 79, 101, 108, 153, 159-161, 189, 191, 192
goldstino 105, 107, 110
Goldstone fermion 101
grand unified theory 81-82, 106, 110, 131, 146, 154, 168-170, 182, 190, 192-195
graviton 80, 82, 84, 93, 189
gravitino 105, 107, 110
    massive 110
gravity
    Einstein's theory of 74, 75, 80-81, 82, 84
    unification with other forces 82-95
group multiplets 31

hadron 36, 80, 91-92, 109, 144, 160-162, 189-191, 195
Heisenberg uncertainty principle 13, 14, 29
helium-3 132
Hierarchy problem 106
Higgs
    bosons 47, 50, 101, 105-107, 131, 192, 194
    mechanism 131
Higgsino 107, 108
hypercharge 37
hypernucleus 159-161

interference effects
    between neutral and charged weak currents 130, 141-143
isotopic spin 30, 37, 189

J/Psi 52, 153, 154, 188-189
jets, hadronic 109, 112, 157

Kaluza-Klein theories 83, 84, 87
kaons 158-162
Kobayashi-Maskawa matrix 51, 123

Lagrangian 14, 17, 34, 100
    complex scalar field 55
    electroweak theory 65-68
    quantum chromodynamics 69-70
    quantum electrodynamics 63
    quantum field theory 25
    real vector fields 56-57
    standard model 71
    weak interactions of quarks 70-71
    Yang-Mills theories 64
Lambda 160, 161
lepton 19, 25, 31, 36, 51, 107, 131, 160, 189, 192, 194
    conservation 128, 136, 137, 145, 154
    families 128, 135, 136, 146
    flavor 133, 136
    multiplets 136
local gauge
    invariance 36
    theory 25, 189
    transformations 39
local symmetry 25, 30, 34, 40

magnetic
    monopoles 110, 192
    pinch 134
Majorana Fields 34
mass
    neutrino 128, 130, 133, 135, 146
    scale 13, 15, 16, 18, 101, 106, 107, 131
massive gravitino 110
meikton 161
meson 111, 135, 136, 161, 189
metabolic rate 5-7
Michel parameters 145, 146, 148
modeling theory 9, 18, 19
Monte Carlo
    calculation 162
    sampling 41
    simulation 137, 140, 191
    modeling 134
multiplets 136
muon 111, 128, 131, 135-141, 143-147, 158-160, 189
    branching ratio 132, 136, 137
    daughter 137
    decay 136, 138-141, 145-148, 155
    discovery 135
    lifetime 138, 144
    Michel parameters 145, 146, 148
    number 34, 136, 137
    prompt 144
    range 136

neutral weak current 46, 48, 130, 131, 141, 192
neutrino 100, 111, 112, 128, 130, 136-146, 155, 158, 159, 162, 192
    appearance mode 143, 144
    astronomy 167, 170
    decay 173-174
    electron neutrino 137, 142, 143, 145
    flavors 130
    mass 122, 128, 130-133, 135, 159
    oscillations 122, 133, 143, 148, 155, 167, 176
    physics 154
    right-handed 131
    solar 155, 167
        flux measurements of 172-177
neutrino-electron scattering 130, 136, 141-145, 159, 162
    detector 144
    background 144
neutron decay. *See* proton decay
non-Abelian gauge theories 16, 18, 40, 188
nucleon decay 111, 146

octets 37
$\Omega^-$ 37, 150

parity
    conservation 128, 146
    violation 49, 69, 146
Pauli exclusion principle 28, 104
photinos 106, 109, 112
photon 14, 16, 19, 29, 36, 76-78, 100, 135
pion 100, 110, 136, 137, 143, 145, 159, 160, 162
    decay 140, 143, 145
    pion dynamics 59, 152, 155
Planck mass, length 80, 82, 83
positrons 29, 109-111, 138, 139, 143, 182
preons 53, 157
propagator 14-19
proton
    beam 137, 158, 160-162
    decay 81-82, 110, 111, 148, 155, 168, 192
        searches for 166-171
pyrgons 84-87, 95

quantum chromodynamics 18, 19, 39-42, 69-70, 79-80, 130, 131, 159, 161, 190, 191, 193. *See also* forces and interactions, basic
quantum electrodynamics 14, 16-19, 28-30, 31, 34, 36, 40, 55, 62-63, 76, 79, 130, 189, 190. *See also* forces and interactions, basic
quantum field theory, 4, 11, 13, 14, 16, 18, 25, 27, 28, 30, 24, 64, 141, 182, 187
quark 3, 17, 19, 25, 31, 38, 39, 42, 51, 100, 107-110, 112, 131, 136, 150, 152-155, 159-161, 180, 182-184, 188-192, 194
    confinement of 42
    families 51
    flavors 39
    masses of 69-71
    mixing of 51, 71
    transitions between states 136

rare decays 128
    limits 138
    of the muon 135, 137
Rayleigh-Ruabouchinsky Paradox 12-13
Regge
    recurrences 95
    trajectories 92
renormalization 11, 14-18, 28, 189-191
    group 11, 16, 18, 19, 189, 191
    group equation 13, 14, 17
rotation group 32
rowing 9, 10

scalar 146
    pseudo 146
    particles 100, 103, 105, 107, 108
scale 2-21, 101, 107, 131, 183, 190, 191
    energy 19
    invariance 11, 13, 190
scaling 4-21, 28, 42, 190
    classical 4
    curve 9

scattering experiments 130, 189, 191
    inelastic 17, 18, 42, 79-80, 190
selectron 100, 109
similitude 5,7
singlets 131
slepton 107
S-matrix theory 189-191
sneutrino 100
solar energy-production models 166, 167, 171-172, 173, 174
solar neutrino 145
    physics 148
space-time manifold 82
    extension to higher dimensions, 76, 83, 84, 86
special relativity 27
spin 30, 87, 100, 131, 189
spin-polarized hydrogen 133
spin-statistics theorem 100
squark 100, 107-109, 11, 113
standard model 18, 19, 23, 25, 30, 42, 50-51, 53, 54, 71, 74, 76-80, 100, 106, 107, 109, 110, 115, 130, 141, 142, 145, 159, 182, 188-190, 192
    minimal 130, 131, 136, 145, 148
strangeness 30, 37
supergap 101
supergravity 76, 88-91, 93, 94, 101, 110
superstring theories 76, 82, 91-95
superspaces 93
supersymmetry 74, 76, 88-90, 98, 100-113, 157
    in quantum mechanics 102-105
    interaction 119
    spontaneous breaking 100, 105, 107, 110
supersymmetry rotation 106, 108
symmetries; symmetry groups, multiplets, and operations 30-36, 38-39, 45-47, 61, 64, 75, 88, 90, 100, 101, 110, 111, 136, 142, 189
    Eightford Way 69, 70
    of electroweak interactions 65, 77
    of Lagrangians 56-57
    of quark-lepton interactions 81
    of strong interactions 69-70
    strong isospin 60-61, 69
    weak isospin 61

symmetry 30, 39, 128, 145, 146, 182, 188-192
  boson-fermion 74
  broken 31, 32, 33, 39, 45, 48, 100,
    105-107, 131, 182
  continuous 31, 56
  CP 33, 131
  discrete 31
  exact 34
  external 100-102
  left-handed 145
  local 25, 30, 34-36, 39-40, 46-47, 54-56
    74-75, 188, 192
  Lorentz in variance 56,59
  phase invariance 56, 59, 62-63, 76
  Poincaré 56, 80
  right-handed 145
  spontaneous breaking of 47-48, 54, 58-59
    62-63, 66-68, 78, 81, 83, 88

tau 30, 128, 136
  particle number 34, 131
time projection chamber (TPC) 146, 147
tritium 128, 132-135
  beta decay 128, 130, 148
  beta-decay spectrometer 128, 132-135
    resolution function 132-134
  end-point energy 132, 133
  final state spectrum 132
  molecular 133-135
  recombination 133
  source 133-135

ultraviolet laser technology 133
uncertainty principle 13, 14, 29
underground science facilities 178-179
unification 26, 101, 141, 188, 190, 192-195
Υ 53, 153-154

vacuum state 58
vector-axial 131, 145
  currents 143
vector potential 28

$W^\pm$. See boson.
weak charge 131
weak force 19, 23-25, 28
  currents 49, 131, 141, 192
weak interaction 42, 107, 110, 112, 128, 136,
    141-143, 145, 146, 192
  constructive and destructive interference
    142, 143
  coupling constant 131, 145, 146
weak mixing angle 48, 66, 67, 110, 111, 131,
    142, 143
weak scale 101
Weinberg angle 48, 110, 111, 131, 142, 143
Weinberg-Salam-Glashow model 130

Yang-Mills theories 40, 45, 188, 189
Yukawa's theory 135

$Z^0$. See boson.
zero modes 85, 86, 95